Laura Costa, Rosa María Miró-Roig, and Joan Pons-Llopis

Ulrich Bundles

De Gruyter Studies in Mathematics

Volume 77

Laura Costa, Rosa María Miró-Roig, and
Joan Pons-Llopis

Ulrich Bundles

From Commutative Algebra to Algebraic Geometry

DE GRUYTER

Mathematics Subject Classification 2020
Primary: 14F06,14J60; Secondary: 13C14, 13D02

Authors
Prof. Dr. Laura Costa
Universitat de Barcelona
Facultat de Matemàtiques i Informàtica
Gran Via des les Corts Catalanes 585
08007 Barcelona
Spain
costa@ub.edu

Dr. Joan Pons-Llopis
Politecnico di Torino
Dipartimento di Scienze Matematiche
Corso Duca degli Abruzzi 24
10129 Torino
Italy
juan.ponsllopis@polito.it

Prof. Dr. Rosa María Miró-Roig
Universitat de Barcelona
Facultat de Matemàtiques i Informàtica
Gran Via des les Corts Catalanes 585
08007 Barcelona
Spain
miro@ub.edu

Despite careful production of the book, sometimes mistakes happen. Unfortunately, the citations numbers in the text after citation [71] were incorrect. The incorrect citation numbers have been replaced by corrected numbers in the book. We apologize for the mistake.

ISBN 978-3-11-064540-8
e-ISBN (PDF) 978-3-11-064768-6
e-ISBN (EPUB) 978-3-11-064580-4
ISSN 0179-0986

Library of Congress Control Number: 2021932020

Bibliographic information published by the Deutsche Nationalbibliothek
The Deutsche Nationalbibliothek lists this publication in the Deutsche Nationalbibliografie; detailed bibliographic data are available on the Internet at http://dnb.dnb.de.

Typesetting: VTeX UAB, Lithuania
Printing and binding: CPI books GmbH, Leck

www.degruyter.com

To Joan Jordi and to my children Anna, Mariona and Bernat
Laura Costa

To my husband Joan
Rosa María Miró-Roig

To my mother and to my wife Hiroko
Joan Pons-Llopis

Contents

Introduction

Since Schwarzenberger's initial works on the existence of holomorphic structures on topological vector bundles over the projective plane, the theory of vector bundles on projective varieties has become one of the mainstreams in algebraic geometry. During the 1970s and 1980s, it received a major boost due to, among other reasons, its relation to mathematical physics. For instance, a class of stable vector bundles, the so-called instanton bundles on $\mathbb{P}^3$, was defined as the algebraic counterpart that permitted to solve a central problem in Yang–Mills theory—the existence of self-dual Yang–Mills $S_p(1)$-connections over the 4-dimensional sphere S^4. This deep link between two a priori unrelated subjects, such as algebraic geometry and theoretical physics, provided a major breakthrough in the theory of vector bundles. We recommend [169] for an account on this topic.

On the other hand, even without leaving the realm of algebraic geometry, vector bundles on their own offer a valuable point of view in order to better understand the geometry and topology of algebraic varieties. For instance, according to the general tenet that moduli spaces of stable vector bundles on a projective variety X inherit from the underlying variety X many of their algebraic-geometric properties, a careful study of the local and global structure of certain moduli spaces of vector bundles led to important advances in the classification problem of projective varieties.

Despite huge achievements in vector bundle theory, researchers in this area should confess that some of the most (illusively) naif questions remain open. A blatant case is the following: although the case of vector bundles of rank one, the so-called line bundles, on (smooth) projective varieties is well understood via the theory of Cartier divisors, once we move just one step forward to the rank-two case, we still lack a clear picture. Indeed, a weaker form of the famous Hartshorne's conjecture states that a smooth projective variety of codimension two in $\mathbb{P}^n$ should be a complete intersection as soon as its dimension is greater than four. In terms of vector bundles, through Serre's correspondence, the conjecture states that a rank-two vector bundle supported on a projective space of dimension greater than six should split as a direct sum of line bundles. Despite many efforts, at the time of writing this monograph this question remains open. In other words, the possible structures of rank-two bundles on the projective space are not known, yet.

In general, since the whole category of vector bundles on a projective variety is usually unwieldy, one has to impose some restrictions on the families of vector bundles under consideration. In this sense, basically two approaches have been proposed: the geometric one is based on the notion of stability developed by many mathematicians as Gieseker, Maruyama, Mumford, and Takemoto. This approach has been very successful, providing the right notion of a bounded family, namely the moduli space

The original version of this chapter was revised: the text on p. XIII, line 39 and p. XIV, line 21, has been corrected.
An Erratum is available at DOI: https://doi.org/10.1515/9783110647686-011
https://doi.org/10.1515/9783110647686-201

of (semi)stable vector bundles with fixed discrete invariants. We suggest as an introduction to the reader interested in this approach the monograph [119].

The second approach, the subject of this book, deals with the cohomological properties of vector bundles. Namely, one is interested in studying and classifying vector bundles which a large number of vanishing cohomology groups. Somehow, the result guiding this approach is Horrocks' theorem which tells that on the projective space $(\mathbb{P}^n, \mathcal{O}_{\mathbb{P}^n}(1))$ of arbitrary dimension, the only indecomposable vector bundles without intermediate cohomology, the so-called arithmetically Cohen–Macaulay (aCM for short) bundles, are exactly the line bundles. Paraphrasing this result, one can say that the simplest projective variety supports only trivial aCM bundles. However, one could still expect a good understanding of aCM bundles either when the complexity of the underlying projective variety is not too high or when one pays attention to aCM bundles satisfying some extra requirements. For example, along the first line of reasoning, it was proved that all projective varieties supporting a finite number of aCM bundles should have minimal degree. A paradigmatic instance of the second line of reasoning will be the study of homogeneous aCM bundles on some rational homogeneous varieties where, as we will see, a characterization has been achieved.

This second approach to the study of vector bundles receives also a very strong input from commutative algebra. Indeed, we will see that there exists a precise bijection between aCM bundles on projective varieties and maximally Cohen–Macaulay (MCM for short) graded modules with respect to the associated coordinated graded ring of the projective variety. MCM-modules have been a central topic of research in commutative algebra and whose numerous open questions have been spurring, through the aforementioned bijection, the research in the algebraic geometric side. Just to have a hint of the rich interplay between the algebraic and geometric point of view, notice that over the coordinate ring of smooth quadrics, the only nontrivial indecomposable MCM modules are closely related to the spinor varieties of planes of maximal dimension lying on the quadric.

It is along this line of thought that Ulrich vector bundles enter into the picture. Among aCM bundles, Ulrich bundles form a subfamily that recently has been the object of intense research. They can be roughly defined as those aCM bundles on a projective variety that have the largest permitted number of global sections. This is just one among several equivalent characterizations, each of them interesting on its own. Although just recently they received a deserved attention from the geometric point of view, they were an important object of study in the algebraic setting already from the 1980s under the name of linear MCM modules or maximally generated MCM modules. Perhaps more enthralling for the algebraic geometers, somehow Ulrich bundles were unawarely the object of an intense research under the disguise of linear determinantal representations of projective hypersurfaces. Indeed, we will see that a very active area of research for algebraists and geometers from the mid-nineteenth century, as it can be the linear determinantal representation of a hypersurface $X \subset \mathbb{P}^n$, turns out to be equivalent to the search of Ulrich line bundles on $X \subset \mathbb{P}^n$.

The existence of an Ulrich bundle on a d-dimensional projective variety has strong consequences for the underlying variety. To mention just a couple, it implies that its cone of cohomology tables of vector bundles on X is the same as that of the projective space $\mathbb{P}^d$. On the other hand, it also implies that its associated Cayley–Chow form has a particularly simple presentation. Despite all of these constraints, it has been asked by Eisenbud and Schreyer whether any projective variety supports an Ulrich bundle. Prompted by this question, the theory of Ulrich bundles has became a very active area of research within algebraic geometry with a constant supply of new achievements. Moreover, due to its Janus-faced inception, it is not surprising that a very broad range of techniques, scattered in the literature, are been used in the development of their theory. All these facts make it difficult for the interested mathematician to keep track of the current state-of-the-art in this subject.

This is one of the reasons that convinced us of the necessity of writing a book about Ulrich bundles on projective varieties. Writing it, the authors had in mind different goals: firstly, to offer an introduction to the topic, trying to underline the main motivations of their inception, the wide range of areas that are involved in the development of this subject, as well as the central questions that remain open in this area. In any case, it should be pointed out that despite our intent to offer a picture of this theory as broad as possible, we do not claim to have exhausted all the results that can be found in the literature.

As a matter of fact, however, we also expect to fulfill a second goal writing this monograph, that can be considered more formative. As it has been mentioned, in order to construct Ulrich bundles on a projective variety, one needs to have a complete acquaintance with a large bunch of techniques that should belong to the toolkit of any mathematician interested in vector bundles. Therefore, we believe that this book can be useful not only for established researchers with a particular interest in this topic, but also for young researchers as a motivation to learn these basic techniques in a practical way and with a definite aim—the construction of Ulrich bundles—in mind.

This book has been structured as follows: the aim of Chapter 1 is to fix notation and introduce general objects and tools that form the background in the theory of vector bundles and that will be used liberally throughout the book without further mention. Despite the existence of excellent references covering these topics, we considered for the reader's convenience to gather here these results. We start from the scratch, recalling the geometric definition of a vector bundle, as well as the equivalence between the category of vector bundles and the category of locally free sheaves on a smooth projective variety. Then we go through the basic results on cohomology of vector bundles, Serre duality theorem, Chern classes, Riemann–Roch theorem, and other results, concluding the chapter with the different notions of stability of coherent sheaves and their moduli spaces.

Chapter 2 is devoted to an introduction of the arithmetically Cohen–Macaulay bundles, namely, those that have no intermediate cohomology groups. As it has been mentioned, Ulrich bundles are a particular subset of aCM bundles. Nevertheless, the study of aCM bundles has a rich and independent history behind them that would deserve an entire monograph on its own. Here we should restrain ourselves to presenting the main results that will be valuable afterwards to work with Ulrich bundles. After giving the definition and several equivalent presentations of aCM bundles, we stress their relation with their algebraic counterpart of maximally Cohen–Macaulay modules. Then, we present the crucial Horrocks' theorem that basically states that on the projective space the only indecomposable aCM bundles are the line bundles. This is the starting point to study aCM bundles on projective varieties of increasing complexity. The first step, smooth quadric varieties, is well understood. Besides the structural sheaf, they support only one (or two, depending on the parity of the dimension of the quadric) aCM bundle(s), and they have a clear geometric meaning. Indeed, projective varieties supporting a finite number of aCM bundles have been completely classified. After developing these topics, we study thoroughly an example where we add an extra structure to the bundles under consideration: we determine all aCM irreducible homogeneous bundles on Grassman varieties. Indeed, this case will represent a general strategy to tackle this kind of problems: one has to restrain oneself to a particular well-behaved family of vector bundles, in this case homogeneous vector bundles on a particular rational homogenous variety, for which a powerful theorem about cohomology groups is known, namely the Borel–Bott–Weil theorem, to obtain relevant information about the aCM property. Finally, we work with smooth hypersurfaces. We will explain how the theory of matrix factorizations permits finding indecomposable aCM bundles of high rank on any smooth hypersurface. This triggers the intriguing question about the smallest rank of an aCM bundle supported on a fixed smooth hypersurface.

Eventually, in Chapter 3, we introduce the main object of this book—Ulrich bundles on a projective variety. Due to the rich history they carry, we start this chapter with a brief historical introduction. Next, we highlight that Ulrich bundles can be introduced from different points of view with distinct flavors. Perhaps it is this malleability that makes them susceptible to be constructed using a wide range of techniques. We continue the chapter gathering the main properties of Ulrich bundles and, in particular, underlying their behavior with respect to the stability conditions. As it will be explained, arguably the leading question in this area, due to Eisenbud and Schreyer, is the following: Does any projective variety support Ulrich bundles? And if it does, which is the lowest rank of such a vector bundle? Although this question is open in general, we show in the last part of this chapter that for projective curves the answer is affirmative.

The rest of the book tries to give an outlook on the state-of-the-art of the aforementioned questions, stressing the wide range of techniques involved in answering them. We start the inquiry about the existence of Ulrich bundles on projective varieties in Chapter 4 with the case of complete intersection varieties. For hypersurfaces, we make explicit the relationship between Ulrich bundles and linear determinantal representations of the homogeneous polynomial associated to the hypersurface. Then we introduce the concept of linear matrix factorization of a homogeneous form which can be understood as a natural generalization of a two-factor matrix factorization. Associated to these concepts, we define the notions of Waring and Chow rank of a homogeneous form, in terms of which it is possible to give an upper bound for the lowest rank of an Ulrich bundle supported on the hypersurface associated to the form. Then, we move forward to complete intersections. We explain how, given Ulrich bundles on each of the hypersurfaces cutting out the complete intersection variety, one can construct an Ulrich bundle (of much larger rank) on it. Gathering the previous results, one obtains that complete intersection varieties always are the support of Ulrich bundles and, moreover, their rank can be bounded in terms of the degrees and Chow ranks of the associated hypersurfaces. Then we apply the previous results to surfaces in the three-dimensional projective space.

We have already mentioned that the construction of Ulrich bundles has received inputs from a priori unrelated areas. To enlighten this point of view, we explain how tools and techniques coming from representation theory allow us to construct them. We conclude the chapter showing that the theory of linear determinantal representations of hypersurfaces has a natural generation in terms of Cayley–Chow forms. Indeed, the search of an explicit presentation of the Cayley–Chow form (in the Plücker coordinates of the related Grassman variety) of a projective variety has been one of the main motivations that prompted the study of Ulrich bundles.

Chapter 5 is completely devoted to the study of Ulrich bundles on smooth projective surfaces. Although, as it has been explained, projective curves always support Ulrich bundles, already in dimension two we still do not have a complete understanding of the situation. It is perhaps not surprising that our knowledge about Ulrich bundles on a projective surface fades away alongside the increase of the Kodaira dimension of the surface. So we decided to frame the known results about Ulrich bundles on surfaces in the setting of the Enriques–Kodaira classification of projective surfaces. We start with minimal rational surfaces, namely Veronese and Hirzebruch surfaces. To extend the existence results to the nonminimal case, we should explain the behavior of Ulrich bundles under blow-up and blow-down maps. As a classically relevant case, we pay particular attention to del Pezzo surfaces. Dealing with the case of projective surfaces of Kodaira dimension zero will provide us with the excuse to introduce the important technique of elementary transformations of a vector bundle along a suitable effective divisor. For positive Kodaira dimension, the picture is still far from being clear. We will show that a large family of elliptic regular surfaces (with Kodaira dimension one) support special Ulrich bundles of rank two. However, for Kodaira dimension two, we can just offer a few sparse results. On the other hand, we conclude the chapter with the remarkable fact that any projective surface can be embedded in a projective space (perhaps of large dimension) in such a way that it supports Ulrich bundles. The material in Section 5.6 is based on the paper *Brill–Noether Problems, Ulrich Bundles and the Cohomology of Moduli Spaces of Sheaves* by Izzet Coskun and Jack Huizenga, Matemática Contemporânea 47 (2020), 21–72 [63].

Chapter 6 deals with Ulrich bundles on Fano threefolds. Indeed, we focus most of our attention in the particular case of the smooth intersection of two quadrics in $\mathbb{P}^5$ for which we provide a complete classification of its Ulrich bundles. In order to justify our choice, notice first that this case lies at the intersection of two approaches that we have been pursuing in the previous chapters: it is the (nontrivial) three-dimensional complete intersection of lowest possible degree and, at the same time, it is the natural generalization of del Pezzo surfaces studied in Chapter 5. But perhaps more important is the fact that a detailed study of Ulrich bundles on this threefold allows us to illustrate one the leitmotif leading the writing of this book: how classical and modern technics from Algebraic Geometry are synergetically used in the theory of Ulrich bundles. In particular, we will see how Orlov's semiorthogonal decomposition of the derived category of coherent sheaves on this variety can be combined with the classical Raynaud bundles on its associated hyperelliptic curve.

Once we increase the dimension of the underlying variety, one is forced to restrain the study of Ulrich bundles to particularly well-defined families. The last chapter is devoted to three particular cases. We start with Segre varieties, namely with the product of projective spaces embedded in a larger projective space with the lowest possible degree. Then we move on to completely classify irreducible homogeneous Ulrich bundles on Grassmann varieties refining the results obtained in Chapter 2 for aCM bundles. Finally, we extend these results to the larger framework of flag varieties. Remarkably, we will show that, thanks to the Borel–Bott–Weil theorem, the problem about the existence of indecomposable irreducible homogeneous Ulrich bundles can be rephrased in a pure combinatorial setting. Using this approach, we will show how to fully answer this problem for most of the flag varieties and propose a conjecture for the rest of them. The authors wish to acknowledge the contribution of Professor Izzet Coskun, Professor Jack Huizenga and Dr. Matthew Woolf for their collaboration on the paper *Ulrich Schur bundles on flag varieties*, J. of Algebra, 474 (2017), 49–96 [61]. The material in Section 7.3 is a reprint of this material. In addition, the authors thank Professor Jack Huizenga for creating all of the original illlustrations. We gratefully acknowledge the permission from the publisher Elsevier for including this material.

In order to try to make this monograph as self-contained as possible, we conclude with an appendix where we have gathered some important results from the theory of derived categories and Fourier–Mukai transforms used throughout the rest of the chapters.

Each chapter will end with a short section where we give further references, historical remarks, as well as a list of additional readings that will complement and offer a better understanding of the material presented.

We have tried hard to keep the text as self-contained as possible. The basics of algebraic geometry supplied by [108] suffices as a foundation for this text. Some familiarity with commutative algebra, as developed, for instance, in [153] and [36], will be helpful. The rudiments on Ext and Tor contained in every introduction to homological algebra will be used freely.

As we already mentioned, despite our intent to offer a picture of the theory of Ulrich bundles as broad as possible, in no case do we claim to have exhausted all the results that can be found in the literature. A large number of mathematicians have made important contributions to this area without even being mentioned in this monograph, and we apologize to those whose work we might have failed to cite properly.

Acknowledgment

Parts of this book were planned while we were guests of the Mathematisches Forschungsinstitut Oberwolfach. We thank the Forschungsinstitut for its generous hospitality. This monograph arrives after years of experience on vector bundles. During this time we have had the opportunity to learn and discuss all this theory with several of our colleagues and we would like to thank for all their contributions.

https://doi.org/10.1515/9783110647686-202

Notation

Throughout this book, $\mathbf{k}$ is an algebraically closed field of characteristic 0 (which is assumed to be uncountable at any instance of Noether–Lefschetz theorem). Given an $(n+1)$-dimensional $\mathbf{k}$-vector space V, we set $\mathbb{P}^n = \mathbb{P}(V)$ to be the projective space of hyperplanes of V, or equivalently, of lines of the dual vector space $V^\vee$. The associated coordinate ring of $\mathbb{P}^n$ is $R := \bigoplus_{i\geq 0} S^iV \cong \mathbf{k}[x_0,\dots,x_n]$. All the schemes will be over $\mathbf{k}$. By an algebraic variety we mean an integral proper scheme of finite type over $\mathbf{k}$, and by points we always mean closed points. Unless otherwise specified in this book, we will only deal with smooth varieties. A projective scheme (respectively a projective variety) will be a couple $(X,\mathcal{L})$, where X is a scheme (respectively an algebraic variety) and $\mathcal{L}$ is a very ample line bundle on it. Analogously, tacitly we assume that the very ample line bundle associated to an embedded projective variety $X \subset \mathbb{P}^n$ is $\mathcal{L} := \mathcal{O}_{\mathbb{P}^n}(1)_{|X}$. Moreover, when the very ample line bundle $\mathcal{L}$ on X is clear from the context, we are going to denote it by $\mathcal{O}_X(1)$ and, for any coherent sheaf $\mathcal{E}$ on X, we are going to denote the twisted sheaf $\mathcal{E}\otimes\mathcal{L}^{\otimes\ell}$ by $\mathcal{E}(\ell)$.

If $X \subset \mathbb{P}^n$ is a closed subscheme of codimension c, we denote the ideal sheaf of X by $\mathcal{I}_X$ and the homogeneous saturated ideal by $I(X) = \bigoplus_{t\in\mathbb{Z}} \mathrm{H}^0(\mathcal{I}_X(t))$; R_X stands for the homogeneous coordinate ring of X defined as $\mathbf{k}[x_0,\dots,x_n]/I(X)$ and $K_X = \mathrm{Ext}^c_R(R_X,R)(-n-1)$ its canonical module. For any coherent sheaf $\mathcal{E}$ on a projective variety $(X,\mathcal{O}_X(H))$, $\mathcal{E}^\vee$ will stand for $\mathcal{H}om_{\mathcal{O}_X}(\mathcal{E},\mathcal{O}_X)$ and $\mathcal{E}nd(\mathcal{E}) := \mathcal{H}om_{\mathcal{O}_X}(\mathcal{E},\mathcal{E})$ denotes the sheaf of endomorphisms of $\mathcal{E}$ while $\mathrm{End}(\mathcal{E}) := \mathrm{Hom}(\mathcal{E},\mathcal{E})$ denotes the group of endomorphisms. As usual, $\mathrm{H}^i(X,\mathcal{E})$ stands for the cohomology groups and $\mathrm{h}^i(X,\mathcal{E})$ for their dimension. If there is no confusion, we write $\mathrm{H}^i(\mathcal{E})$ instead of $\mathrm{H}^i(X,\mathcal{E})$ and denote by $\mathrm{h}^i(\mathcal{E})$ its dimension. In addition, we denote by $\mathrm{H}^i_*\,\mathcal{E} := \bigoplus_{t\in\mathbb{Z}} \mathrm{H}^i(\mathcal{E}(t))$.

We denote by $\mathrm{Gr}(s,n)$ the Grassmann variety (sometimes called Grassmannian) which parameterizes linear subspaces $\mathbb{P}^s$ of dimension s of $\mathbb{P}^n$ and by $\mathrm{Gr}^a(s,n)$ the Grassmann variety which parameterizes vector subspaces $\mathbf{k}^s$ of dimension s of $\mathbf{k}^n$.

Let X and Y be projective varieties. Given coherent sheaves $\mathcal{F}$ on X and $\mathcal{E}$ on Y, we set $\mathcal{F}\boxtimes\mathcal{E} := p^*\mathcal{F}\otimes q^*\mathcal{E}$ where p and q are the natural projections:

$$\begin{array}{ccccc} & & X\times Y & & \\ & \overset{p}{\swarrow} & & \overset{q}{\searrow} & \\ X & & & & Y \end{array} \tag{1}$$

Given an irreducible space V, we say that *a property P holds for a general element of V* if P holds for any element from a nonempty open set U of V. On the other hand, we say that *P holds for a very general element of V* if P holds for any element from $V\setminus T$, where $T \neq V$ is a countable union of proper closed subsets of V. Notice in particular that if P holds for a general element of V, then P also holds for a very general element of V.

https://doi.org/10.1515/9783110647686-203

1 Background

As we pointed out in the introduction, the main purpose of this first chapter is to review basic facts on vector bundles and, in particular, those results regarding the cohomology of vector bundles, Serre duality theorem, Chern classes, Riemann–Roch theorem, stability, and moduli spaces. This chapter contains only a few proofs, and we refer the reader for more details to the book of Hartshorne [108] and the book of Okonek, Schneider, and Spindler [169].

1.1 Definition and examples of vector bundles

Let us start introducing the main object of study in this book.

Definition 1.1.1. Let X be an algebraic variety. A *linear fibration* of rank r on X is an algebraic variety E and a surjective morphism $p : E \longrightarrow X$ such that for each $x \in X$, $E_x := p^{-1}(x)$ is a $\mathbf{k}$-vector space of dimension r.

Given two fibrations $p : E \longrightarrow X$ and $p' : E' \longrightarrow X$, a morphism of varieties $f : E \longrightarrow E'$ is a map of linear fibrations if it is compatible with the projections p and p'. This means that $p'f = p$, and, for each $x \in X$, the induced map $f_x : E_x \longrightarrow E'_x$ is linear. The fibration $\pi : X \times \mathbf{k}^r \longrightarrow X$ given by projection to the first factor is called the *trivial fibration* of rank r. For each open set $U \subset X$, we write $E_{|U}$ for the fibration $p^{-1}(U) \longrightarrow U$ given by restriction to U.

A *vector bundle* of rank r on X is a linear fibration $f : E \longrightarrow X$ which is locally trivial, that is, for any $x \in X$ there exists an open neighborhood U of x and an isomorphism of fibrations $\varphi : E_{|U} \longrightarrow U \times \mathbf{k}^r$. The isomorphism φ is called a *local chart* of the vector bundle E. A *line bundle* is a rank-one vector bundle.

Definition 1.1.2. Let $\varphi_i : E_{|U_i} \longrightarrow U_i \times \mathbf{k}^r$ and $\varphi_j : E_{|U_j} \longrightarrow U_j \times \mathbf{k}^r$ be local charts on the open subsets U_i and U_j, respectively. The change of chart is defined over $U_{ij} := U_i \cap U_j$ by

$$\varphi_i \circ \varphi_j^{-1} : U_{ij} \times \mathbf{k}^r \longrightarrow U_{ij} \times \mathbf{k}^r, \quad (x, v) \mapsto (x, g_{ij}(x)v),$$

where $g_{ij} : U_{ij} \longrightarrow \mathrm{GL}(r)$ is a morphism of varieties; we call it *transition function*. The transition functions verify:

(i) $g_{ii} = \mathrm{id}_{\mathbf{k}^r}$ on U_i.
(ii) $g_{it} = g_{ij}g_{jt}$ on $U_{ijt} := U_i \cap U_j \cap U_t$.

Let X be an algebraic variety and $(U_i)_{i\in I}$ an open cover of X. For any $i, j \in I$, let $g_{ij} : U_{ij} \longrightarrow \mathrm{GL}(r)$ be a morphism satisfying the above conditions (i) and (ii). We consider the quotient

$$E = \Big(\coprod_{i\in I} U_i \times \mathbf{k}^r\Big)/ \sim$$

https://doi.org/10.1515/9783110647686-001

where ~ is the equivalence relation which identifies $(x, v) \in U_i \times \mathbf{k}^r$ with $(x', v') \in U_j \times \mathbf{k}^r$ if $x = x'$ and $v' = g_{ij}(x)v$. We endow E with the quotient topology. The projection $p : E \longrightarrow X$ is a continuous map and, in addition, we have homeomorphisms $\varphi_i : E_{|U_i} \longrightarrow U_i \times \mathbf{k}^r$. In a natural way we can endow the topological quotient space E with the structure of a separated algebraic variety of finite type induced from the algebraic structure of $U_i \times \mathbf{k}^r$. Over U_{ij} the structures induced from $U_i \times \mathbf{k}^r$ and $U_j \times \mathbf{k}^r$ coincide. We can also carry over the $\mathbf{k}$-vector space structure of the fibers. Therefore, we obtain a vector bundle $E \longrightarrow X$ with local charts the above isomorphisms φ_i.

Let E and F be vector bundles of rank r and s, respectively, on X. A morphism of vector bundles $f : E \longrightarrow F$ is a morphism of the underlying linear fibrations. If $\varphi : E_{|U} \longrightarrow U \times \mathbf{k}^r$ and $\psi : F_{|U} \longrightarrow U \times \mathbf{k}^s$ are local charts for E and F over the open subset U then the local expression $\psi f \varphi^{-1} : U \times \mathbf{k}^r \longrightarrow U \times \mathbf{k}^s$ of f in the charts ψ and φ is given by $(x, v) \mapsto (x, g(x)v)$ where $g : U \longrightarrow L(\mathbf{k}^r, \mathbf{k}^s)$ is a morphism of algebraic varieties.

It is worthwhile to point out that if $f : E \longrightarrow F$ is a morphism of vector bundles on X, $x \in X$ a point such that f_x is invertible, then there exists an open neighborhood U of x such that $f_{|U} : E_{|U} \longrightarrow F_{|U}$ is an isomorphism.

Using transition functions it is possible to define in a natural way the direct sum, tensor product, the wedge or symmetric power of a vector bundle, the dual of a vector bundle, etc. Let us give as an example the direct sum of two vector bundles, and the reader can check how the transition functions are constructed in the other cases.

Let E and F be vector bundles of rank r and s, respectively, on X with local charts $\varphi_i : E_{|U_i} \longrightarrow U_i \times \mathbf{k}^r$ and $\psi_i : F_{|U_i} \longrightarrow U_i \times \mathbf{k}^s$ over the same open subset U_i. The direct sum of E and F is defined as a set by

$$E \oplus F = \coprod_{x \in X} (E_x \oplus F_x).$$

We use the bijections $(E \oplus F)_{|U_i} \longrightarrow U_i \times (\mathbf{k}^r \oplus \mathbf{k}^s)$ associated to φ_i and ψ_i to carry over the algebraic structure from $U_i \times (\mathbf{k}^r \oplus \mathbf{k}^s)$ to $(E \oplus F)_{|U_i}$. This structure is independent of the choice of charts. Indeed, over the open set U_{ij} the change of charts for $E \oplus F$ is given by the transition functions $U_{ij} \longrightarrow \mathrm{GL}(r + s)$ defined by the matrix

$$\begin{pmatrix} g_{ij} & 0 \\ 0 & h_{ij} \end{pmatrix}$$

where g_{ij} and h_{ij} are the transition functions of E and F, respectively.

Example 1.1.3. (1) Let $\mathbb{P}^n$ be the n-dimensional projective space over the field $\mathbf{k}$ and denote by $(x_0, \ldots, x_n)$ the homogeneous coordinates of a point $x \in \mathbb{P}^n$. We consider the closed subvariety $H_{-1} := \{(x, v) \in \mathbb{P}^n \times \mathbf{k}^{n+1} \mid v \in x\} \subset \mathbb{P}^n \times \mathbf{k}^{n+1}$. Over the open subset $U_i \subset \mathbb{P}^n$ defined by $x_i \neq 0$, H_{-1} is defined by the equations $v_j = \frac{x_j}{x_i} v_i$, $j = 0, \ldots, n$. The fiber $(H_{-1})_x$ of $H_{-1} \longrightarrow \mathbb{P}^n$ over a point $x \in \mathbb{P}^n$ can be identified with the line $x \subset \mathbf{k}^{n+1}$.

Therefore, we have an isomorphism of fibrations:

$$U_i \times \mathbf{k} \longrightarrow (H_{-1})_{|U_i}$$

given by $(x,t) \mapsto (x,v)$ where v is defined by $v_j = \frac{x_j}{x_i}t$ for $j = 0,\dots,n$. It immediately follows that H_{-1} is a rank-one vector bundle on $\mathbb{P}^n$. It is the so-called *tautological line bundle* and its dual is denoted by H_1.

More generally, H_0 denotes the trivial line bundle, $H_d \otimes H_e \cong H_{e+d}$ and $H_d^{-1} \cong H_{-d}$.

We observe that the transition functions of H_{-1} are $g_{ij}(x) = \frac{x_i}{x_j}$. If as a transition function we take $g_{ij}(x) = \frac{x_i^d}{x_j^d}$, we get the line bundle H_{-d} for any $d \in \mathbb{Z}$.

(2) We will define the cotangent bundle of $\mathbb{P}^n$ in terms of its transition matrices with respect to the open sets $U_i = \{(x_0,\dots,x_n) \in \mathbb{P}^n \mid x_i \neq 0\}$. We first define a differential form on U_i as a formal expression of the form:

$$\omega = f_0 d\left(\frac{x_0}{x_i}\right) + \dots + f_{i-1} d\left(\frac{x_{i-1}}{x_i}\right) + f_{i+1} d\left(\frac{x_{i+1}}{x_i}\right) + \dots + f_n d\left(\frac{x_n}{x_i}\right)$$

where $f_0,\dots,f_{i-1},f_{i+1},\dots,f_n$ are regular functions depending on the affine coordinates $\frac{x_0}{x_i},\dots,\frac{x_{i-1}}{x_i},\frac{x_{i+1}}{x_i},\dots,\frac{x_n}{x_i}$. For any $t \neq i$, we have

$$d\left(\frac{x_t}{x_i}\right) = d\left(\frac{x_t x_j}{x_i x_j}\right) = \frac{x_j}{x_i} d\left(\frac{x_t}{x_j}\right) - \frac{x_t x_j}{x_i^2} d\left(\frac{x_i}{x_j}\right).$$

Therefore,

$$\omega = \frac{x_j}{x_i}\left(f_0 d\left(\frac{x_0}{x_j}\right) + \dots + \left(-\frac{x_0}{x_i}f_0 - \dots - \frac{x_{i-1}}{x_i}f_{i-1} - \frac{x_{i+1}}{x_i}f_{i+1} - \dots - \frac{x_n}{x_i}f_n\right) d\left(\frac{x_i}{x_j}\right) + \dots + f_n d\left(\frac{x_n}{x_j}\right)\right).$$

In other words, the transition matrix from U_i to U_j is given by

$$A_{ij} = \frac{x_j}{x_i}\begin{pmatrix} 1 & 0 & \cdots & 0 & \cdots & 0 \\ 0 & 1 & \cdots & 0 & \cdots & 0 \\ \vdots & \vdots & & \vdots & & \vdots \\ -\frac{x_0}{x_i} & -\frac{x_1}{x_i} & \cdots & -\frac{x_j}{x_i} & \cdots & -\frac{x_n}{x_i} \\ \vdots & \vdots & & \vdots & & \vdots \\ 0 & 0 & \cdots & 0 & \cdots & 1 \end{pmatrix}.$$

The rank n vector bundle on $\mathbb{P}^n$ associated to the transition matrix is called the *cotangent bundle* and denoted by $\Omega^1_{\mathbb{P}^n}$. It fits into the so-called Euler exact sequence:

$$0 \longrightarrow \Omega^1_{\mathbb{P}^n} \longrightarrow \mathcal{O}_{\mathbb{P}^n}(-1)^{n+1} \longrightarrow \mathcal{O}_{\mathbb{P}^n} \longrightarrow 0. \tag{1.1}$$

Its dual is the *tangent bundle* $T_{\mathbb{P}^n}$. The p-exterior power $\Omega^p_{\mathbb{P}^n} := \wedge^p \Omega^1_{\mathbb{P}^n}$ is the vector bundle of p-differential forms which has rank $\binom{n}{p}$.

The values of $\mathrm{h}^q(\mathbb{P}^n, \Omega^p_{\mathbb{P}^n}(k))$ are given by the *Bott formula*:

$$\mathrm{h}^q(\mathbb{P}^n, \Omega^p_{\mathbb{P}^n}(k)) = \begin{cases} \binom{k+n-p}{k}\binom{k-1}{p} & \text{for } q = 0, 0 \le p \le n, k > p, \\ 1 & \text{for } k = 0, 0 \le p = q \le n, \\ \binom{-k+p}{-k}\binom{-k-1}{n-p} & \text{for } q = n, 0 \le p \le n, k < p - n, \\ 0 & \text{otherwise.} \end{cases} \tag{1.2}$$

Definition 1.1.4. Let $p : E \longrightarrow X$ be a rank r vector bundle on X. A *subbundle* of rank s of E is a subvariety $F \subset E$ such that for any point $x \in X$, $F \cap E_x \subset E_x$ is a vector subspace of dimension s and the induced fibration $p_{|F} : F \longrightarrow X$ is locally trivial.

Given a vector subbundle $F \subset E$, we consider the family of vector bundles on X:

$$E/F = \coprod_{x \in X} (E_x/F_x).$$

Any map of vector bundles $f : E \longrightarrow G$ which vanishes on F factorizes uniquely through a map $\overline{f} : E/F \longrightarrow G$. Even more, there exists a unique vector bundle structure of $E/F \longrightarrow X$ which satisfies the following universal property. For any morphism $f : E \longrightarrow G$ which vanishes on F, the map $\overline{f}$ is a map of vector bundles. The vector bundle E/F is called the *quotient bundle*.

Example 1.1.5. Let $f : E \longrightarrow F$ be a morphism of vector bundles on X. Assume that for any point $x \in X$ the rank of f_x is constant equal to n. Then $\ker(f)$ and $\mathrm{im}(f)$ are subbundles of E and F, respectively.

1.2 Vector bundles vs locally free sheaves

Definition 1.2.1. Let $p : E \longrightarrow X$ be a vector bundle of rank r over an algebraic variety X. We define a *regular section* of E over an open subset $U \subset X$ to be a morphism $s : U \longrightarrow E$ of algebraic varieties such that $ps = \mathrm{Id}_U$.

The set $\Gamma(U, E)$ of regular sections of E over U is an $\mathcal{O}_X(U)$-module. Indeed, for any $s, t \in \Gamma(U, E)$ and $\alpha \in \mathcal{O}_X(U)$, we define $(s + t)(x) = s(x) + t(x)$ and $(\alpha s)(x) = \alpha(x)s(x)$. So we obtain a sheaf of $\mathcal{O}_X$-modules $\mathcal{E} = \mathcal{O}_X(E)$ over X, locally isomorphic to $\mathcal{O}_X^r$. This means that $\mathcal{E}$ is a locally free sheaf of rank r.

Proposition 1.2.2. *The functor which associates a locally free sheaf $\mathcal{E} = \mathcal{O}_X(E)$ on X to a vector bundle E on X is an equivalence of categories between the category of rank r vector bundles over X and the category of rank r locally free sheaves on X.*

Proof. See [108, Chapter II, Exercise 5.18]. □

Remark 1.2.3. From now on, we will not distinguish between a vector bundle E on X and its locally free sheaf $\mathcal{E}$ of sections and we will also denote with calligraphic letters all vector bundles. For instance, via the identification of vector bundles and locally free sheaves, the vector bundle H_d introduced in Example 1.1.3 (1) is the line bundle associated to the invertible sheaf $\mathcal{O}_{\mathbb{P}^n}(d)$.

1.3 Theorems A and B of Serre and Serre duality

In this section, we summarize the basic results about cohomology of coherent sheaves needed in the sequel. We refer the reader to [108, Chapter III] for more details.

Recall that a sheaf $\mathcal{F}$ on an algebraic variety X with structure sheaf $\mathcal{O}_X$ is said to be *coherent* if for any $x \in X$ there exists an open neighborhood U of x, locally free sheaves $\mathcal{O}_U^r$ and $\mathcal{O}_U^s$, and an exact sequence of sheaves over U,

$$\mathcal{O}_U^r \longrightarrow \mathcal{O}_U^s \longrightarrow \mathcal{F}_{|U} \longrightarrow 0.$$

If X is a projective variety embedded in $\mathbb{P}^n$ and $\mathcal{O}_X(1)$ is the restriction to X of $\mathcal{O}_{\mathbb{P}^n}(1)$, we set $\mathcal{F}(n) = \mathcal{F} \otimes_{\mathcal{O}_X} \mathcal{O}_X(n)$ for any coherent sheaf $\mathcal{F}$ on X.

Theorem 1.3.1. *Let $X \subset \mathbb{P}^N$ be a projective variety and let $\mathcal{F}$ be a coherent sheaf. It holds:*
(1) $\mathrm{H}^i(X, \mathcal{F}) = 0$ *if* $i > \dim X$.
(2) *(Theorem A of Serre) There exists $n_0 \in \mathbb{Z}$ such that for all $n \geq n_0$, $\mathcal{F}(n)$ is generated by global sections.*
(3) $\mathrm{H}^i(X, \mathcal{F})$ *is a* **k***-vector space of finite dimension.*
(4) *(Theorem B of Serre) There exists $n_0 \in \mathbb{Z}$ such that for all $i > 0$ and all $n \geq n_0$,* $\mathrm{H}^i(X, \mathcal{F}(n)) = 0$.

Proof. (1) This vanishing theorem is due to Grothendieck, and the reader can see [108, Chapter III, Theorem 2.7] for a proof.

(2) See [108, Chapter II, Theorem 5.17] for the proof of Theorem A of Serre.

(3)–(4) For the finiteness theorem and Theorem B of Serre, the reader can see [108, Chapter III, Theorem 5.2]. □

Let us now introduce Serre's duality.

Theorem 1.3.2. *Let X be a smooth irreducible projective variety of dimension n and let ω_X be the canonical line bundle of X. Let $\mathcal{E}$ be a locally free sheaf on X. Then, for any $1 \leq i \leq n$,*

$$\mathrm{H}^i(X, \mathcal{E}^\vee \otimes \omega_X) \cong \mathrm{H}^{n-i}(X, \mathcal{E})^\vee.$$

Proof. See [108, Chapter III, Theorem 7.6]. □

Definition 1.3.3. Let $\mathcal{F}$ be a coherent sheaf on a projective variety $X \subset \mathbb{P}^N$ of dimension n. We say that $\mathcal{F}$ is *m-regular* if $\mathrm{H}^i(X, \mathcal{F}(m-i)) = 0$ for $i \geq 1$. We say that $\mathcal{F}$ is *minimal regular* if $\mathcal{F}$ is 0-regular but not (−1)-regular.

It is worthwhile to point out that

(1) If $\mathcal{F}$ is a 0-regular sheaf then $\mathcal{F}$ is globally generated;
(2) If $\mathcal{F}$ is m-regular then $\mathcal{F}$ if p-regular for any $p \geq m$.

We define the *Euler–Poincaré* characteristic of $\mathcal{F}$ as

$$\chi(X, \mathcal{F}) := \sum_{i \geq 0} (-1)^i \, \mathrm{h}^i(\mathcal{F}),$$

and the *Hilbert polynomial* of $\mathcal{F}$ as (see [119, Lemma 1.2.1])

$$\mathrm{HP}_{\mathcal{F}}(t) := \chi(X, \mathcal{F}(t)) \in \mathbb{Q}[t].$$

It follows from Serre's duality (see Theorem 1.3.2) that for any locally free sheaf $\mathcal{E}$ on a smooth irreducible projective variety X of dimension n,

$$\chi(X, \mathcal{E}) = (-1)^n \chi(X, \mathcal{E}^\vee \otimes \omega_X)$$

where ω_X is the canonical line bundle on X.

In the next section, we will compute the Euler–Poincaré characteristic of a coherent sheaf in terms of its Chern classes.

1.4 Chern classes

In this section, we outline the theory of Chern classes, and the reader can look at [105] and [116] for more details.

Let X be a smooth irreducible projective variety of dimension n. By a *cycle* on X we mean an element of the free abelian group generated by all closed irreducible subvarieties of X. If all the components of a cycle have codimension r, we say that the cycle has codimension r. There is an equivalence relation, called *rational equivalence*, on codimension r cycles which coincides with linear equivalence for $r = 1$. The group of equivalence classes of cycles of codimension r is denoted by $A^r(X)$. There exists an intersection pairing $A^r(X) \times A^s(X) \longrightarrow A^{r+s}(X)$ making

$$A(X) := \bigoplus_{i=0}^{n} A^i(X)$$

into a commutative graded ring, called the *Chow ring* of X.

If $\mathcal{E}$ is a vector bundle on X, we can define certain classes $c_s(\mathcal{E}) \in A^s(X)$, $0 \leq s \leq n$, called the *Chern classes of* $\mathcal{E}$ such that

$$c_0(\mathcal{E}) = 1 \quad \text{and} \quad c_s(\mathcal{E}) = 0 \quad \text{for } s > \operatorname{rank}(\mathcal{E}).$$

By the *total Chern class* of $\mathcal{E}$ we mean

$$c(\mathcal{E}) = c_0(\mathcal{E}) + c_1(\mathcal{E}) + c_2(\mathcal{E}) + \cdots + c_n(\mathcal{E}) \in A(X)$$

and by the *Chern polynomial* we mean

$$c_t(\mathcal{E}) = 1 + c_1(\mathcal{E})t + c_2(\mathcal{E})t^2 + \cdots + c_n(\mathcal{E})t^n.$$

One can show that this theory of Chern classes is uniquely determined by the following three properties:

(C1) If $\mathcal{L}$ is a line bundle on X, then it is of the form $\mathcal{L} = \mathcal{O}_X(D)$ for some divisor D on X and then $c_1(\mathcal{L})$ is the class of D in $A^1(X)$. In particular, $c_1(\mathcal{L}_1 \otimes \mathcal{L}_2) = c_1(\mathcal{L}_1) + c_1(\mathcal{L}_2)$ and $c_1(\mathcal{L}^\vee) = -c_1(\mathcal{L})$.

(C2) If

$$0 \longrightarrow \mathcal{E} \longrightarrow \mathcal{F} \longrightarrow \mathcal{G} \longrightarrow 0$$

is an exact sequence of vector bundles on X, then $c_t(\mathcal{F}) = c_t(\mathcal{E}) \cdot c_t(\mathcal{G})$.

(C3) If $f : X \longrightarrow Y$ is a morphism of projective varieties and $\mathcal{E}$ a vector bundle on Y then

$$c(f^*\mathcal{E}) = f^*c(\mathcal{E}).$$

Example 1.4.1. If X is the projective space $\mathbb{P}^n$, then $A(X) = \mathbb{Z}[h]/(h^{n+1}) \cong \mathbb{Z}$ where $h \in A^1(X)$ is the class of a hyperplane. So we can view the Chern classes of a vector bundle $\mathcal{E}$ over $\mathbb{P}^n$ as integers. In particular, $c_1(\mathcal{O}_{\mathbb{P}^n}(s)) = s$. Using Euler sequence (1.1)

$$0 \longrightarrow \Omega^1_{\mathbb{P}^n} \longrightarrow \mathcal{O}_{\mathbb{P}^n}(-1)^{n+1} \longrightarrow \mathcal{O}_{\mathbb{P}^n} \longrightarrow 0,$$

we can compute the Chern polynomial of $\Omega^1_{\mathbb{P}^n}$ and get $c_t(\Omega^1_{\mathbb{P}^n}) = (1 - ht)^{n+1}$.

There are also formalisms for computing the Chern classes of exterior powers, tensor products, etc., all of them based on the following *splitting principle*. Given a vector bundle $\mathcal{E}$ on X, there exists a morphism $f : Y \longrightarrow X$ such that $f^* : A(X) \longrightarrow A(Y)$ is injective and $\mathcal{E}' := f^*\mathcal{E}$ splits. This means that $\mathcal{E}'$ has a filtration

$$\mathcal{E}' = \mathcal{E}'_0 \supseteq \mathcal{E}'_1 \supseteq \cdots \supseteq \mathcal{E}'_r = 0$$

whose successive quotients are all invertible sheaves. It holds:

(C4) If $\mathcal{E}$ splits and the filtration has the invertible sheaves $\mathcal{L}_1, \ldots, \mathcal{L}_r$ as quotients then

$$c_t(\mathcal{E}) = \prod_{i=1}^{r} c_t(\mathcal{L}_i).$$

(C5) If $\mathcal{E}$ and $\mathcal{F}$ are vector bundles of rank r and s on X, respectively, and we formally write

$$c_t(\mathcal{E}) = \prod_{i=1}^{r}(1 + a_i t) \quad \text{and} \quad c_t(\mathcal{F}) = \prod_{i=1}^{s}(1 + b_i t),$$

then we have

$$c_t(\mathcal{E} \otimes \mathcal{F}) = \prod_{i,j}(1 + (a_i + b_j)t),$$

$$c_t(\wedge^p \mathcal{E}) = \prod_{1 \le i_1 < \cdots < i_p \le r} (1 + (a_{i_1} + \cdots + a_{i_p})t).$$

(C6) Let $\mathcal{E}$ be a rank r vector bundle on X, $s \in \Gamma(X, \mathcal{E})$ a global section, and $Y \subset X$ the scheme of zeros of s. We also denote by Y the associated cycle of X of codimension r. Then, if Y has codimension r, $c_r(\mathcal{E}) = Y$ in $A^r(X)$.

Example 1.4.2. If $\mathcal{E}$ is a rank r vector bundle on a smooth projective variety X of dimension n and $\mathcal{L}$ is a line bundle on X, then

$$c_k(\mathcal{E} \otimes \mathcal{L}) = \sum_{i=0}^{k} \binom{r-i}{k-i} c_i(\mathcal{E}) c_1(\mathcal{L})^{k-i}.$$

We end this section with Riemann–Roch theorem which allows us to express the Euler–Poincaré characteristic of a vector bundle in terms of its Chern classes. Given a rank r vector bundle $\mathcal{E}$ on X, we define the *Todd class* of $\mathcal{E}$ as follows:

$$\mathrm{td}(\mathcal{E}) = \prod_{i=1}^{r} \frac{a_i}{1 - e^{-a_i}} = 1 + \frac{1}{2}a_i + \frac{1}{12}a_i^2 + \cdots.$$

The Riemann–Roch theorem can be formulated as follows.

Theorem 1.4.3. *Let $\mathcal{E}$ be a rank r vector bundle on a smooth projective variety X of dimension n. Then*

$$\chi(\mathcal{E}) = \sum_{i=0}^{n} (-1)^i \dim \mathrm{H}^i(X, \mathcal{E}) = \deg((\mathrm{ch}(\mathcal{E})\mathrm{td}(T_X))_n) \tag{1.3}$$

where $(\cdot)_n$ denotes the component of degree n in $A(X) \otimes \mathbb{Q}$ and T_X the tangent bundle on X.

In particular, if X is a curve and d is the degree of a rank r vector bundle $\mathcal{E}$ on X, we have

$$\chi(\mathcal{E}) = \dim \mathrm{H}^0(X, \mathcal{E}) - \dim \mathrm{H}^1(X, \mathcal{E}) = r(1 - g(X)) + d \tag{1.4}$$

where $g(X)$ stands for the genus of X.

If X is a surface and c_1, c_2 are the Chern classes of a rank r vector bundle $\mathcal{E}$ on X, we have

$$\chi(\mathcal{E}) = \sum_{i=0}^{2} (-1)^i \dim \mathrm{H}^i(X, \mathcal{E}) = r(1 + p_a(X)) - \frac{c_1 K}{2} + \frac{c_1^2 - 2c_2}{2} \tag{1.5}$$

where K denotes the class of the canonical divisor on X and $p_a(X) := \chi(\mathcal{O}_X) - 1$ the arithmetic genus of X.

Remark 1.4.4. The Chern polynomial $c_t(\mathcal{E})$ of a vector bundle $\mathcal{E}$ on a smooth projective variety X can be defined on the Grothendieck group $K(X)$ of vector bundles on X (see [108, p. 435]). For smooth varieties, $K(X)$ coincides with the Grothendieck group $K^0(X)$ of all coherent sheaves on X (see [108, III, Exercise 6.9]). Therefore $c_t(\mathcal{F})$ is defined for all coherent sheaves $\mathcal{F}$ on X. For a rank r vector bundle $\mathcal{E}$ on X, we know that $c_i(\mathcal{E}) = 0$ for $i > r$, but this does not need to hold for coherent sheaves.

1.5 Stability and moduli spaces

Roughly speaking, moduli spaces are schemes whose closed points are in bijection with sets of equivalence classes of certain objects. In this book, given a projective variety $X \subset \mathbb{P}^n$, we are interested in parameterizing vector bundles on X. More precisely, we would like to endow the set of vector bundles on X with a natural algebraic structure but, in general, we cannot expect that such structure exists even if we fix the rank $r \in \mathbb{Z}$ and the Chern classes $c_i \in A^i(X)$. The necessity of restricting to some subset of the set of all vector bundles can be understood by the following well-known example.

Example 1.5.1. We consider the set of vector bundles $\{\mathcal{O}_{\mathbb{P}^1}(n) \oplus \mathcal{O}_{\mathbb{P}^1}(-n)\}_{n\in\mathbb{Z}}$ over $\mathbb{P}^1$. Since $\operatorname{rank}(\mathcal{O}_{\mathbb{P}^1}(n)\oplus\mathcal{O}_{\mathbb{P}^1}(-n)) = 2$ and $c_1(\mathcal{O}_{\mathbb{P}^1}(n)\oplus\mathcal{O}_{\mathbb{P}^1}(-n)) = 0$, the rank and the Chern classes are fixed, but we have infinitely many points which cannot form a nice algebraic variety. Moreover, for any $n > 0$, there exists a family of vector bundles $\{\mathcal{E}_t\}_{t\in\mathbf{k}}$ parameterized by an affine line inside $\operatorname{Ext}^1(\mathcal{O}_{\mathbb{P}^1}(n), \mathcal{O}_{\mathbb{P}^1}(-n)) \cong \mathrm{H}^1(\mathcal{O}_{\mathbb{P}^1}(-2n)) \cong \mathbf{k}^{2n-1}$ such that $\mathcal{E}_t \cong \mathcal{O}_{\mathbb{P}^1} \oplus \mathcal{O}_{\mathbb{P}^1}$ for $t \neq 0$ and $\mathcal{E}_0 \cong \mathcal{O}_{\mathbb{P}^1}(n) \oplus \mathcal{O}_{\mathbb{P}^1}(-n)$. So, the point corresponding to $\mathcal{O}_{\mathbb{P}^1} \oplus \mathcal{O}_{\mathbb{P}^1}$ would not be a closed point in the moduli space.

The natural class of vector bundles which admits a nice natural algebraic structure comes from Mumford's Geometric Invariant Theory. The corresponding vector bundles are called stable vector bundles. Let us recall their definition.

Definition 1.5.2. Let X be a smooth irreducible projective variety of dimension d and let H be an ample divisor on X. For a rank r torsion-free sheaf $\mathcal{F}$ on X, one sets

$$\deg_H(\mathcal{F}) = \deg(\mathcal{F}) =: c_1(\mathcal{F})H^{d-1}.$$

In particular, $\deg(\mathcal{F}) = \deg(\det(\mathcal{F})) := \deg(\bigwedge^r \mathcal{F})$.

We define the *slope* of $\mathcal{F}$ (with respect to H) as

$$\mu_H(\mathcal{F}) = \frac{\deg_H(\mathcal{F})}{\operatorname{rank}(\mathcal{F})} = \frac{c_1(\mathcal{F})H^{d-1}}{\operatorname{rank}(\mathcal{F})}$$

and the *reduced Hilbert polynomial* of $\mathcal{F}$ as

$$P_{\mathcal{F}}(m) := \frac{\chi(\mathcal{F} \otimes \mathcal{O}_X(mH))}{\operatorname{rank}(\mathcal{F})} = \frac{\mathrm{HP}_{\mathcal{F}}(m)}{\operatorname{rank}(\mathcal{F})}.$$

Example 1.5.3. The slope of the cotangent bundle $\Omega^1_{\mathbb{P}^n}$ with respect to $\mathcal{O}_{\mathbb{P}^n}(1)$ is $\frac{-n-1}{n}$.

Remark 1.5.4. (1) Dealing with a smooth projective curve X, $\mathcal{O}_X(H)$ an ample line bundle of degree k on X, and $\mathcal{F}$ a rank r torsion-free sheaf on X, we identify $c_1(\mathcal{F})$ with $\deg(\mathcal{F})$ and, in particular, $c_1(\mathcal{F}(nH)) = c_1(\mathcal{F}) + rnk = \deg(\mathcal{F}) + rnk$.

(2) For torsion-free sheaves $\mathcal{F}$ and $\mathcal{G}$ on X, from the definition of slope and the properties of Chern classes it follows that

$$\mu_H(\mathcal{F} \otimes \mathcal{G}) = \mu_H(\mathcal{F}) + \mu_H(\mathcal{G}).$$

Definition 1.5.5. Let X be a smooth irreducible projective variety of dimension d and let H be an ample divisor on X. A torsion-free sheaf $\mathcal{F}$ on X is *μ-semistable* (respectively *semistable*) with respect to the ample divisor H if and only if

$$\mu_H(\mathcal{E}) \leq \mu_H(\mathcal{F}) \quad (\text{respectively } P_{\mathcal{E}}(m) \leq P_{\mathcal{F}}(m) \text{ for } m \gg 0)$$

for all nonzero subsheaves $\mathcal{E} \subset \mathcal{F}$ with $\mathrm{rank}(\mathcal{E}) < \mathrm{rank}(\mathcal{F})$; if strict inequality holds then $\mathcal{F}$ is *μ-stable* (respectively *stable*) with respect to H.

The notion of μ-stability was introduced for vector bundles on curves by Mumford and later generalized to sheaves on higher dimensional varieties by Takemoto, Gieseker, Maruyama, and Simpson.

The definition of stability depends on the choice of the ample divisor H. We will simply say μ-(semi)stable (resp (semi)stable) when there is no confusion on H. One can check the implications

$$\mu\text{-stable} \Rightarrow \text{stable} \Rightarrow \text{semistable} \Rightarrow \mu\text{-semistable}.$$

Example 1.5.6. As an example of μ-stable bundle, we have $\Omega^1_{\mathbb{P}^n}$. An example of μ-semistable but not μ-stable vector bundle is $\mathcal{O}_{\mathbb{P}^n} \oplus \mathcal{O}_{\mathbb{P}^n}$. Finally, for a general morphism $\phi : \mathcal{O}_{\mathbb{P}^3}(-1)^2 \longrightarrow \mathcal{O}^4_{\mathbb{P}^3} \oplus \mathcal{O}_{\mathbb{P}^3}(2)$, $\mathcal{E} := \mathrm{coker}(\phi)$ is a rank 3 not μ-semistable vector bundle on $\mathbb{P}^3$.

Definition 1.5.7. A vector bundle $\mathcal{E}$ on a projective variety X is said to be *simple* if its only endomorphisms are the homotheties, i. e., $\mathrm{End}(\mathcal{E}) \cong \mathrm{Hom}(\mathcal{E}, \mathcal{E}) \cong \mathrm{H}^0(X, \mathcal{E} \otimes \mathcal{E}^\vee) = \mathbf{k}$.

Example 1.5.8. As a first example of simple vector bundle, we have the tangent bundle $T_{\mathbb{P}^n}$. Indeed, we consider the Euler sequence

$$0 \longrightarrow \mathcal{O}_{\mathbb{P}^n}(-1) \longrightarrow \mathcal{O}^{n+1}_{\mathbb{P}^n} \longrightarrow T_{\mathbb{P}^n}(-1) \longrightarrow 0.$$

We tensor it with $\Omega^1_{\mathbb{P}^n}(1)$ and get

$$0 \longrightarrow \Omega^1_{\mathbb{P}^n} \longrightarrow \Omega^1_{\mathbb{P}^n}(1)^{n+1} \longrightarrow \Omega^1_{\mathbb{P}^n} \otimes T_{\mathbb{P}^n} \longrightarrow 0.$$

Taking cohomology and using Bott's formula, we obtain

$$\mathrm{H}^0(\Omega^1_{\mathbb{P}^n} \otimes T_{\mathbb{P}^n}) \cong \mathrm{H}^1(\Omega^1_{\mathbb{P}^n}) = \mathbf{k}.$$

Therefore, $T_{\mathbb{P}^n}$ is simple.

Remark 1.5.9. (1) A decomposable vector bundle $\mathcal{E} \oplus \mathcal{F}$ on a projective variety X certainly has a nontrivial endomorphism. Therefore, simple vector bundles are always indecomposable but, in general, the converse is not true.

(2) Any μ-stable vector bundle $\mathcal{E}$ is simple and hence indecomposable.

(3) We will see in Chapter 3 that all Ulrich bundles are μ-semistable but this does not need to hold for aCM bundles, that is, for vector bundles without intermediate cohomology. Indeed, $\mathcal{O}^4_{\mathbb{P}^3} \oplus \mathcal{O}_{\mathbb{P}^3}(2)$ is an aCM bundle on $\mathbb{P}^3$ and it is not μ-semistable.

We have the following property relating μ-stability and simplicity.

Proposition 1.5.10. *Let X be a smooth projective variety and*

$$0 \longrightarrow \mathcal{E} \longrightarrow \mathcal{F} \longrightarrow \mathcal{G} \longrightarrow 0$$

be a nonsplitting short exact sequence of vector bundles where $\mathcal{E}$ and $\mathcal{G}$ are μ-stable of the same slope. Then, $\mathcal{F}$ is a simple vector bundle.

Proof. See [41, Lemma 4.2]. □

To go further, we need to recall the notion of Jordan–Hölder filtration.

Definition 1.5.11. Let $\mathcal{E}$ be a rank r semistable sheaf on a d-dimensional projective variety $X \subset \mathbb{P}^n$. A *Jordan–Hölder filtration* of $\mathcal{E}$ is a filtration

$$0 = \mathcal{E}_0 \subset \mathcal{E}_1 \subset \cdots \subset \mathcal{E}_l = \mathcal{E}$$

such that the factors $\mathcal{E}_i/\mathcal{E}_{i-1}$ are stable with reduced Hilbert polynomial $P_{\mathcal{E}}$.

It must be mention that Jordan–Hölder filtrations always exist (see [119, Proposition 1.5.2]).

We are interested in providing the set of isomorphic classes of μ-stable vector bundles with fixed rank and Chern classes on a smooth, irreducible, projective variety with a natural structure of scheme. This leads us to consider the following contravariant moduli functor.

Let X be a smooth, irreducible, projective variety of dimension n and let H be an ample divisor on X. For a fixed polynomial $P \in \mathbb{Q}[z]$, we consider the contravariant moduli functor $\mathcal{M}_X^{s,H,P}(-)$:

$$\begin{aligned} \mathcal{M}_X^{s,H,P}(-) : (\mathrm{Sch}/\mathbf{k}) &\longrightarrow (\mathrm{Sets}) \\ S &\longmapsto \mathcal{M}_X^{s,H,P}(S), \end{aligned}$$

where $\mathcal{M}_X^{s,H,P}(S) = \{S\text{-flat families } \mathbb{J} \longrightarrow X \times S$ of vector bundles on X all whose fibers are μ-stable with respect to H and have Hilbert polynomial $P\}/\sim$, with $\mathbb{J} \sim \mathbb{J}'$ if and only if $\mathbb{J} \cong \mathbb{J}' \otimes p^*\mathcal{L}$ for some $\mathcal{L} \in \mathrm{Pic}(S)$, and $p : S \times X \longrightarrow S$ being the natural projection. In 1977, Maruyama proved the following result.

Theorem 1.5.12. *Let X be a smooth, irreducible, projective variety of dimension n and let H be an ample divisor on X. Then, the moduli functor $\mathcal{M}_X^{s,H,P}(-)$ has a coarse moduli scheme $M_X^{s,H,P}$ which is a separated scheme and locally of finite type over* $\mathbf{k}$. *This means that*

(1) *There is a natural transformation*

$$\Psi : \mathcal{M}_X^{s,H,P}(-) \longrightarrow \mathrm{Hom}(-, M_X^{s,H,P}),$$

which is bijective for any reduced point x_0.

(2) *For any scheme N and any natural transformation*

$$\Phi : \mathcal{M}_X^{s,H,P}(-) \longrightarrow \mathrm{Hom}(-, N),$$

there is a unique morphism $\varphi : M_X^{s,H,P} \longrightarrow N$ for which the diagram

$$\begin{array}{ccc} \mathcal{M}_X^{s,H,P}(-) & \xrightarrow{\quad\Psi\quad} & \mathrm{Hom}(-, M_X^{s,H,P}) \\ & {\scriptstyle\Phi}\searrow \quad \swarrow{\scriptstyle\varphi_*} & \\ & \mathrm{Hom}(-, N) & \end{array}$$

commutes.

In addition, $M_X^{s,H,P}$ decomposes into a disjoint union of schemes $M_{X,H}^s(r; c_1, \ldots, c_{\min(r,n)})$ and $M_{X,H}^s(r; c_1, \ldots, c_{\min(r,n)})$ is the moduli space of rank r μ-stable, with respect to H, vector bundles on X with Chern classes $(c_1, \ldots, c_{\min(r,n)})$ up to numerical equivalence.

Proof. See [151, Theorem 5.6] and [152]. □

Definition 1.5.13. If the natural transformation Ψ from Theorem 1.5.12 is an isomorphism, then it is said that the moduli functor $\mathcal{M}_X^{s,H,P}(-)$ is *representable* and that $M_X^{s,H,P}$ is a *fine moduli space.*

Remark 1.5.14. (1) If a coarse moduli space exists for a given classification problem, then it is unique (up to isomorphism). So $M_X^{s,H,P}$ is unique (up to isomorphism).

(2) A fine moduli space for a given classification problem is always a coarse moduli space for this problem but, in general, not vice versa. In fact, there is no a priori reason why the map $\Psi(S) : \mathcal{M}_X^{s,H,P}(S) \to \mathrm{Hom}(S, M_X^{s,H,P})$ should be bijective for varieties S other than a point $\{pt\}$.

(3) Analogously, we define $\overline{M}_{X,H}(r; c_1, \dots, c_{\min(r,n)})$ as the moduli space of S-equivalence classes of semistable (with respect to H) rank r sheaves on X with Chern classes $(c_1, \dots, c_{\min(r,n)})$ up to numerical equivalence.

We refer to [119, Section 4.5] for general facts on the infinitesimal structure of the moduli space $M^s_{X,H}(r; c_1, \dots, c_{\min(r,n)})$. Let us recall the results which are basic for this book.

Proposition 1.5.15. *Let X be a smooth, irreducible, projective variety of dimension n and let $\mathcal{E}$ be a μ-stable vector bundle on X with Chern classes $c_i(\mathcal{E}) = c_i \in A^i(X)$, represented by a point $[\mathcal{E}] \in M^s_{X,H}(r; c_1, \dots, c_{\min(r,n)})$. Then, the Zariski tangent space of $M^s_{X,H}(r; c_1, \dots, c_{\min(r,n)})$ at $[\mathcal{E}]$ is canonically given by*

$$T_{[\mathcal{E}]} M^s_{X,H}(r; c_1, \dots, c_{\min(r,n)}) \cong \operatorname{Ext}^1(\mathcal{E}, \mathcal{E}).$$

If $\operatorname{Ext}^2(\mathcal{E}, \mathcal{E}) = 0$ then $M^s_{X,H}(r; c_1, \dots, c_{\min(r,n)})$ is smooth at $[\mathcal{E}]$. In general, we have the following bounds:

$$\begin{aligned} \dim \operatorname{Ext}^1(\mathcal{E}, \mathcal{E}) &\geq \dim_{[\mathcal{E}]} M^s_{X,H}(r; c_1, \dots, c_{\min(r,n)}) \\ &\geq \dim \operatorname{Ext}^1(\mathcal{E}, \mathcal{E}) - \dim \operatorname{Ext}^2(\mathcal{E}, \mathcal{E}). \end{aligned}$$

Proof. See, for instance [119, Corollary 4.5.2]. □

Indeed, the estimates from the previous proposition can be improved. For this, notice that, by the universal property of the moduli space, there exists a map

$$M^s_{X,H}(r; c_1, \dots, c_{\min(r,n)}) \to \operatorname{Pic}(X)$$

sending the class $[\mathcal{E}]$ of a vector bundle to the determinant line bundle $\det(\mathcal{E}) := \bigwedge^r \mathcal{E} \in \operatorname{Pic}(X)$. On the other hand, the trace map $\operatorname{tr} : \mathcal{E}nd(\mathcal{E}) \to \mathcal{O}_X$ induces maps $\operatorname{tr}_i : \operatorname{Ext}^i(\mathcal{E}, \mathcal{E}) \to \mathrm{H}^i(\mathcal{O}_X)$. Let us denote by $\operatorname{Ext}^i_0(\mathcal{E}, \mathcal{E})$ the kernel of tr_i. Then we have:

Proposition 1.5.16. *Let X be a smooth, irreducible, projective variety of dimension n and let $\mathcal{E}$ be a μ-stable vector bundle on X with Chern classes $c_i(\mathcal{E}) = c_i \in A^i(X)$, represented by a point $[\mathcal{E}] \in M^s_{X,H}(r; c_1, \dots, c_{\min(r,n)})$. Let $\mathcal{Q} \cong \bigwedge^r \mathcal{E}$ be its determinant line bundle. Let $M(\mathcal{Q})$ be the fiber of the map*

$$M^s_{X,H}(r; c_1, \dots, c_{\min(r,n)}) \to \operatorname{Pic}(X)$$

at the point $[\mathcal{Q}]$. Then, the Zariski tangent space $T_{[\mathcal{E}]} M(\mathcal{Q}) \cong \operatorname{Ext}^1_0(\mathcal{E}, \mathcal{E})$. If $\operatorname{Ext}^2_0(\mathcal{E}, \mathcal{E}) = 0$ then $M^s_{X,H}(r; c_1, \dots, c_{\min(r,n)})$ and $M(\mathcal{Q})$ are smooth at $[\mathcal{E}]$. In general, we have the following bounds:

$$\dim \operatorname{Ext}^1_0(\mathcal{E}, \mathcal{E}) \geq \dim_{[\mathcal{E}]} M(\mathcal{Q}) \geq \dim \operatorname{Ext}^1_0(\mathcal{E}, \mathcal{E}) - \dim \operatorname{Ext}^2_0(\mathcal{E}, \mathcal{E}).$$

Proof. See [119, Theorem 4.5.4]. □

Remark 1.5.17. (1) In case X is a smooth projective curve, we can make the above dimension bounds more explicit. Indeed, for any μ-stable vector bundle $\mathcal{E}$ on X, $\mathrm{Ext}^2_0(\mathcal{E},\mathcal{E}) = 0$. Thus, the moduli space of μ-stable locally free sheaves of rank r and fixed determinant bundle on X is smooth and its dimension is given by

$$\dim \mathrm{Ext}^1_0(\mathcal{E},\mathcal{E}) = -\chi(\mathcal{E}\otimes\mathcal{E}^\vee) + \chi(\mathcal{O}_X) = (r^2-1)(g-1)$$

where the last equality follows from Riemann–Roch theorem.

(2) In case X is a smooth surface, we can also make the above dimension bounds more explicit. Given a rank r vector bundle $\mathcal{E}$ on X with $c_i(\mathcal{E}) = c_i$, define the *discriminant* of $\mathcal{E}$ by

$$\Delta(\mathcal{E}) = 2rc_2 - (r-1)c_1^2.$$

Then, for any μ-stable vector bundle $\mathcal{E} \in M^s_{X,H}(r; c_1, c_2)$, we have

$$\begin{aligned} &\dim \mathrm{Ext}^1_0(\mathcal{E},\mathcal{E}) - \dim \mathrm{Ext}^2_0(\mathcal{E},\mathcal{E}) \\ &= \chi(\mathcal{O}_X) - \sum_{i=0}^{2}(-1)^i \dim \mathrm{Ext}^i_0(\mathcal{E},\mathcal{E}) = \Delta(\mathcal{E}) - (r^2-1)\chi(\mathcal{O}_X) \\ &= 2rc_2 - (r-1)c_1^2 - (r^2-1)\chi(\mathcal{O}_X) \end{aligned}$$

where the last equality follows from Riemann–Roch theorem. The number $2rc_2 - (r-1)c_1^2 - (r^2-1)\chi(\mathcal{O}_X)$ is called the *expected dimension* of $M(\mathcal{Q})$.

1.6 Final comments and additional reading

This introductory chapter intends to refresh the reader's memory with the main objects (vector bundles, Chern classes, stability and moduli spaces, etc.) that will be all-pervasive during the rest of the book. By no means it can be considered as a complete study of these objects and, in particular, it does not substitute standard references on these topics as they can be [103, 119, 169], among others.

Sections 1.1 and 1.2 deal with the intertwined notions of vector bundle and locally free sheaf on a projective variety. Vector bundles' inception lies on Schwarzenberger's work (see [193]) on the existence of holomorphic structures on topological vector bundles over the projective plane. During the 1970s and 1980s, it developed into a major branch of algebraic geometry. A major boost to the theory was given by its relation to mathematical physics (see [15]). On the other hand, [195] rooted firmly the use of coherent sheaves in algebraic geometry. Proposition 1.2.2 shows that we can exchange freely the use of vector bundles and locally free sheaves. In Sections 1.3 and 1.4, we introduce some basic results concerning coherent sheaves. Also [108, Chapter III and Appendix A] can be used as an excellent introduction.

Finally, Section 1.5 deals with the theory of moduli spaces of vector bundles. The notion of stability was coined to identify a family of vector bundles (or, more generally, torsion-free sheaves) that could be gathered into a malleable family. Stability was first introduced for bundles on curves by Mumford (see [163]) and later generalized to sheaves on higher dimensional varieties by Takemoto, Gieseker, Maruyama, and Simpson. It has developed as a very active area of research in algebraic geometry. A complete account on moduli spaces of vector bundles can be found, for instance, in [119], [151] and [152].

2 Vector bundles without intermediate cohomology

Given a projective variety $X \subset \mathbb{P}^n$, it is usual to try to understand the complexity of X in terms of the associated category of vector bundles that it supports. Since, in general, this category is unwieldy, one usually restricts attention to the category of (semi)stable vector bundles which is known to behave well since there exist nice moduli spaces parameterizing (semi)stable bundles. Whereas this approach has been largely and fruitfully exploited, it is also possible to pay attention to another property of a vector bundle $\mathcal{E}$ on X, that is, the fact of having the simplest possible cohomology, i. e., $\mathrm{H}^i(X, \mathcal{E}(t)) = 0$ for all $t \in \mathbb{Z}$ and $1 \leq i \leq \dim X - 1$. The vector bundles holding this property are called arithmetically Cohen–Macaulay (aCM for short) bundles. A seminal result due to Horrocks asserts that, on the projective space $\mathbb{P}^n$, any aCM bundle splits into a direct sum of line bundles (see Theorem 2.2.2). This complies with the general philosophy that a simple variety should have a simple associated category of aCM bundles. Following these lines, a cornerstone result is the classification of varieties that support only a finite number of indecomposable aCM bundles. These varieties fall into a very short list: $\mathbb{P}^n$, three or less reduced points on $\mathbb{P}^2$, a smooth quadric hypersurface $Q \subset \mathbb{P}^n$, a cubic scroll in $\mathbb{P}^4$, the Veronese surface in $\mathbb{P}^5$, or a rational normal curve.

For the rest of varieties, the classification of *all* aCM bundles seems out of reach, and we usually restrict our attention to certain subfamilies like irreducible homogeneous aCM bundles in the case of Grassmann varieties or aCM bundles of low rank in the case of hypersurfaces. In next chapters we will focus our attention on a special subfamily of aCM bundles, mainly we will focus our attention on aCM bundles such that their module of global sections has the maximal number of generators.

This chapter is entirely devoted to the study of aCM bundles and it has been organized as follows. In Section 2.1, we fix the definition of aCM bundle and the basic facts needed in the sequel. In Section 2.2, we prove Horrocks' theorem and classify all aCM bundles on projective spaces $\mathbb{P}^n$ while in Section 2.3 we state Knörrer theorem which classifies all aCM bundles on quadrics $Q_n \subset \mathbb{P}^{n+1}$. In Section 2.4, we deal with Grassmann varieties $\mathrm{Gr}(k, n)$ and determine all irreducible homogeneous aCM bundles on $\mathrm{Gr}(k, n)$, and, finally, in Section 2.5 we prove that low rank aCM bundles on general hypersurfaces of high degree $X \subset \mathbb{P}^n$, $n \geq 3$, split into a direct sum of line bundles.

2.1 Definition and examples of aCM bundles

The aim of this section is to provide an account of basic results on aCM sheaves (respectively vector bundles) and on the classification of aCM varieties according to the complexity of the families of aCM bundles that they support. Recall that, if there is no confusion, when we consider a variety $X \subset \mathbb{P}^n$ we mean that we embed it into

https://doi.org/10.1515/9783110647686-002

the corresponding projective space by means of the very ample line bundle $\mathcal{O}_X(1)$ or, equivalently, we consider the couple $(X, \mathcal{O}_X(1))$.

Definition 2.1.1. Let $X \subset \mathbb{P}^n$ be a variety. A coherent sheaf $\mathcal{E}$ on X is *arithmetically Cohen–Macaulay* (aCM for short) if it is locally Cohen–Macaulay (i. e., $\operatorname{depth}(\mathcal{E}_x) = \dim \mathcal{O}_{X,x}$ for any point $x \in X$) and has no intermediate cohomology:

$$\mathrm{H}^i(X, \mathcal{E}(t)) = 0 \quad \text{for all } t \in \mathbb{Z} \text{ and } 1 \le i \le \dim X - 1.$$

A coherent sheaf $\mathcal{E}$ on a variety $X \subset \mathbb{P}^n$ is said to be *initialized* if

$$\mathrm{H}^0(X, \mathcal{E}) \neq 0 \quad \text{and} \quad \mathrm{H}^0(X, \mathcal{E}(t)) = 0 \quad \text{for all } t \in \mathbb{Z}_{<0}.$$

Notice that for a coherent sheaf $\mathcal{E}$ with support of positive dimension on a variety $X \subset \mathbb{P}^n$ there is always an integer t_0 such that $\mathcal{E}(t_0)$ is initialized. We usually denote it by $\mathcal{E}_{\mathrm{init}}$. In the following sections and chapters we will be interested in initialized aCM sheaves with the maximal number of global sections.

Lemma 2.1.2. *Let $X \subset \mathbb{P}^n$ be a smooth variety. Then any locally Cohen–Macaulay sheaf $\mathcal{E}$ on X is locally free.*

Proof. Let $x \in X$ be any point. Then by hypothesis, $\operatorname{depth}(\mathcal{E}_x) = \dim \mathcal{O}_{X,x}$. But now, since $\mathcal{O}_{X,x}$ is a regular local ring because $x \in X$ is a smooth point, the projective dimension of $\mathcal{E}_x$ is finite, and we can apply the *Auslander–Buchsbaum formula* (see [36, Theorem 1.3.3])

$$\operatorname{pd}(\mathcal{E}_x) = \operatorname{depth}(\mathcal{O}_{X,x}) - \operatorname{depth}(\mathcal{E}_x)$$

to see that $\operatorname{pd}(\mathcal{E}_x) = 0$, i. e., $\mathcal{E}_x$ is $\mathcal{O}_{X,x}$-free and $\mathcal{E}$ locally free. □

Example 2.1.3. (1) For any $a \in \mathbb{Z}$, $\mathcal{O}_{\mathbb{P}^n}(a)$ is an aCM bundle on $\mathbb{P}^n$. Indeed, for all $t \in \mathbb{Z}$ and $1 \le i \le n-1$, we have $\mathrm{H}^i(\mathbb{P}^n, \mathcal{O}_{\mathbb{P}^n}(a+t)) = 0$ by Bott formula (see (1.2)).

(2) The tangent bundle $T_{\mathbb{P}^n}$ of $\mathbb{P}^n$ is not aCM because, using Serre's duality (see Theorem 1.3.2) and the dual of Euler sequence (1.1),

$$0 \longrightarrow \mathcal{O}_{\mathbb{P}^n} \longrightarrow \mathcal{O}_{\mathbb{P}^n}(1)^{n+1} \longrightarrow T_{\mathbb{P}^n} \longrightarrow 0, \tag{2.1}$$

we get $\mathrm{H}^{n-1}(\mathbb{P}^n, T_{\mathbb{P}^n}(-n-1)) = \mathrm{H}^1(\mathbb{P}^n, \Omega^1_{\mathbb{P}^n}) = \mathbf{k}$.

(3) Let $S \subset \mathbb{P}^3$ be a smooth cubic surface embedded by the very ample line bundle $\mathcal{O}_S(H)$. Then S is isomorphic to the blow-up of $\mathbb{P}^2$ at six general points $p_1, \dots, p_6 \in \mathbb{P}^2$, $\operatorname{Pic}(S) \cong \mathbb{Z}^7 = \langle h, e_1, \dots, e_6 \rangle$ where h is the pull-back under the blow-down map of any line l in $\mathbb{P}^2$ and e_i are the exceptional divisors corresponding to the points p_i. The embedding in $\mathbb{P}^3$ is given by the anticanonical divisor $H := -K_S = 3h - \sum_{i=1}^6 e_i$ corresponding to the plane cubic curves passing through the six points p_i. The line bundle $\mathcal{O}_S(h - e_1)$ is aCM while $\mathcal{O}_S(5h - 2e_1 - 2e_2 - 2e_3 - 2e_4 - 2e_5 - e_6)$ is not aCM. In

fact, since S is a surface, it is enough to see that $\mathrm{H}^1(\mathcal{O}_S(h - e_1 - tH)) = 0$ for any $t \in \mathbb{Z}$ meanwhile $\mathrm{H}^1(\mathcal{O}_S(5h - 2e_1 - 2e_2 - 2e_3 - 2e_4 - 2e_5 - e_6 - t_0H)) \neq 0$ for some $t_0 \in \mathbb{Z}$ which can be seen using the Riemann–Roch theorem (Theorem 1.4.3) in its form for surfaces (1.5) and the rules for the intersection product on S: $h^2 = 1$, $he_i = 0$, and $e_ie_j = -\delta_{ij}$.

(4) Consider the Segre variety $\mathbb{P}^n \times \mathbb{P}^m \subset \mathbb{P}^{nm+n+m}$. Using the *Künneth formula* for coherent sheaves $\mathcal{F}_1$ and $\mathcal{F}_2$ on projective varieties X_1 and X_2,

$$\mathrm{H}^i(X_1 \times X_2, \mathcal{F}_1 \boxtimes \mathcal{F}_2) = \bigoplus_{p+q=i} \mathrm{H}^p(X_1, \mathcal{F}_1) \otimes \mathrm{H}^q(X_2, \mathcal{F}_2), \tag{2.2}$$

we can see that the line bundles $\mathcal{O}_{\mathbb{P}^n\times\mathbb{P}^m}(0, n)$ and $\mathcal{O}_{\mathbb{P}^n\times\mathbb{P}^m}(m, 0)$ are aCM bundles while the line bundle $\mathcal{O}_{\mathbb{P}^n\times\mathbb{P}^m}(-1, n)$ is not aCM.

A particularly meaningful case of aCM sheaf is the structural sheaf $\mathcal{O}_X$ of an *arithmetically Cohen–Macaulay variety* $X \subset \mathbb{P}^n$:

Definition 2.1.4. A variety $X \subset \mathbb{P}^n$ of dimension d is said to be *arithmetically Cohen–Macaulay* (aCM for short) if its homogeneous coordinate ring R_X is a Cohen–Macaulay ring or, equivalently, $\dim R_X = \operatorname{depth}(R_X)$.

Using the graded version of the *Auslander–Buchsbaum formula,* which asserts that for any finitely generated R-module M,

$$\operatorname{pd}(M) = n + 1 - \operatorname{depth}(M),$$

we immediately get that a variety $X \subsetneq \mathbb{P}^n$ is aCM if and only if the $\operatorname{pd}(R_X) = \operatorname{codim}(X)$. Therefore, if $X \subset \mathbb{P}^n$ is a codimension c aCM variety, a graded minimal free R-resolution of $I(X)$ looks like

$$0 \longrightarrow F_c \longrightarrow F_{c-1} \longrightarrow \cdots \longrightarrow F_1 \longrightarrow I(X) \longrightarrow 0 \tag{2.3}$$

where $F_i = \bigoplus_{j\in\mathbb{Z}} R(-j)^{b_{i,j}(X)}$, $i = 1, \ldots, c$. The $b_{i,j}(X) \in \mathbb{Z}_{\geq 0}$ are called the *graded Betti numbers* of X. They are uniquely determined by X and defined as follows:

$$b_{i,j}(X) = \dim_{\mathbf{k}} \operatorname{Tor}_i^R(I(X), \mathbf{k})_j.$$

Example 2.1.5. (1) Let $g_1, \ldots, g_c$, $c \leq n$, be a regular sequence of elements of R. Then the variety $X \subset \mathbb{P}^n$ associated to the (saturated) ideal $I = (g_1, \ldots, g_c)$ is called a *complete intersection.* Since $\operatorname{depth}(R_X) = \dim R_X$, we have that complete intersections are examples of aCM varieties.

(2) As an example of aCM curve which is not a complete intersection, we have the *rational normal cubic curve* $C \subset \mathbb{P}^3$ defined as the image of the Veronese map μ : $\mathbb{P}^1 \longrightarrow \mathbb{P}^3$ which maps $(a : b)$ to $(a^3 : a^2b : ab^2 : b^3)$. Its minimal free resolution has the following shape:

$$0 \longrightarrow R(-3)^2 \xrightarrow{\phi} R(-2)^3 \longrightarrow R \longrightarrow R_C \longrightarrow 0, \tag{2.4}$$

where the map ϕ can be represented by the matrix of linear forms

$$M := \begin{pmatrix} x_0 & x_1 & x_2 \\ x_1 & x_2 & x_3 \end{pmatrix}$$

and the ideal $I(C)$ is generated by the determinants of maximal minors of M, i. e., $I(C) = (x_0x_2 - x_1^2, x_1x_3 - x_2^2, x_0x_3 - x_1x_2)$. This presentation shows that $C \subset \mathbb{P}^3$ is a *standard determinantal variety*. Since $\mathrm{pd}(R_C) = 2 = \mathrm{codim}(C)$, C is an aCM curve, but it is not a complete intersection since $I(C)$ is minimally generated by three quadrics. Indeed, it is known that any determinantal variety is aCM (see [155]).

(3) The *smooth rational quartic curve* $C \subset \mathbb{P}^3$ given as the image of the map

$$\begin{array}{ccc} \mathbb{P}^1 & \longrightarrow & \mathbb{P}^3 \\ (u:t) & \mapsto & (u^4 : u^3t : ut^3 : t^4), \end{array}$$

is not aCM. In fact, C has minimal free resolution

$$0 \longrightarrow R(-5) \longrightarrow R(-4)^4 \longrightarrow R(-3)^3 \oplus R(-2) \longrightarrow R \longrightarrow R_C \longrightarrow 0$$

from where it is seen that $\mathrm{pd}(R_C) = 3 > \mathrm{codim}(C) = 2$. Indeed, it is immediate from the definition that a curve $C \subset \mathbb{P}^n$ is aCM if and only if it is *projectively normal*, equivalently, for any $d \geq 1$, any section of $\mathcal{O}_C(d)$ is the intersection of C with a hypersurface of $\mathbb{P}^n$ of degree d.

Remark 2.1.6. It is worthwhile to point out that if $X \subset \mathbb{P}^n$ is an aCM variety of codimension c, then the dual of the minimal free resolution of R_X,

$$0 \longrightarrow F_c \longrightarrow F_{c-1} \longrightarrow \cdots \longrightarrow F_1 \longrightarrow R \longrightarrow R_X \longrightarrow 0,$$

provides a resolution of a twist of the *canonical module*,

$$0 \longrightarrow R \longrightarrow F_1^\vee \longrightarrow \cdots \longrightarrow F_c^\vee \longrightarrow K_X(n+1) := \mathrm{Ext}_R^c(R_X, R) \longrightarrow 0.$$

Definition 2.1.7. Let $X \subset \mathbb{P}^n$ be a variety. A graded R_X-module M is a *maximal Cohen–Macaulay* module (MCM for short) if $\mathrm{depth}(M) = \dim M = \dim R_X$.

MCM modules and aCM sheaves are closely related. To state this relationship, recall that to any R-graded module E we can canonically associate a coherent $\mathcal{O}_{\mathbb{P}^n}$-sheaf $\widetilde{E}$ (see [108, p. 116]). Then, we have:

Proposition 2.1.8. *Let $X \subset \mathbb{P}^n$ be a variety. There exists a bijection between aCM sheaves $\mathcal{E}$ on X and MCM R_X-modules given by the functors*

$$E \longrightarrow \widetilde{E} \quad \text{and} \quad \mathcal{E} \longrightarrow \mathrm{H}^0_*(X, \mathcal{E}) := \bigoplus_{t \in \mathbb{Z}} \mathrm{H}^0(X, \mathcal{E}(t)).$$

Proof. First of all, recall that, given a graded finitely generated R_X-module M, there exist an exact sequence

$$0 \longrightarrow \mathrm{H}^0_{\mathfrak{m}_X}(M) \longrightarrow M \longrightarrow \mathrm{H}^0_*(X, \widetilde{M}) \longrightarrow \mathrm{H}^1_{\mathfrak{m}_X}(M) \longrightarrow 0$$

and isomorphisms $\mathrm{H}^i_*(X, \widetilde{M}) := \bigoplus_{t\in\mathbb{Z}} \mathrm{H}^i(X, \widetilde{M}(t)) \cong \mathrm{H}^{i+1}_{\mathfrak{m}_X}(M)$ for $i \geq 1$, where $\mathfrak{m}_X$ denotes the irrelevant ideal of R_X.

So let $\mathcal{E}$ be an aCM sheaf on X. Then $E := \mathrm{H}^0_*(X, \mathcal{E})$ will be a finitely generated R_X-module that verifies $E \cong \mathrm{H}^0_*(X, \widetilde{E})$ and therefore $\mathrm{H}^i_{\mathfrak{m}_X}(E) = 0$ for $i \leq \dim X = \dim R_X - 1$. This allows us to conclude that E is MCM by the local cohomological criterion of depth.

On the other hand, let E be an MCM R_X-module. Then $\mathcal{E} := \widetilde{E}$ will be a locally Cohen–Macaulay sheaf, and again by the previous isomorphisms we will have $\mathrm{H}^i_*(X, \mathcal{E}) = 0$ for $i = 1, \ldots, \dim X - 1$. □

The following proposition relates the notions previously introduced in this chapter.

Proposition 2.1.9. *Let $X \subset \mathbb{P}^n$ be a variety of dimension $d \geq 1$. The following statements are equivalent:*
(1) *X is an aCM variety.*
(2) *$\mathcal{O}_X$ is an aCM sheaf and $\bigoplus_{t\in\mathbb{Z}} \mathrm{H}^1(\mathcal{I}_X(t)) = 0$.*
(3) *$\bigoplus_{t\in\mathbb{Z}} \mathrm{H}^i(\mathcal{I}_X(t)) = 0$ for all $1 \leq i \leq d$.*

Proof. From the long exact sequence of cohomology groups associated to the structural sequence,

$$0 \longrightarrow \mathcal{I}_X(t) \longrightarrow \mathcal{O}_{\mathbb{P}^n}(t) \longrightarrow \mathcal{O}_X(t) \longrightarrow 0, \tag{2.5}$$

we immediate obtain the equivalence between (2) and (3). Notice that also using (2.5) we can see that $R_X \cong \bigoplus_{t\in\mathbb{Z}} \mathrm{H}^0(\mathcal{O}_X(t))$ is equivalent to $\bigoplus_{t\in\mathbb{Z}} \mathrm{H}^1(\mathcal{I}_X(t)) = 0$. Therefore, the fact that (1) is a consequence of (2) follows from Theorem 2.1.8.

Finally, let us suppose (1). Then applying the functor $E \longrightarrow \widetilde{E}$ to the free minimal resolution (2.3) of R_X and splitting it in short exact sequences, one can see that $\bigoplus_{t\in\mathbb{Z}} \mathrm{H}^i(\mathcal{I}_X(t)) = 0$ for all $1 \leq i \leq d$. □

Definition 2.1.10. A codimension c variety $X \subset \mathbb{P}^n$ is *arithmetically Gorenstein* (or aG for short) if its homogeneous coordinate ring R_X is a Gorenstein ring or, equivalently, its saturated homogeneous ideal, $I(X)$, has a minimal free graded R-resolution of the following type:

$$0 \longrightarrow R(-t) \longrightarrow F_{c-1} \longrightarrow \cdots \longrightarrow F_1 \longrightarrow I(X) \longrightarrow 0.$$

In other words, an aG variety is an aCM variety with Cohen–Macaulay type 1 (i. e., the rank of the last graded free R-module in the minimal free resolution of $I(X)$ is one).

We have the following equivalent definitions of aG varieties:

Proposition 2.1.11. *Let $X \subset \mathbb{P}^n$ be an aCM variety of codimension c. Then the following conditions are equivalent:*
(1) *X is aG.*
(2) *$R_X \cong K_X(s)$ for some $s \in \mathbb{Z}$.*
(3) *The minimal free resolution of R_X is self-dual, up to a twist by $n+1$.*

Proof. See [155, Proposition 1.1.27]. □

Example 2.1.12. (1) The minimal free R-resolution of a codimension c *complete intersection* X with $I(X) = (g_1, \dots, g_c)$ and $d_i = \deg(g_i)$ is given by the *Koszul resolution*:

$$0 \longrightarrow R\left(-\sum_{i=1}^{c} d_i\right) \cong \bigwedge^{c} F_1 \longrightarrow \bigwedge^{c-1} F_1 \longrightarrow \cdots \longrightarrow \bigwedge^{2} F_1 \longrightarrow F_1 \longrightarrow I(X) \longrightarrow 0,$$

where $F_1 = \bigoplus_{i=1}^{c} R(-d_i)$. Therefore, complete intersections are examples of aG varieties.

(2) According to the resolution (2.4), as an example of aCM curve that is not aG we have the *rational normal cubic curve* $C \subset \mathbb{P}^3$.

(3) By Buchsbaum–Eisenbud theorem (see [37, Theorem 2.1]), given a generic skew-symmetric matrix A of odd size l with entries homogenous polynomials in R, the ideal (usually called $\mathrm{Pf}_{l-1}(A)$) generated by the pfaffians of the matrices of size $l-1$ obtained by deleting the ith row and the ith column of A, $i = 1, \dots, l$, is an aG subvariety of $\mathbb{P}^n$ of codimension 3 which is not a complete intersection whenever $l > 3$. Conversely, any aG variety of codimension 3 arises in this way. For instance, a set of five generic points $Z \subset \mathbb{P}^3$ is an aG variety with minimal free resolution

$$0 \longrightarrow R(-5) \longrightarrow R(-3)^5 \xrightarrow{A} R(-2)^5 \longrightarrow R \longrightarrow R_Z \longrightarrow 0$$

where A is a skew-symmetric matrix of size 5 with linear entries. Notice, however, that any codimension 2 aG variety is indeed a complete intersection.

In the next proposition we summarize basic properties of aCM sheaves.

Proposition 2.1.13. *Let $X \subset \mathbb{P}^n$ be a variety of dimension d. It holds:*
(1) *The extension of aCM sheaves is an aCM sheaf. In particular, the direct sum of aCM sheaves is an aCM sheaf.*
(2) *Let H be a hyperplane section of X. The restriction $\mathcal{E}_{|H}$ to H of an aCM sheaf $\mathcal{E}$ on X is an aCM sheaf on H.*
(3) *If X is smooth and $\omega_X \cong \mathcal{O}_X(a)$ for some $a \in \mathbb{Z}$ then the dual $\mathcal{E}^\vee$ of an aCM bundle $\mathcal{E}$ is again an aCM bundle.*

Proof.

(1) Let $\mathcal{E}_1$ and $\mathcal{E}_2$ be two aCM sheaves on X and let $\mathcal{E}$ be an extension given by $e \in \operatorname{Ext}^1(\mathcal{E}_2, \mathcal{E}_1)$. By construction, $\mathcal{E}$ fits into an exact sequence

$$0 \longrightarrow \mathcal{E}_1 \longrightarrow \mathcal{E} \longrightarrow \mathcal{E}_2 \longrightarrow 0,$$

and, taking cohomology, we get $\mathrm{H}^i(X, \mathcal{E}(t)) = 0$ for all $t \in \mathbb{Z}$ and $1 \le i \le d-1$. Therefore, $\mathcal{E}$ is aCM.

(2) It immediately follows from the exact cohomology sequence associated to

$$0 \longrightarrow \mathcal{E}(-1) \longrightarrow \mathcal{E} \longrightarrow \mathcal{E}_{|H} \longrightarrow 0.$$

(3) By Serre's duality (see Theorem 1.3.2), we have

$$\mathrm{H}^i(X, \mathcal{E}^\vee(t)) \cong \mathrm{H}^{d-i}(X, \mathcal{E} \otimes \omega_X(-t))^\vee \cong \mathrm{H}^{d-i}(X, \mathcal{E}(a-t))^\vee = 0$$

for all $t \in \mathbb{Z}$ and $1 \le i \le d-1$. □

A possible way to classify aCM varieties is according to the complexity of the families of indecomposable aCM vector bundles that they support. To this end, the following definitions were proposed (see [79] for the case of curves and [40] for the higher-dimensional case):

Definition 2.1.14. Let $X \subset \mathbb{P}^n$ be a variety.

(1) X is of *finite representation type* if it has, up to twist and isomorphism, only a finite number of indecomposable aCM sheaves.

(2) X is of *tame representation type* if for each rank r, the indecomposable aCM sheaves of rank r form a finite number of families of dimension at most one.

(3) X is of *wild representation type* if there exist ℓ-dimensional families of nonisomorphic indecomposable aCM sheaves for arbitrary large ℓ.

This trichotomy was inspired by an analogous classification for quivers and for k-algebras of finite type.

Remark 2.1.15. (1) Auslander proved that if a graded Cohen–Macaulay ring S supports only a finite number of MCM modules up to isomorphism and degree shift, then S is an isolated singularity (see [205, Theorem 4.22]). This means, from the geometric point of view, that aCM varieties of finite type are smooth. Therefore, by Lemma 2.1.2, the family of aCM bundles coincides with the family of aCM sheaves.

(2) ACM varieties of finite representation type have been classified. In fact, if $X \subset \mathbb{P}^n$ is an aCM variety of finite representation type, then X is a set of either three or less reduced points on $\mathbb{P}^2$, a projective space (see Theorem 2.2.2), a smooth quadric $X \subset \mathbb{P}^n$ (see Theorem 2.3.4), a cubic scroll in $\mathbb{P}^4$, the Veronese surface in $\mathbb{P}^5$ or a rational normal curve (see [38, Theorem C] and [85, p. 348]).

(3) A smooth aCM curve $C \subset \mathbb{P}^n$ of genus g is of finite (respectively tame, wild) representation type if and only if $g = 0$ (respectively $g = 1$, $g > 1$) (see [157, Example 4.4]).

(4) Buchweitz, Greuel, and Schreyer proved that a hypersurface $X \subset \mathbb{P}^n$ of degree $\deg(X) \geq 3$ is not of finite representation type by constructing a surjective map from the set of isomorphism classes of MCM modules over R_X either to $\mathbb{P}^n$ (if $\deg(X) = 3$) or to a cubic hypersurface in $\mathbb{P}^n$ (if $\deg(X) > 3$). Nevertheless, it remains open to prove, in general, that they are of wild representation type (see [38, Theorem C]).

In the next chapters we will give plenty of examples of varieties of wild representation type. Often the aCM sheaves of the families that we will construct will share another important property, namely, the associated MCM will be maximally generated.

We end the section with a cohomological notion that will appear later on.

Definition 2.1.16. Let $X \subset \mathbb{P}^n$ be a variety. We say that a vector bundle $\mathcal{E}$ on X has *natural cohomology* if for any integer i, at most one cohomology group $\mathrm{H}^j(X, \mathcal{E}(i)) \neq 0$ for $0 \leq j \leq \dim X$.

Example 2.1.17. As an example of vector bundle with natural cohomology, we have the cotangent bundle $\Omega^1_{\mathbb{P}^n}$.

2.2 aCM bundles on $\mathbb{P}^n$: Horrocks' theorem

The goal of this section is to classify all aCM bundles on projective spaces. This will be achieved by means of the splitting criterion of Horrocks. We start with the case of the projective line.

Theorem 2.2.1 (Grothendieck's theorem). *Every rank r vector bundle $\mathcal{E}$ on $\mathbb{P}^1$ splits as a direct sum*

$$\mathcal{E} \cong \mathcal{O}_{\mathbb{P}^1}(a_1) \oplus \cdots \oplus \mathcal{O}_{\mathbb{P}^1}(a_r)$$

for uniquely determined integers $a_1 \leq \cdots \leq a_r$.

Proof. See [169, Theorem 2.1.1]. □

Theorem 2.2.2 (Horrocks' theorem). *A vector bundle $\mathcal{E}$ on $\mathbb{P}^n$ splits as a sum of line bundles if and only if $\mathcal{E}$ is aCM. In particular, $\mathbb{P}^n$ is a variety of finite representation type.*

Proof. If $\mathcal{E} = \bigoplus_{j=1}^r \mathcal{O}_{\mathbb{P}^n}(a_j)$ then $\mathrm{H}^i(\mathcal{E}(t)) = \mathrm{H}^i(\oplus_{j=1}^r \mathcal{O}_{\mathbb{P}^n}(a_j + t)) = 0$ for all $t \in \mathbb{Z}$ and $1 \leq i \leq n-1$ by Bott formula (1.2). Therefore, $\mathcal{E}$ is an aCM bundle. Conversely, assume that $\mathcal{E}$ is an aCM bundle of rank r. Using induction on n, we will check that $\mathcal{E}$ splits into a sum of line bundles. For $n = 1$, the result is clear because any vector bundle on $\mathbb{P}^1$ is aCM and, on the other hand, by Grothendieck's theorem any vector bundle on $\mathbb{P}^1$

is a direct sum of line bundles (see Theorem 2.2.1). Assume that the result is true for $m < n$. Using the exact cohomology sequence

$$\cdots \longrightarrow \mathrm{H}^i(\mathbb{P}^n, \mathcal{E}(t)) \longrightarrow \mathrm{H}^i((\mathbb{P}^{n-1}, \mathcal{E}_{|\mathbb{P}^{n-1}}(t)) \longrightarrow \mathrm{H}^{i+1}(\mathbb{P}^n, \mathcal{E}(t-1)) \longrightarrow \cdots$$

associated to the exact sequence

$$0 \longrightarrow \mathcal{E}(-1) \longrightarrow \mathcal{E} \longrightarrow \mathcal{E}_{|\mathbb{P}^{n-1}} \longrightarrow 0, \tag{2.6}$$

we obtain $\mathrm{H}^i(\mathbb{P}^{n-1}, \mathcal{E}_{|\mathbb{P}^{n-1}}(t)) = 0$ for all $t \in \mathbb{Z}$ and $1 \le i \le n-2$, i. e., $\mathcal{E}_{|\mathbb{P}^{n-1}}$ is an aCM bundle on $\mathbb{P}^{n-1}$, and hence by induction hypothesis we have

$$\mathcal{E}_{|\mathbb{P}^{n-1}} = \bigoplus_{j=1}^{r} \mathcal{O}_{\mathbb{P}^{n-1}}(a_j).$$

Set $\mathcal{F} = \bigoplus_{j=1}^{r} \mathcal{O}_{\mathbb{P}^n}(a_j)$ and let $\varphi : \mathcal{F}_{|\mathbb{P}^{n-1}} \longrightarrow \mathcal{E}_{|\mathbb{P}^{n-1}}$ be an isomorphism. We will prove:
(1) φ extends to a morphism $\psi : \mathcal{F} \longrightarrow \mathcal{E}$, and
(2) ψ is an isomorphism.

(1) We tensor the exact sequence (2.6) with $\mathcal{F}^\vee$ and obtain the cohomology long exact sequence

$$\cdots \longrightarrow \mathrm{H}^0(\mathbb{P}^n, \mathcal{F}^\vee \otimes \mathcal{E}) \longrightarrow \mathrm{H}^0(\mathbb{P}^{n-1}, \mathcal{F}^\vee \otimes \mathcal{E}_{|\mathbb{P}^{n-1}}) \longrightarrow \mathrm{H}^1(\mathbb{P}^n, \mathcal{F}^\vee \otimes \mathcal{E}(-1)) \longrightarrow \cdots.$$

Since $\mathrm{H}^1(\mathbb{P}^n, \mathcal{F}^\vee \otimes \mathcal{E}(-1)) = \bigoplus_{j=1}^{r} \mathrm{H}^1(\mathbb{P}^n, \mathcal{E}(-1-a_j)) = 0$, every global section of $\mathcal{F}^\vee \otimes \mathcal{E}_{|\mathbb{P}^{n-1}}$ extends to a global section of $\mathcal{F}^\vee \otimes \mathcal{E}$ and, therefore, φ extends to a morphism $\psi : \mathcal{F} \longrightarrow \mathcal{E}$.
(2) Since $\mathrm{rank}(\mathcal{F}) = \mathrm{rank}(\mathcal{E}) = r$ and $c_1 := c_1(\mathcal{F}) = c_1(\mathcal{E})$, the morphism ψ induces a morphism $\det(\psi) : \wedge^r \mathcal{F} \cong \mathcal{O}_{\mathbb{P}^n}(c_1) \longrightarrow \wedge^r \mathcal{E} \cong \mathcal{O}_{\mathbb{P}^n}(c_1)$ which is constant and whose restriction to $\mathbb{P}^{n-1}$ is nonzero. Thus, ψ is an isomorphism which proves what we want. □

Example 2.2.3. As an immediate application of Horrocks' theorem, we get that the cotangent bundle $\Omega^1_{\mathbb{P}^2}$ of $\mathbb{P}^2$ does not split as a sum of line bundles because the exact cohomology sequence associated to the Euler sequence (1.1) gives us $\mathrm{H}^1(\mathbb{P}^2, \Omega^1_{\mathbb{P}^2}) \neq 0$.

2.3 aCM bundles on Q_n: Knörrer's theorem

In this section we are going to prove that on smooth n-dimensional quadrics $Q_n \subset \mathbb{P}^{n+1}$ is also possible to give a complete description of all aCM bundles. To this end, we start introducing spinor bundles on quadrics $Q_n \subset \mathbb{P}^{n+1}$. They are related to linear spaces of maximal dimension on Q_n. In Section 4.2, we will give an explicit construction of spinor bundles in terms of linear matrix factorization of the homogeneous polynomial defining $Q_n \subset \mathbb{P}^{n+1}$.

First suppose that $n = 2k + 1$ is odd. In this case, k is the maximal dimension of a linear space contained in Q_n. Let $\mathrm{Gr}(k, n + 1)$ be the Grassmann variety which parameterizes linear subspaces $\mathbb{P}^k$ of dimension k of $\mathbb{P}^{n+1}$. We define the closed subset S_{k+1} of $\mathrm{Gr}(k, n + 1)$ as follows:

$$S_{k+1} := \{\mathbb{P}^k \in \mathrm{Gr}(k, n + 1) \mid \mathbb{P}^k \subset Q_n\} \subset \mathrm{Gr}(k, n + 1).$$

This is a smooth projective variety of dimension $(k+1)(k+2)/2$, called the *spinor variety*. It turns out that $\mathrm{Pic}(S_{k+1}) \cong \mathbb{Z}$ and that the ample generator of $\mathrm{Pic}(\mathrm{Gr}(k, n+1))$ restricts to twice the generator $\mathcal{O}_{S_{k+1}}(1)$ of $\mathrm{Pic}(S_{k+1})$.

Let us consider the incidence variety $\mathbb{F} := \{(x, \mathbb{P}^k) \mid x \in \mathbb{P}^k\} \subset Q_n \times S_{k+1}$ and the following diagram:

$$\begin{array}{ccc} & \mathbb{F} & \\ {}^{p}\swarrow & & \searrow^{q} \\ Q_n & & S_{k+1} \end{array} \tag{2.7}$$

where p and q are the natural projections. We define the spinor bundle

$$\mathcal{S} := p_* q^* (\mathcal{O}_{S_{k+1}}(1))^\vee.$$

In case $n = 2k$, it turns out that the variety of linear spaces of maximal dimension of Q_n in $\mathrm{Gr}(k, n+1)$ has two isomorphic connected components S'_k and S''_k each of them of dimension $k(k + 1)/2$. A similar incidence relation as in (4.11) allows us to construct two spinor bundles $\mathcal{S}'$ and $\mathcal{S}''$ on Q_n. In any case, they are vector bundles of rank $2^{\lfloor \frac{n-1}{2} \rfloor}$. In Example 4.2.6 we will explicitly construct the spinor bundles using the matrix factorization approach.

If $Q_n \subset Q_{n+1}$ is a smooth hyperplane section, their spinor bundles are closely related.

Theorem 2.3.1.
(1) *Let $\mathcal{S}'$ and $\mathcal{S}''$ be the spinor bundles on Q_{2k} and let $i : Q_{2k-1} \longrightarrow Q_{2k}$ be a smooth hyperplane section. Then $i^*\mathcal{S}' \cong i^*\mathcal{S}'' \cong \mathcal{S}$, where $\mathcal{S}$ is the spinor bundle on Q_{2k-1}.*
(2) *Let $\mathcal{S}$ be the spinor bundle on Q_{2k+1} and let $i : Q_{2k} \longrightarrow Q_{2k+1}$ be a smooth hyperplane section. Then $i^*\mathcal{S} \cong \mathcal{S}' \oplus \mathcal{S}''$, where $\mathcal{S}'$ and $\mathcal{S}''$ are the spinor bundles on Q_{2k}.*

Proof. See [172, Theorem 1.4]. □

If there is no confusion, $\mathcal{S}$ denotes either the unique spinor bundle on Q_{2k+1} or one of the two spinor bundles on Q_{2k}.

Theorem 2.3.2. *Let $\mathcal{S}$ be a spinor bundle on Q_n. Then:*
(1) $\mathrm{H}^i(Q_n, \mathcal{S}(t)) = 0$ *for* $0 < i < n$ *and* $t \in \mathbb{Z}$.
(2) $\mathrm{H}^0(Q_n, \mathcal{S}(t)) = 0$ *for* $t \le 0$ *and* $\mathrm{h}^0(Q_n, \mathcal{S}(1)) = 2^{\lfloor \frac{n+1}{2} \rfloor}$.

In other words, $\mathcal{S}$ is an initialized aCM bundle on Q_n.

Proof. We will prove it by induction on n. For $n = 1$, we have that $Q_1 \cong \mathbb{P}^1$ and $\mathcal{S} \cong \mathcal{O}_{\mathbb{P}^1}$. For $n = 2$, $Q_2 \cong \mathbb{P}^1 \times \mathbb{P}^1 \subset \mathbb{P}^3$ and the spinor bundles are $\mathcal{O}_{\mathbb{P}^1\times\mathbb{P}^1}(-1,0)$ and $\mathcal{O}_{\mathbb{P}^1\times\mathbb{P}^1}(0,-1)$. Therefore for these two cases the statement is well-known. Let $\mathcal{S}$ be a spinor bundle on Q_{n+1}, $n \geq 2$. Consider the exact hyperplane sequence

$$0 \longrightarrow \mathcal{S}(t-1) \longrightarrow \mathcal{S}(t) \longrightarrow \mathcal{S}_{|Q_n}(t) \longrightarrow 0.$$

Then the result follows from Serre's Theorem B (see Theorem 1.3.1 (4)), Serre duality (see Theorem 1.3.2), and the induction hypothesis. □

Theorem 2.3.3. *Let $\mathcal{E}$ be a coherent sheaf on Q_n such that $\mathrm{H}^i(\mathcal{E}(-i-1)) = 0$ for $0 \leq i \leq n-2$. It holds:*

(1) *If $\mathrm{H}^{n-1}(\mathcal{E} \otimes \mathcal{S}(-n+1)) = 0$ for any spinor bundle $\mathcal{S}$ on Q_n, then $\mathcal{E}$ has $\mathcal{O}_{Q_n}^{\mathrm{h}^0(\mathcal{E})}$ as direct summand.*

(2) *Assume that $\mathrm{H}^{n-1}(\mathcal{E}(-n)) = 0$. Then for n odd, $\mathcal{E}$ has $(\mathcal{S}^\vee)^{\mathrm{h}^{n-1}(\mathcal{E}\otimes\mathcal{S}(-n+1))}$ as direct summand and for n even, $\mathcal{E}$ has $(\mathcal{S}'^\vee)^{\mathrm{h}^{n-1}(\mathcal{E}\otimes\mathcal{S}'(-n+1))} \oplus (\mathcal{S}''^\vee)^{\mathrm{h}^{n-1}(\mathcal{E}\otimes\mathcal{S}''(-n+1))}$ as a direct summand.*

Proof. We are going to give the details of the proof for n odd (the proof in the even case follows the same lines). By Example A.0.28 (2), there exists an exact sequence

$$0 \longrightarrow \mathcal{S}^a \longrightarrow \mathcal{O}_{Q_n}^b \longrightarrow \mathcal{F} \longrightarrow 0$$

with $\mathcal{F}$ a direct summand of $\mathcal{E}$, $a := \mathrm{h}^{n-1}(\mathcal{E} \otimes \mathcal{S}(-n+1)) = \mathrm{h}^{n-1}(\mathcal{F} \otimes \mathcal{S}(-n+1))$, and $b := \mathrm{h}^0(\mathcal{E}) = \mathrm{h}^0(\mathcal{F})$. Therefore, in the situation (1) of the statement of the theorem, we immediately obtain the result. In situation (2), we can assume that $\mathcal{E}$ has no direct sums of copies of $\mathcal{O}_{Q_n}$, and hence by (1) we have $a \neq 0$. We will prove that $\mathcal{F}$ contains $\mathcal{S}^\vee$ as a direct summand.

We construct the following diagram:

$$\begin{array}{ccccccccc}
0 & \longrightarrow & \mathcal{S}^a & \xrightarrow{\alpha} & \mathcal{O}_{Q_n}^b & \xrightarrow{\beta} & \mathcal{F} & \longrightarrow & 0 \\
 & & \downarrow{\scriptstyle p} & & \downarrow{\scriptstyle t} & & \downarrow{\scriptstyle z} & & \\
0 & \longrightarrow & \mathcal{S} & \xrightarrow{\gamma} & \mathcal{O}_{Q_n}^{2^{\frac{n+1}{2}}} & \xrightarrow{\delta} & \mathcal{S}^\vee & \longrightarrow & 0
\end{array}$$

where the existence of the short exact sequence on the second row is due to [172, Theorem 2.8], p is the projection on the first summand, and t exists because $\mathrm{Ext}^1(\mathcal{F}, \mathcal{O}_{Q_n}) \subset \mathrm{Ext}^1(\mathcal{E}, \mathcal{O}_{Q_n}) \cong \mathrm{H}^{n-1}(\mathcal{E}(-n)) = 0$, and therefore the map

$$\alpha^* : \mathrm{Hom}(\mathcal{O}_{Q_n}^b, \mathcal{O}_{Q_n}^{2^{\frac{n+1}{2}}}) \longrightarrow \mathrm{Hom}(\mathcal{S}^a, \mathcal{O}_{Q_n}^{2^{\frac{n+1}{2}}})$$

is surjective and γp has a preimage. Analogously, since $\mathcal{S}$ is an aCM bundle and, in particular, $\mathrm{H}^1(\mathcal{S}) = 0$, we can construct the diagram (i is the immersion on the first summand):

$$\begin{array}{ccccccccc}
0 & \longrightarrow & \mathcal{S}^a & \xrightarrow{\alpha} & \mathcal{O}_{Q_n}^b & \xrightarrow{\beta} & \mathcal{F} & \longrightarrow & 0 \\
 & & \uparrow{\scriptstyle i} & & \uparrow{\scriptstyle t'} & & \uparrow{\scriptstyle z'} & & \\
0 & \longrightarrow & \mathcal{S} & \xrightarrow{\gamma} & \mathcal{O}_{Q_n}^{2^{\frac{n+1}{2}}} & \xrightarrow{\delta} & \mathcal{S}^\vee & \longrightarrow & 0.
\end{array}$$

"Diagram chasing" shows that tt' is the identity on $\operatorname{im}(\gamma)$. In particular, we can suppose that there is a trivial line bundle $\mathcal{O}_{Q_n} \cong \mathcal{L} \subset \mathcal{O}_{Q_n}^{2^{\frac{n+1}{2}}}$ such that $tt'_{|\mathcal{L}}$ is an automorphism. Now, since $\mathrm{H}^0(\mathcal{S}) = 0$ (see Theorem 2.3.2), there exists $l \in \mathcal{L}$ such that $\delta(l) \neq 0$. Let $l' \in \mathcal{L}$ such that $tt'(l') = l$. Another diagram chasing shows that $zz'\delta(l') = \delta(l) \neq 0$. Since spinor bundles are defined by irreducible representations (see [172, p. 305]), by a theorem of Ramanan (see [201]) $\mathcal{S}^\vee$ is stable, and so in particular simple, which implies that zz' is a nonzero multiple of the identity and therefore $\mathcal{S}^\vee$ is a direct summand of $\mathcal{F}$. □

As a consequence of this result, we classify all aCM bundles on smooth quadrics.

Theorem 2.3.4. *The only indecomposable aCM bundles on a smooth quadric $Q_n \subset \mathbb{P}^{n+1}$ are the line bundles and the spinor bundles up to a twist. In particular, Q_n is of finite representation type.*

Proof. Let $\mathcal{E}$ be an indecomposable aCM bundle. Since $\mathcal{E}$ is locally free, there exists t_0 such that $0 = \mathrm{h}^0(\mathcal{E}(t_0 - 1)) < \mathrm{h}^0(\mathcal{E}(t_0))$. Then we apply Theorem 2.3.3 to conclude. □

We have seen that smooth quadrics $Q_n \subset \mathbb{P}^{n+1}$ admit only a finite number of aCM bundles (up to isomorphism and twist). We end the chapter with an example that shows that this is not necessarily the case if $Q_n \subset \mathbb{P}^{n+1}$ is not smooth.

Example 2.3.5. Let $Q \subset \mathbb{P}^3$ be a quadric cone and $\mathcal{O}_Q$ its structural sheaf. Then, as for any other hypersurface, $\mathcal{O}_Q$ is an aCM line bundle. On the other hand, let $L \subset Q$ be any line. The ideal sheaf $\mathcal{I}_{L|Q}$ is a Cartier divisor which is not a Weil divisor. Indeed, $2L$ is the hyperplane section of Q. We have the short exact sequence

$$0 \longrightarrow \mathcal{I}_L \longrightarrow \mathcal{I}_Q \longrightarrow \mathcal{I}_{L|Q} \longrightarrow 0.$$

Since L and Q are aCM subvarieties of $\mathbb{P}^3$, we have that $\mathrm{H}^1(\mathcal{I}_{L|Q}(t)) = 0$ for all $t \in \mathbb{Z}$. On the other hand, the structural exact sequence

$$0 \longrightarrow \mathcal{I}_{L|Q} \longrightarrow \mathcal{O}_Q \longrightarrow \mathcal{O}_L \longrightarrow 0$$

and [36, Proposition 1.2.9] show that $\operatorname{depth}(\mathcal{I}_{L|Q,x}) = 2$ at any point $x \in Q$. Namely, $\mathcal{I}_{L|Q}$ is an aCM $\mathcal{O}_Q$-sheaf of rank one. On the other hand, we have

$$\operatorname{Ext}^1(\mathcal{I}_{L|Q}, \mathcal{I}_{L|Q}(-l+1)) = \begin{cases} \mathbf{k} & \text{for } l > 0, \\ 0 & \text{otherwise.} \end{cases}$$

Therefore, we have a sequence of indecomposable aCM rank-two sheaves on Q indexed by $l \in \mathbb{Z}_{>0}$:

$$0 \longrightarrow \mathcal{I}_{L|Q} \longrightarrow \mathcal{E}_l \longrightarrow \mathcal{I}_{L|Q}(-l+1) \longrightarrow 0.$$

Now Knörrer's periodicity theorem (see [132, Theorem 3.1]) and [38, Theorem 4.1] tell us that this is the complete list of aCM indecomposable sheaves on Q.

2.4 aCM bundles on Grassmann varieties

In the previous sections we managed to completely characterize aCM bundles on the projective spaces and smooth quadrics. They are particular cases of *rational homogeneous varieties*, namely, varieties of the form $X = G/P$ where G is a simple algebraic group and $P \subset G$ is a parabolic subgroup. In this setting, one can restrict oneself to the family of *homogenous bundles* $\mathcal{E}$, for which the action of G on X extends to $\mathcal{E}$ (see [173]). It was proved in [30] that the category of homogeneous vector bundles on X is equivalent to the category of representations of the parabolic group P. Since this group is not reductive, a full description of its representations seems out of reach in general. However, things become handier when one deals with *irreducible representations of* P (equivalently, *irreducible homogeneous bundles* on X).

The goal of this section is to characterize irreducible homogeneous aCM bundles on a particular family of rational homogeneous varieties, the Grassmann varieties. To this end, let us start fixing some notation.

As usual $\mathbb{P}^n$ denotes the n-dimensional projective space associated to an $(n+1)$-dimensional $\mathbf{k}$-vector space V and $\mathrm{Gr}(k,n)$ the *Grassmann variety* of linear subspaces of $\mathbb{P}^n$ of dimension k together with the *Plücker embedding*

$$\mathrm{Gr}(k,n) \longrightarrow \mathbb{P}\Big(\bigwedge^{k+1} V\Big) \cong \mathbb{P}^{\binom{n+1}{k+1}-1}.$$

Since $\mathrm{Gr}(k,n) \cong \mathrm{Gr}(n-k-1,n)$, we can assume that $k \leq \frac{n-1}{2}$. Then $\mathrm{Gr}(k,n)$ is an aCM variety of dimension $(k+1)(n-k)$ and

$$\deg(\mathrm{Gr}(k,n)) = \frac{((k+1)(n-k))!k!(k-1)!\cdots 2!}{n!(n-1)!\cdots(n-k)!} \tag{2.8}$$

(see, for instance, [162, Proposition 1.10]). On $\mathrm{Gr}(k,n)$ (G for short), there is the universal exact sequence

$$0 \longrightarrow \mathcal{S}^\vee \longrightarrow V \otimes \mathcal{O}_G \longrightarrow \mathcal{Q} \longrightarrow 0 \tag{2.9}$$

defining *the universal vector subbundle* $\mathcal{S}^\vee$ and *the universal quotient bundle* $\mathcal{Q}$ of ranks $n-k$ and $k+1$, respectively. Note that $\bigwedge^{k+1}\mathcal{Q} \cong \bigwedge^{n-k}\mathcal{S}$ is the ample generator $\mathcal{O}_G(1)$

of Pic(Gr(k,n)). The tangent bundle of Gr(k,n) is $\mathcal{Q}\otimes\mathcal{S}$ and hence the canonical line bundle is $\mathcal{O}_G(-n-1)$.

Throughout this section the weight lattice of the group GL(V) is identified with $\mathbb{Z}^{n+1}$ and this identification takes the kth fundamental weight to the vector $(1,\dots,1,0,\dots,0)$ where the first k entries are 1 and the last $n+1-k$ are 0. Under this correspondence, the cone of dominant weights of GL(V) gets identified with the set of nonincreasing sequences $\alpha=(\alpha_1,\dots,\alpha_{n+1})\in\mathbb{Z}^{n+1}$ of integers. For such α, we denote by $\Sigma^\alpha V=\Sigma^{(\alpha_1,\dots,\alpha_{n+1})}V$ the corresponding representation of GL(V). Analogously, given a vector bundle $\mathcal{E}$ of rank r on a projective variety X, we denote by $\Sigma^\alpha\mathcal{E}$ the vector bundle associated with the GL(r)-representation of highest weight α. We choose the conventions such that

$$\Sigma^{(\lambda_1,0,\dots,0)}\mathcal{E}=S^{\lambda_1}\mathcal{E}\quad(\text{respectively }\Sigma^{(\lambda_1,0,\dots,0)}V=S^{\lambda_1}V)$$

is the λ_1-symmetric power of $\mathcal{E}$ (respectively V) and

$$\Sigma^{(\overbrace{1,\dots,1}^{m},0,\dots,0)}\mathcal{E}=\bigwedge^{m}\mathcal{E}\quad\left(\text{respectively }\Sigma^{(\overbrace{1,\dots,1}^{m},0,\dots,0)}V=\bigwedge^{m}V\right)$$

is the mth exterior power of $\mathcal{E}$ (respectively V). We also fix the convention

$$\Sigma^{(\lambda_1,\dots,\lambda_s)}\mathcal{E}=\Sigma^{(\lambda_1,\dots,\lambda_s,0,\dots,0)}\mathcal{E}$$

whenever $s<r$.

Remark 2.4.1. (1) Recall (see [173, Proposition 10.13] and [124, Section 2.5]) that every irreducible homogeneous bundle on Gr(k,n) is isomorphic to $\Sigma^\beta\mathcal{Q}\otimes\Sigma^\gamma\mathcal{S}^\vee$ for some nonincreasing $\beta=(b_1,\dots,b_{k+1})\in\mathbb{Z}^{k+1}$ and $\gamma=(a_1,\dots,a_{n-k})\subset\mathbb{Z}^{n-k}$ and according to the above conventions, for any $t\in\mathbb{Z}$, we have $\Sigma^\beta\mathcal{Q}\otimes\mathcal{O}_G(t)=\Sigma^{\beta+t}\mathcal{Q}$ and $\Sigma^\gamma\mathcal{S}^\vee\otimes\mathcal{O}_G(-t)=\Sigma^{\gamma+t}\mathcal{S}^\vee$ where $\beta+t:=\beta+(t,\dots,t)=(b_1+t,\dots,b_{k+1}+t)$ and $\gamma+t:=\gamma+(t,\dots,t)=(a_1+t,\dots,a_{n-k}+t)$.

(2) It is worthwhile to mention that Ramanan proved (see [178, Theorem 1]) that irreducible homogeneous bundles $\mathcal{E}=\Sigma^{(b_1,\dots,b_{k+1})}\mathcal{Q}\otimes\Sigma^{(a_1,\dots,a_s)}\mathcal{S}^\vee$ are simple. Therefore, a fortiori, they are indecomposable on Gr(k,n).

From the previous remark we have:

Remark 2.4.2. For any integer $l\in\mathbb{Z}$, it holds

$$\Sigma^{(b_1,\dots,b_{k+1})}\mathcal{Q}\otimes\Sigma^{(a_1,\dots,a_{n-k})}\mathcal{S}^\vee\cong\Sigma^{(b_1+l,\dots,b_{k+1}+l)}\mathcal{Q}\otimes\Sigma^{(a_1+l,\dots,a_{n-k}+l)}\mathcal{S}^\vee.$$

Thus, from now on, while dealing with irreducible homogeneous bundles $\Sigma^{(b_1,\dots,b_{k+1})}\mathcal{Q}\otimes\Sigma^{(a_1,\dots,a_{n-k})}\mathcal{S}^\vee$ on Gr(k,n), we will assume that $a_{n-k}=0$.

For a later use, we compute here the rank of the Schur power of a rank r vector bundle on $\mathrm{Gr}(k,n)$. To this end, we need to fix some notation. Given a partition $\lambda = (\lambda_1, \ldots, \lambda_r) \in \mathbb{Z}^r$ with r parts, i. e., $\lambda_1 \geq \lambda_2 \geq \cdots \geq \lambda_r \geq 0$, we will draw the Young diagram corresponding to λ by putting λ_i boxes in the ith row and left justifying the picture. We will denote also by λ the corresponding Young diagram and by λ^t its transpose. For example, the partition $\lambda = (8,5,3,2,2,0,0)$ corresponds to

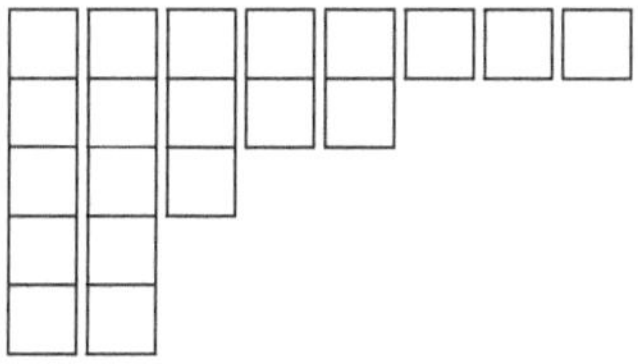

and $\lambda^t = (5,5,3,2,2,1,1,1)$. We have:

Lemma 2.4.3. *Let $\mathcal{E}$ be a rank r vector bundle on $\mathrm{Gr}(k,n)$ (respectively let W be any r-dimensional vector space) and $\lambda = (\lambda_1, \ldots, \lambda_r) \in \mathbb{Z}^r$ a partition with r parts. Then*

$$\mathrm{rank}(\Sigma^\lambda \mathcal{E}) = \dim_{\mathbf{k}}(\Sigma^\lambda W) = \prod_{(i,j)\in\mathcal{F}_\lambda} \frac{(r+j-i)}{d_{ij}}$$

where $\mathcal{F}_\lambda = \{(i,j) \mid (i,j)$ is a position in the Young diagram $\lambda\}$ and

$$d_{ij} = \lambda_i + (\lambda^t)_j - (i+j) + 1.$$

For example, if W is a vector space of dimension 6 and we consider the partition $\lambda = (6,5,2,2,1,1)$, we get $\dim_{\mathbf{k}}(\Sigma^\lambda W) = 17640$.

The Borel–Bott–Weil theorem is a powerful tool which computes the cohomology of irreducible homogeneous bundles on $\mathrm{Gr}(k,n)$ and plays an important role in our goal of characterizing aCM bundles on $\mathrm{Gr}(k,n)$. To state it, recall that for every $\alpha \in \mathbb{Z}^{n+1}$ there exists an element σ of the Weyl group $\mathfrak{S}_{n+1}$ of $\mathrm{GL}(V)$ (the group of permutations) such that $\sigma(\alpha)$ is nonincreasing, and it is unique if and only if all the entries of α are distinct. Denote by $\rho = (n+1, n, n-1, \ldots, 2, 1)$ the half-sum of the positive roots of $\mathrm{GL}(V)$ and by $l : \mathfrak{S}_{n+1} \longrightarrow \mathbb{Z}$ the standard length function.

Theorem 2.4.4. *Let $\beta = (b_1, \ldots, b_{k+1}) \in \mathbb{Z}^{k+1}$ and $\gamma = (a_1, \ldots, a_{n-k}) \in \mathbb{Z}^{n-k}$ be two nonincreasing sequences and let $\alpha = (\beta, \gamma) \in \mathbb{Z}^{n+1}$ be their concatenation.*

(1) *Assume that all entries of $\alpha+\rho$ are distinct and let σ be the unique permutation such that $\sigma(\alpha+\rho)$ is strictly decreasing. Then*

$$\mathrm{H}^m(\mathrm{Gr}(k,n), \Sigma^\beta \mathcal{Q} \otimes \Sigma^\gamma \mathcal{S}^\vee) = \begin{cases} \Sigma^{\sigma(\alpha+\rho)-\rho} V^* & \text{if } m = l(\sigma), \\ 0 & \text{otherwise.} \end{cases}$$

(2) *If at least two entries of $\alpha+\rho$ coincide then, for any $i \geq 0$,*

$$\mathrm{H}^i(\mathrm{Gr}(k,n), \Sigma^\beta \mathcal{Q} \otimes \Sigma^\gamma \mathcal{S}^\vee) = 0.$$

Proof. See [76, Theorem 2]. □

Remark 2.4.5. Theorem 2.4.4 implies that for any twist $\mathcal{E}(t)$ of an irreducible homogeneous bundle $\mathcal{E}$ on $\mathrm{Gr}(k,n)$ there exists at most one i_t, $0 \leq i_t \leq (k+1)(n-k)$ such that $\mathrm{H}^{i_t}(\mathrm{Gr}(k,n), \mathcal{E}(t)) \neq 0$. Hence, according to Definition 2.1.16, any irreducible homogeneous bundle $\mathcal{E}$ on the Grassmann variety $\mathrm{Gr}(k,n)$ has natural cohomology.

Lemma 2.4.6. *Let $\mathcal{E} = \Sigma^{(b_1,\dots,b_{k+1})}\mathcal{Q} \otimes \Sigma^{(a_1,\dots,a_s)}\mathcal{S}^\vee$ be an initialized irreducible homogeneous bundle on the Grassmann variety $\mathrm{Gr}(k,n)$ with $s < n-k$ and $a_s \neq 0$. We have*

$$b_1 \geq \cdots \geq b_{k+1} = a_1 \geq \cdots \geq a_s > 0.$$

Proof. Since $\mathcal{E}$ is initialized, $\mathrm{H}^0(\mathrm{Gr}(k,n), \mathcal{E}) \neq 0$ and $\mathrm{H}^0(\mathrm{Gr}(k,n), \mathcal{E}(-1)) = 0$. By Theorem 2.4.4, the first condition implies that the entries of

$$(b_1,\dots,b_{k+1},a_1,\dots,a_s,0,\dots,0) + (n+1,n,\dots,2,1)$$

are strictly decreasing. Thus

$$b_1 \geq \cdots \geq b_{k+1} \geq a_1 \geq \cdots \geq a_s > 0. \tag{2.10}$$

Since $\mathrm{H}^0(\mathrm{Gr}(k,n), \mathcal{E}(-1)) = 0$, by Theorem 2.4.4 we have at least two coincident entries in

$$(b_1-1,\dots,b_{k+1}-1,a_1,\dots,a_s,0,\dots,0) + (n+1,n,\dots,2,1),$$

and this forces $b_{k+1} = a_1$. Therefore,

$$b_1 \geq \cdots \geq b_{k+1} = a_1 \geq \cdots \geq a_s > 0.$$ □

Up to now, we have seen few examples of aCM bundles. The goal of this section is to characterize irreducible homogeneous aCM bundles on $\mathrm{Gr}(k,n)$. In particular, we will see that there are infinitely many of them. The key point is to associate to any irreducible homogeneous bundle $\mathcal{E}$ on $\mathrm{Gr}(k,n)$ a $(n-k)(k+1)$-matrix of integers which encodes the cohomology of $\mathcal{E}$. So, let us start fixing the definition of these matrices.

Definition 2.4.7. Let $T = (t_i^j)$, $1 \leq i \leq p$, $1 \leq j \leq q$, be a $(p \times q)$-matrix with coefficients $t_i^j \in \mathbb{Z}_{>0}$. We say that T is a *$(p \times q)$-step matrix* (or a step matrix for short) if the differences between columns and rows are constant, i. e.,

$$\begin{aligned} t_i^{j+1} - t_i^j &= \epsilon_j \geq 0 \quad \text{for } 1 \leq i \leq p, 1 \leq j \leq q-1, \\ t_{i+1}^j - t_i^j &= \gamma_i \geq 0 \quad \text{for } 1 \leq i \leq p-1, 1 \leq j \leq q. \end{aligned}$$

A step matrix $T = (t_i^j)$ is said to be *normalized* if $t_1^1 = 1$.

To any initialized irreducible homogeneous bundle $\mathcal{E}$ on $\mathrm{Gr}(k,n)$, we associate a normalized $(n-k)\times(k+1)$-step matrix. Indeed, given any initialized irreducible homogeneous bundle $\mathcal{E} = \Sigma^{(b_1,\dots,b_{k+1})}\mathcal{Q}\otimes\Sigma^{(a_1,\dots,a_{n-k})}\mathcal{S}^\vee$ with $b_1 \ge \cdots \ge b_{k+1} = a_1 \ge \cdots \ge a_{n-k} = 0$, we define its associated $(n-k)\times(k+1)$-step matrix $T_\mathcal{E} = (t_i^j)$ as follows:

$$t_i^j := b_{k+2-j} - a_i + i + j - 1, \quad \text{for } 1 \le i \le n-k, \quad 1 \le j \le k+1.$$

This $T_\mathcal{E}$ is a step matrix. In fact, we have

$$t_i^{j+1} - t_i^j = b_{k+2-j-1} - b_{k+2-j} + 1 \ge 1 \quad \text{for } 1 \le i \le n-k, \quad 1 \le j \le k,$$
$$t_{i+1}^j - t_i^j = a_i - a_{i+1} + 1 \ge 1 \quad \text{for } 1 \le i \le n-k-1, \quad 1 \le j \le k+1.$$

Moreover, $T_\mathcal{E}$ is normalized since $t_1^1 = 1$.

Conversely, given a normalized $(n-k)\times(k+1)$-step matrix $T = (t_i^j)$, we associate to it an initialized irreducible homogeneous bundle $\mathcal{E}_T = \Sigma^{(b_1,\dots,b_{k+1})}\mathcal{Q}\otimes\Sigma^{(a_1,\dots,a_{n-k})}\mathcal{S}^\vee$ where

$$a_i = t_{n-k}^1 - t_i^1 - (n-k-i),$$
$$b_{k+2-j} = t_1^j + t_{n-k}^1 - (n-k) - j.$$

Remark 2.4.8. (1) It holds that $T_{\mathcal{E}_T} = T$ and $\mathcal{E}_{T_\mathcal{E}} = \mathcal{E}$.

(2) Given any initialized irreducible homogeneous bundle $\mathcal{E} = \Sigma^{(b_1,\dots,b_{k+1})}\mathcal{Q}\otimes\Sigma^{(a_1,\dots,a_{n-k})}\mathcal{S}^\vee$ on $\mathrm{Gr}(k,n)$, the entries t_i^j of its associated step matrix $T_\mathcal{E}$ are bounded as follows:

$$1 \le t_i^j \le b_1 + n.$$

Now we are ready to state the characterization of aCM irreducible homogeneous bundles on $\mathrm{Gr}(k,n)$ in terms of their step matrix.

Theorem 2.4.9. *Let $\mathcal{E} = \Sigma^{(b_1,\dots,b_{k+1})}\mathcal{Q}\otimes\Sigma^{(a_1,\dots,a_{n-k})}\mathcal{S}^\vee$ be any irreducible initialized homogeneous bundle on $\mathrm{Gr}(k,n)$ with $b_1 \ge \cdots \ge b_{k+1} = a_1 \ge \cdots \ge a_{n-k} = 0$ and let $T_\mathcal{E} = (t_i^j)$ be its associated normalized step matrix. Denote by $n_l := \#\{t_i^j \mid t_i^j = l\}$. Then, $\mathcal{E}$ is an aCM bundle if and only if*

$$\sum_{l=1}^{b_1+n} (n_l - 1)_+ = (k+1)(n-k) - (b_1+n)$$

where, for an integer $a \in \mathbb{Z}$, $a_+ := \max\{a, 0\}$.

Proof. Let $\mathcal{E} = \Sigma^{(b_1,\dots,b_{k+1})}\mathcal{Q}\otimes\Sigma^{(a_1,\dots,a_{n-k})}\mathcal{S}^\vee$ be any initialized homogeneous bundle on $\mathrm{Gr}(k,n)$ with $b_1 \ge \cdots \ge b_{k+1} = a_1 \ge \cdots \ge a_{n-k} = 0$. First of all, recall that $\mathcal{E}$ has natural cohomology (Remark 2.4.5). Since by assumption $\mathcal{E}$ is initialized, for any integer $t \ge 0$,

$\mathcal{E}(t)$ has cohomology concentrated in degree 0 and for $t < 0$, $\mathrm{H}^0(\mathcal{E}(t)) = 0$. On the other hand, by Borel–Bott–Weil theorem (Theorem 2.4.4), for any $t \le -b_1 - n - 1$,

$$\mathrm{H}^i(\mathcal{E}(t)) \neq 0 \quad \text{if and only if} \quad i = \dim \mathrm{Gr}(k,n) = (k+1)(n-k)$$

and for any $t > -b_1 - n - 1$, $\mathrm{H}^{(k+1)(n-k)}(\mathcal{E}(t)) = 0$.

Thus, $\mathcal{E}$ is an aCM bundle if and only if for any $t \in [1, b_1 + n]$ and all $i \ge 0$,

$$\mathrm{H}^i(\mathcal{E}(-t)) = 0.$$

By Borel–Bott–Weil theorem, this is equivalent to the fact that for any $t \in [1, b_1 + n]$, at least two entries of

$$(b_1 - t, b_2 - t, \ldots, b_{k+1} - t, a_1, a_2, \ldots, a_{n-k}) + (n+1, n, \ldots, 2, 1)$$

coincide. That is, $\mathcal{E}$ is an aCM bundle if and only if for any $t \in [1, b_1 + n]$ there exists a pair $(i,j) \in [1, n-k] \times [1,k]$ such that

$$b_{k+2-j} + n - (k+2-j) + 2 - t = a_i + (n-k) - (i-1). \tag{2.11}$$

Finally, recall that by definition the entries of the step matrix $T_{\mathcal{E}} = (t_i^j)$ associated to $\mathcal{E}$ are given by

$$t_i^j = b_{k+2-j} - a_i + i + j - 1.$$

Now assume that $\mathcal{E}$ is an aCM bundle. According to the equality (2.11), there are at least $b_1 + n$ entries of $T_{\mathcal{E}}$ taking exactly the values between 1 and $b_1 + n$. Moreover, by Remark 2.4.8 (2), for any entry t_i^j of $T_{\mathcal{E}}$ we have

$$1 \le t_i^j \le b_1 + n.$$

Therefore, $T_{\mathcal{E}}$ must contain $(k+1)(n-k) - (b_1 + n)$ repeated entries, that is,

$$\sum_{l=1}^{b_1+n} (n_l - 1)_+ = (k+1)(n-k) - (b_1 + n).$$

Conversely, if $\sum_{l=1}^{b_1+n}(n_l - 1)_+ = (k+1)(n-k) - (b_1+n)$ then $T_{\mathcal{E}}$ has exactly $b_1 + n$ different entries that take value between 1 and $b_1 + n$. Therefore, for any $t \in [1, b_1 + n]$ there exists a t_i^j such that

$$t = t_i^j = b_{k+2-j} - a_i + i + j - 1,$$

which by equality (2.11) is equivalent to the fact that $\mathcal{E}$ is an aCM bundle. □

In the last theorem we have characterized irreducible homogeneous aCM bundles on $\mathrm{Gr}(k,n)$. If we delete the hypothesis of being irreducible and homogeneous, much more examples of aCM bundles on $\mathrm{Gr}(k,n)$ can be built. In fact, by [69, Theorem 4.6], we have that for all integers $0 < k < n-1$ and $(k,n) \neq (1,3)$, the Grassmann variety $\mathrm{Gr}(k,n)$ is of wild representation type (see Definition 2.1.14). For $k = 0$ or $k = n-1$ or $(k,n) = (1,3)$, it is of finite representation type.

We finish this section with a couple of examples which illustrate Theorem 2.4.9.

Example 2.4.10. (1) On the Grassmann variety $\mathrm{Gr}(3,8)$ we consider the vector bundles $\mathcal{E} = \Sigma^{(10,10,3,2)}\mathcal{Q}\otimes\Sigma^{(2,2,1,1,0)}\mathcal{S}^\vee$ and $\mathcal{F} = \Sigma^{(8,8,2,2)}\mathcal{Q}\otimes\Sigma^{(2,2,1,0,0)}\mathcal{S}^\vee$. Let us see that $\mathcal{F}$ is an aCM bundle but $\mathcal{E}$ is not an aCM bundle. Indeed, the associated normalized step matrices are

$$T_{\mathcal{E}} = \begin{pmatrix} 1 & 2 & 4 & 5 & 7 \\ 3 & 4 & 6 & 7 & 9 \\ 11 & 12 & 14 & 15 & 17 \\ 12 & 13 & 15 & 16 & 18 \end{pmatrix}$$

and

$$T_{\mathcal{F}} = \begin{pmatrix} 1 & 2 & 4 & 6 & 7 \\ 2 & 3 & 5 & 7 & 8 \\ 9 & 10 & 12 & 14 & 15 \\ 10 & 11 & 13 & 15 & 16 \end{pmatrix}.$$

In the first case, $\sum_{l=1}^{18}(n_l-1)_+ = 4 > 20-18 = (k+1)(n-k)-(b_1+n)$. Thus, by Theorem 2.4.9, $\mathcal{E}$ is not an aCM bundle. On the other hand, according to $T_{\mathcal{F}}$, $\sum_{l=1}^{16}(n_l-1)_+ = 4 = 20-16$. Therefore, $\mathcal{F}$ is an aCM bundle.

(2) We consider the Grassmann variety $\mathrm{Gr}(k,n)$ with $2k \leq n-1$. Let us see that $\mathcal{E} = \Sigma^{(k,\dots,k,0)}\mathcal{Q}$ is an aCM bundle on $\mathrm{Gr}(k,n)$. The associated normalized step matrix is given by

$$T_{\mathcal{E}} = \begin{pmatrix} 1 & 2 & \cdots & n-k \\ k+2 & k+3 & \cdots & n+1 \\ \vdots & \vdots & \ddots & \vdots \\ 2k+1 & 2k+2 & \cdots & n+k \end{pmatrix}.$$

Since $2k \leq n-1$, we have $\sum_{l=1}^{k+n}(n_l-1)_+ = (k+1)(n-k)-(k+n)$ and by Theorem 2.4.9, $\mathcal{E}$ is an aCM bundle.

(3) On the Grassmann variety $\mathrm{Gr}(k,n)$ with $2k \leq n-1$ and $n-k > 2$, the vector bundle $\mathcal{E} = \Sigma^{(2,1)}\mathcal{S}$ is aCM. In fact, by Proposition 2.1.13, $\mathcal{E}$ is an aCM bundle on $\mathrm{Gr}(k,n)$ if and only if $\mathcal{E}^\vee$ is aCM. This is equivalent to saying that $\mathcal{E}^\vee(2)$ is an aCM bundle, which follows from the fact that (notice that the functors Σ^α commute with taking duals)

$$\Sigma^{(2,1)}\mathcal{S}^\vee(2) \cong \Sigma^{(2,2,\dots,2)}\mathcal{Q}\otimes\Sigma^{(2,1)}\mathcal{S}^\vee,$$

and hence it is initialized and its $(n-k)\times(k+1)$-step matrix is given by

$$\begin{pmatrix} 1 & 3 & 5 & 6 & \cdots & n-k+2 \\ 2 & 4 & 6 & 7 & \cdots & n-k+3 \\ 3 & 5 & 7 & 8 & \cdots & n-k+4 \\ \vdots & \vdots & \vdots & \vdots & & \vdots \\ k+1 & k+3 & k+5 & k+6 & \cdots & n+2 \end{pmatrix}.$$

2.5 aCM bundles of low rank on hypersurfaces

We have seen that there is a complete classification of aCM bundles on $\mathbb{P}^n$ and on $Q_n \subset \mathbb{P}^{n+1}$. The same is true for aCM bundles on Grassmann varieties if we restrict our attention to irreducible homogeneous bundles. For other varieties, to obtain a full classification of aCM bundles, one has to restrict the family under consideration and fix some invariants of the vector bundle as, for instance, the rank or the Chern classes. In this section, we will focus our attention on the classification of low rank aCM bundles on smooth hypersurfaces of $\mathbb{P}^n$.

Through this section, X will be a smooth hypersurface of degree $d \geq 2$ in $\mathbb{P}^n$ for $n \geq 3$. By the *Grothendieck–Lefschetz theorem*, if $d \geq 2$ and $n \geq 4$, $\operatorname{Pic}(X) \cong \mathbb{Z}$, and hence any line bundle $\mathcal{L}$ on X can be written as $\mathcal{L} = \mathcal{O}_X(f)$ for some $f \in \mathbb{Z}$. If $n = 3$, *Noether–Lefschetz theorem* asserts that when $d \geq 4$ and X is very general then again $\operatorname{Pic}(X) \cong \mathbb{Z}$, and for any line bundle $\mathcal{L}$ on X there is $f \in \mathbb{Z}$ such that $\mathcal{L} = \mathcal{O}_X(f)$. If $n = 3$, $d = 2$ (respectively $d = 3$) then $\operatorname{Pic}(X) \cong \mathbb{Z}^2$ (respectively $\operatorname{Pic}(X) \cong \mathbb{Z}^7$).

Let us start analyzing the existence of aCM line bundles on X. It holds:

(1) If $n \geq 4$, then $\mathcal{O}_X$ is (up to twist) the only aCM line bundle.
(2) If X is very general, $n = 3$, and $d \geq 4$, then $\mathcal{O}_X$ is (up to twist) the only aCM line bundle.
(3) If X is smooth and $n = d = 3$ we saw in Example 2.1.3 that $\operatorname{Pic}(X) \cong \mathbb{Z}^7 = \langle h; e_1, \dots, e_6\rangle$. Then $\mathcal{O}_X(D)$ is an aCM line bundle if (up to permutation of the exceptional divisors e_i), D is one of the following ten divisors:

$$\begin{array}{ll} h, & 5h-2e_1-e_2-2e_3-2e_4-2e_5-2e_6, \\ h-e_1, & 4h-2e_1-2e_2-2e_3-e_4-e_5-e_6, \\ h-e_2, & 3h-2e_1-e_2-e_3-e_4-e_5-e_6, \\ 2h-e_1-e_2-e_3, & 3h-2e_1-e_2-e_3-e_4-e_5, \\ 2h-e_1-e_2-e_3-e_4, & 2h-e_1-e_2-e_3-e_4-e_5. \end{array} \tag{2.12}$$

Notice that they correspond to rational normal curves on X (see [175] and also Corollary 5.3.22).

(4) If $n = 3$, $d = 2$, and X is smooth, it holds that $X = \mathbb{P}^1 \times \mathbb{P}^1$ and then $\mathcal{O}_X$, $\mathcal{O}_X(0,1)$, and $\mathcal{O}_X(1,0)$ are up to twist the only aCM line bundles.

We now discuss the existence and classification of rank 2 aCM bundles $\mathcal{E}$ on X.

Proposition 2.5.1. *On any smooth surface $(X, \mathcal{O}_X(H))$ in $\mathbb{P}^3$ of degree $d \geq 3$, there exists an indecomposable rank 2 aCM bundle.*

Proof. We choose a point $p \in X$ and set $Z = \{p\}$. Then Z is an aG variety with canonical module K_Z generated in degree 1 by a unique element ε which provides an extension:

$$\varepsilon : \quad 0 \longrightarrow R_X \longrightarrow M \longrightarrow I_{Z/X}(3-d) \longrightarrow 0 \tag{2.13}$$

via the isomorphism $K_Z = \operatorname{Ext}^1(I_{Z/X}(-1), R_X(d-4))$. Applying $\operatorname{Hom}(-, R_X(d-4))$ to the exact sequence (2.13), we get that $\operatorname{Ext}^1(M, K_X) = 0$, i. e., M is an MCM R_X-module. Therefore, by Proposition 2.1.8 its sheafification $\mathcal{E} = \widetilde{M}$ is an aCM bundle. For more details on this construction, the reader should have a look at Theorem 5.1.6.

Now, since $c_1(\mathcal{E}) = (3-d)H$, we see that $\mathcal{E}$ cannot decompose. Otherwise, Z would be the complete intersection of two effective divisors D_1 and D_2 on S such that $\mathcal{E} \cong \mathcal{O}_S(D_1) \oplus \mathcal{O}_S(D_2)$. But then $D_1 + D_2 \equiv (3-d)H$, which is impossible for $d \geq 3$. □

Remark 2.5.2. (1) If in the previous proposition we suppose that $\operatorname{Pic}(X) \cong \mathbb{Z}$, the irreducibility of the aCM bundle $\mathcal{E}$ follows immediately from Proposition 2.1.13 and the fact that the only initialized aCM line bundle on X is $\mathcal{O}_X$.

(2) Notice that for $d = 2$ the situation is very different. Indeed, any point $p \in \mathbb{P}^1 \times \mathbb{P}^1 \subset \mathbb{P}^3$ can be written as the complete intersection of two divisors associated to the aCM line bundles $\mathcal{O}_{\mathbb{P}^1\times\mathbb{P}^1}(1,0)$ and $\mathcal{O}_{\mathbb{P}^1\times\mathbb{P}^1}(0,1)$ (see Example 2.1.3).

Once we know the existence of rank 2 aCM bundles on a smooth surface $X \subset \mathbb{P}^3$ of degree $d \geq 3$, it would be interesting to determine the integers (c_1, c_2) such that on a smooth surface $X \subset \mathbb{P}^3$ of degree $d \geq 2$ there exists a rank 2 (μ-stable) aCM bundle $\mathcal{E}$ on X with Chern classes $c_i(\mathcal{E}) = c_i$. More generally, we would like to determine pairs $(r, d) \in \mathbb{Z}^2$ such that on a smooth surface $X \subset \mathbb{P}^3$ of degree d there exists a rank r (μ-stable) aCM bundle $\mathcal{E}$ on X (see [55] and [174, Remark 4.5.12]).

Let X be a smooth hypersurface of degree $d \geq 2$ in $\mathbb{P}^n$, $n \geq 4$, and $\mathcal{E}$ be an indecomposable rank 2 aCM bundle on X. If $d = 2$, according to Theorem 2.3.4, $\mathcal{E}$ exists only for $n = 4$ and $n = 5$ (see Section 2.3). Hence, in the remaining part of this section, X will be a smooth hypersurface of degree $d \geq 3$ in $\mathbb{P}^n$, $n \geq 4$ and $\mathcal{E}$ a rank 2 aCM bundle on X.

Let us deal first with smooth threefolds $X \subset \mathbb{P}^4$ of low degree where it is easy to construct indecomposable rank-two aCM bundles.

Proposition 2.5.3. *Let $X \subset \mathbb{P}^4$ be a smooth threefold of degree $3 \leq d \leq 5$. Then, X supports indecomposable rank-two aCM bundles.*

Proof. Let $L \subset X$ be a line in X (indeed, it is well-known that, for a fixed d, any smooth threefold $X \subset \mathbb{P}^4$ of degree d contains a line if and only if $2 \leq d \leq 5$, see [133, Chapter V]). Associated to the line L, there exists the short exact sequence

$$0 \to \mathcal{E} \to \mathcal{O}_X(-1)^3 \xrightarrow{\phi} \mathcal{I}_{L|X} \to 0$$

where $\mathcal{I}_{L|X}$ is the ideal sheaf of the line in X, ϕ is the map associated to the linear forms of $\mathbb{P}^4$ defining L, and $\mathcal{E}$ is the kernel of ϕ. It turns out, by a depth computation, that $\mathcal{E}$ is a vector bundle. Since $\mathrm{H}^0_*(\mathcal{O}_X(-1)^3) \to \mathrm{H}^0_*(\mathcal{I}_{L|X})$ is surjective, we have $\mathrm{H}^1_*(\mathcal{E}) = 0$ which automatically implies, by Serre duality, that also $H^2_*(\mathcal{E}) = 0$. In other words, $\mathcal{E}$ is an aCM bundle of rank two. Since $c_1(\mathcal{E}) = -3$, it is immediate that $\mathcal{E}$ cannot split as a direct sum of line bundles. □

For a general threefold $X \subset \mathbb{P}^4$ of degree $d \geq 6$, the situation is completely different.

Proposition 2.5.4. *Let $X \subset \mathbb{P}^4$ be a general threefold of degree $d \geq 6$. Then, any rank-two aCM bundle $\mathcal{E}$ on X splits as the direct sum of two line bundles.*

Proof. See [135, Main Theorem]. The case $d = 6$ was first proved in [53, Theorem 1.3]. The result for arbitrarily $d \geq 6$ was conjectured in the latter paper. □

Now we move forward to the study of rank-two aCM bundles on smooth hypersurfaces $X \subset \mathbb{P}^n$ with $n \geq 5$. We set $e := c_1(\mathcal{E})$ so that $\bigwedge^2 \mathcal{E} \cong \mathcal{O}_X(e)$ and since $\mathcal{E}$ is a rank two vector bundle on X, $\mathcal{E}^\vee \cong \mathcal{E}(-e)$. From the fact that $\mathcal{E}$ is an aCM $\mathcal{O}_{\mathbb{P}^n}$-sheaf of dimension $n-1$, we deduce that it has a minimal resolution of the form (see [22, Theorem A]):

$$0 \longrightarrow \mathcal{F}_1 \longrightarrow \mathcal{F}_0 \longrightarrow \mathcal{E} \longrightarrow 0 \tag{2.14}$$

where the $\mathcal{F}_i$'s are direct sums of r line bundles on $\mathbb{P}^n$.

Restricting the exact sequence (2.14) to X, we get the exact sequence

$$0 \longrightarrow \mathcal{E}(-d) \longrightarrow \overline{\mathcal{F}}_1 \xrightarrow{\overline{\phi}} \overline{\mathcal{F}}_0 \longrightarrow \mathcal{E} \longrightarrow 0,$$

where $\overline{\mathcal{F}}_i := \mathcal{F}_i \otimes \mathcal{O}_X$. Denote by $\mathcal{G}$ the image of $\overline{\phi}$ so that we have the two short exact sequences:

$$0 \longrightarrow \mathcal{E}(-d) \longrightarrow \overline{\mathcal{F}}_1 \longrightarrow \mathcal{G} \longrightarrow 0, \tag{2.15}$$

$$0 \longrightarrow \mathcal{G} \longrightarrow \overline{\mathcal{F}}_0 \longrightarrow \mathcal{E} \longrightarrow 0. \tag{2.16}$$

Finally, taking the second exterior power in (2.14), we get

$$0 \longrightarrow \bigwedge^2 \mathcal{F}_1 \longrightarrow \bigwedge^2 \mathcal{F}_0 \longrightarrow \mathcal{F} \longrightarrow 0 \tag{2.17}$$

for some cokernel sheaf $\mathcal{F}$ with reduced support X.

The following technical result will be the key ingredient that we will use to obtain the classification of rank-two aCM bundles on X.

Proposition 2.5.5. *With the above notations, the following holds:*
(1) *There is an exact sequence*

$$0 \longrightarrow \mathcal{L} \longrightarrow \mathcal{F} \longrightarrow \overline{\mathcal{F}} = \mathcal{F} \otimes \mathcal{O}_X \longrightarrow 0 \tag{2.18}$$

where $\mathcal{L}$ is a line bundle on X and $\overline{\mathcal{F}}$ is a rank $2r-3$ vector bundle on X which fits into an exact sequence

$$0 \longrightarrow \mathcal{E} \otimes \mathcal{G} \longrightarrow \overline{\mathcal{F}} \longrightarrow \mathcal{O}_X(e) \longrightarrow 0 \tag{2.19}$$

with $\mathcal{O}_X(e) \cong \bigwedge^2 \mathcal{E}$.
(2) *For any $k \in \mathbb{Z}$ and i, $2 \le i \le n-3$, $\mathrm{H}^i(\mathcal{E} \otimes \mathcal{G}(k)) = 0$.*
(3) *For any $k \in \mathbb{Z}$ and i, $1 \le i \le n-3$,*

$$\mathrm{H}^i(\mathcal{E}^\vee \otimes \mathcal{G}(k)) = \mathrm{H}^{i+1}(\mathcal{E}^\vee \otimes \mathcal{E}(k-d)),$$
$$\mathrm{H}^i(\mathcal{E}^\vee \otimes \mathcal{E}(k)) = \mathrm{H}^{i+1}(\mathcal{E}^\vee \otimes \mathcal{G}(k)).$$

(4) *If $\mathcal{E}$ is indecomposable, then the finite length module $\bigoplus_k \mathrm{H}^1(\mathcal{E} \otimes \mathcal{G}(k))$ is a nonzero cyclic module generated by an element of degree $-e$.*

Proof. (1) It follows from [134, Lemma 2.1].

(2) From the exact sequence (2.14) and the fact that $\mathcal{F}_j$, $j = 0, 1$, is a direct sum of line bundles on $\mathbb{P}^n$ (see Theorem 2.2.2), we get for $1 \le i \le n-2$,

$$\mathrm{H}^i_*(\mathcal{E}) = 0.$$

By the same argument and using the fact that $\mathcal{F}$ is the cokernel bundle defined in (2.17), we get $\mathrm{H}^i_*(\mathcal{F}) = 0$, for $1 \le i \le n-2$, and also have $\mathrm{H}^i_*(\mathcal{L}) = 0$ for $1 \le i \le n-2$. Hence from the exact sequence (2.18) we deduce

$$\mathrm{H}^i_*(\overline{\mathcal{F}}) = 0 \quad \text{for } 1 \le i \le n-3. \tag{2.20}$$

Finally, from the exact sequence (2.19), for any $k \in \mathbb{Z}$ we get the long exact sequence

$$\longrightarrow \mathrm{H}^{i-1}(\mathcal{O}_X(k+e)) \longrightarrow \mathrm{H}^i(\mathcal{E} \otimes \mathcal{G}(k)) \longrightarrow \mathrm{H}^i(\overline{\mathcal{F}}(k)) \longrightarrow \cdots$$

from which we deduce that $\mathrm{H}^i(\mathcal{E} \otimes \mathcal{G}(k)) = 0$ for $2 \le i \le n-3$.

(3) For any $k \in \mathbb{Z}$, we tensor the exact sequence (2.15) with $\mathcal{E}^\vee(k)$ and, using the fact that $\mathcal{F}_1$ is a direct sum of line bundles, get

$$\mathrm{H}^i(\mathcal{E}^\vee \otimes \mathcal{G}(k)) = \mathrm{H}^{i+1}(\mathcal{E}^\vee \otimes \mathcal{E}(k-d)), \quad \text{for } 1 \le i \le n-3,$$

and similarly using (2.16) we get

$$\mathrm{H}^i(\mathcal{E}^\vee \otimes \mathcal{E}(k)) = \mathrm{H}^{i+1}(\mathcal{E}^\vee \otimes \mathcal{G}(k))$$

for all $k \in \mathbb{Z}$ and $1 \leq i \leq n-3$.

(4) Using the exact sequence (2.19), taking cohomology, and using (2.20), we get that the module $\bigoplus_k \mathrm{H}^1(\mathcal{E} \otimes \mathcal{G}(k))$ is cyclically generated by an element of degree $-e$. If it is zero, then

$$\mathrm{H}^1(\mathcal{E} \otimes \mathcal{G}(-e)) = \mathrm{H}^1(\mathcal{E}^\vee \otimes \mathcal{G}) = \mathrm{Ext}^1(\mathcal{E}, \mathcal{G}) = 0,$$

which implies that the exact sequence (2.16) splits and, since $\overline{\mathcal{F}}_0$ is a direct sum of line bundles, the same holds for $\mathcal{E}$, leading to a contradiction. Therefore, the finite length module $\bigoplus_k \mathrm{H}^1(\mathcal{E} \otimes \mathcal{G}(k))$ is a nonzero cyclic module generated by an element of degree $-e$. □

Using arguments similar to those used in the construction of Quot schemes, one can obtain some results on vector bundles on families of varieties that allow getting the following:

Proposition 2.5.6. *Let $X \subset \mathbb{P}^n$, $n \geq 4$, be a general hypersurface of degree $d \geq 3$ and $\mathcal{E}$ a rank-two aCM bundle on X. Then, for any $g \in \mathrm{H}^0(\mathcal{O}_X(d))$,*

$$\mathrm{H}^2(\mathcal{E}^\vee \otimes \mathcal{E}(-d)) \xrightarrow{g} \mathrm{H}^2(\mathcal{E}^\vee \otimes \mathcal{E})$$

is zero.

Proof. It follows from [134, Corollaries 3.5 and 3.8]. □

Now we are ready to state the main result concerning the existence of rank 2 aCM bundles on hypersurfaces $X \subset \mathbb{P}^n$, $n \geq 5$, of degree $d \geq 3$.

Theorem 2.5.7. *Let $X \subset \mathbb{P}^n$ be a smooth hypersurface of degree $d \geq 3$.*
(1) *If $n \geq 6$, any aCM rank-two bundle $\mathcal{E}$ on X is a direct sum of line bundles.*
(2) *If $n = 5$ and X is general, then any aCM rank-two bundle $\mathcal{E}$ on X is a direct sum of line bundles.*

Proof. (1) Let $\mathcal{E}$ be a rank-two aCM bundle on $X \subset \mathbb{P}^n$, $n \geq 6$. Since $n \geq 6$, by Proposition 2.5.5, (3), and (2),

$$\mathrm{H}^2(\mathcal{E}^\vee \otimes \mathcal{E}(-d)) = \mathrm{H}^3(\mathcal{E}^\vee \otimes \mathcal{G}(-d)) = \mathrm{H}^3(\mathcal{E} \otimes \mathcal{G}(-d-e)) = 0.$$

Hence, applying once more Proposition 2.5.5 (3), we get

$$0 = \mathrm{H}^2(\mathcal{E}^\vee \otimes \mathcal{E}(-d)) = \mathrm{H}^1(\mathcal{E}^\vee \otimes \mathcal{G}) = \mathrm{Ext}^1(\mathcal{E}, \mathcal{G}),$$

which implies that the exact sequence (2.16) splits. Since $\overline{\mathcal{F}}_0$ is a direct sum of line bundles, we deduce that $\mathcal{E}$ also decomposes into a direct sum of line bundles.

(2) We will prove it by contradiction. Assume that a general hypersurface X of degree $d \geq 3$ has an indecomposable aCM bundle $\mathcal{E}$ of rank two. By Proposition 2.5.6, for any $g \in \mathrm{H}^0(\mathcal{O}_X(d))$,

$$\mathrm{H}^2(\mathcal{E}^\vee \otimes \mathcal{E}(-d)) \xrightarrow{g} \mathrm{H}^2(\mathcal{E}^\vee \otimes \mathcal{E})$$

is zero. By Proposition 2.5.5, (3), and (4), $N := \bigoplus_k N_k$ where

$$N_k := \mathrm{H}^2(\mathcal{E}^\vee \otimes \mathcal{E}(k)) = \mathrm{H}^1(\mathcal{E}^\vee \otimes \mathcal{G}(k+d)) = \mathrm{H}^1(\mathcal{E} \otimes \mathcal{G}(k-e+d))$$

is a nonzero cyclic module generated by an element ξ of degree $-d$. Thus, N_0 consist of multiples of ξ by elements $g \in \mathrm{H}^0(\mathcal{O}_X(d))$. Since these are zero, we deduce that $N_i = 0$ for $i \geq 0$.

On the other hand, since $n = 5$, by Serre duality

$$\mathrm{H}^2(\mathcal{E}^\vee \otimes \mathcal{E}(-d))^\vee \cong \mathrm{H}^2(\mathcal{E}^\vee \otimes \mathcal{E}(2d-6)) = N_{2d-6}.$$

Since N is nonzero generated in degree $-d$, $\mathrm{H}^2(\mathcal{E}^\vee \otimes \mathcal{E}(-d)) \neq 0$, and thus $N_{2d-6} \neq 0$, which contradicts the fact that for $d \geq 3$, $2d-6 \geq 0$. Therefore, $\mathcal{E}$ splits. □

Remark 2.5.8. (1) The hypothesis $d \geq 3$ in Theorem 2.5.7 cannot be dropped. Indeed, on a quadric $Q_4 \subset \mathbb{P}^5$, there are exactly two rank 2 indecomposable aCM bundles, namely, the spinor bundles S_2' and S_2'' (see Theorem 2.3.4).

(2) If $n \geq 5$ and $d \geq 2$, there certainly exist indecomposable rank 2 aCM bundles on certain singular hypersurfaces $X \subset \mathbb{P}^n$ of degree d. For instance, we can take a hypersurface defined as the determinant of a suitable homogeneous square matrix. For such examples, the reader could look at [131, Theorem 3.13 and Corollary 3.16].

Similar results have been obtained in [181] for rank-three aCM bundles on hypersurfaces. More concretely, the following holds:

Theorem 2.5.9. *Let $X \subset \mathbb{P}^n$ be a smooth hypersurface of degree $d \geq 3$.*
(1) *If $n = 6$, any rank-three aCM bundle on X is the direct sum of line bundles.*
(2) *If $n = 5$ and X is general, then any rank-three aCM bundle on X is the direct sum of line bundles.*

Proof. See [181, Theorem 1.1]. □

Now we will turn our attention to rank $r \geq 4$ aCM vector bundles $\mathcal{E}$ on general hypersurfaces X of degree $d \geq 3$ on $\mathbb{P}^n$, $n \geq 4$. In fact, we will determine an upper bound for the rank r in terms of d and n such that any aCM bundle of rank less than this bound on a general hypersurface $X \subset \mathbb{P}^n$ of degree d splits into a direct sum of line bundles. This will also be a classification result for higher rank aCM bundles on hypersurfaces in the spirit of a question raised by Hartshorne that morally says that a vector bundle on a projective space splits if the dimension of the projective space is sufficiently larger than the rank of the vector bundle (see [107, Conjecture 6.3]).

To reach our new goal, we need to explain some basic facts on the two-factor matrix factorization introduced by Eisenbud in [80]. In Chapter 4 we will deal with linear matrix factorizations in a more general setting.

Fix $n \geq 3$ and let $f \in R = \mathbf{k}[x_0, \dots, x_n]$ be a homogenous polynomial. Set $S = R/(f)$.

Definition 2.5.10. A *two-factor matrix factorization of* f is a pair (φ, ψ) of square matrices of size l over R which gives rise to homogeneous homomorphisms between graded free R-modules satisfying the following conditions:

(1) All entries of φ are nonunits;

(2) $\varphi\psi = \psi\varphi = fI_l$, where I_l is the identity matrix of size l; and

(3) φ defines a homogeneous homomorphism

$$\bigoplus_{i=1}^{l} R(a_i) \longrightarrow \bigoplus_{i=1}^{l} R(b_i).$$

A two-factor matrix factorization (φ, ψ) is *reduced* if, in addition to (1), all entries of ψ are nonunits.

Example 2.5.11. We consider the smooth quadric $Q_4 \subset \mathbb{P}^5$ defined by the equation $f = x_1x_2 + x_3x_4 + x_5x_0 \in R = \mathbf{k}[x_0, x_1, x_2, x_3, x_4, x_5]$. The 4×4 matrices

$$A = \begin{pmatrix} x_2 & x_4 & x_0 & 0 \\ -x_3 & x_1 & 0 & x_0 \\ -x_5 & 0 & x_1 & -x_4 \\ 0 & -x_5 & x_3 & x_2 \end{pmatrix} \quad \text{and} \quad B = \begin{pmatrix} x_1 & -x_4 & -x_0 & 0 \\ x_3 & x_2 & 0 & -x_0 \\ x_5 & 0 & x_2 & x_4 \\ 0 & x_5 & -x_3 & x_1 \end{pmatrix}$$

define graded R morphisms φ and ψ such that $\varphi\psi = \psi\varphi = fI_4$ and give a reduced graded two-factor matrix factorization of f. See [139] for further details.

The notion of morphism between two two-factor matrix factorizations (φ_1, ψ_1) and (φ_2, ψ_2) of f is defined and consists of a couple (α, β) of morphisms making commutative the natural diagram.

The use of two-factor matrix factorization as a tool for classifying aCM bundles on hypersurfaces is motivated by the following result due to Eisenbud [80, Theorem 6.3]:

Theorem 2.5.12. *Let the notation be as before. Then, there is a one-to-one correspondence between two-factor matrix factorizations of f and graded MCM S-modules. Furthermore, there is a one-to-one correspondence between reduced two-factor matrix factorizations of f and graded MCM S-modules without free summands. This correspondence sends (φ, ψ) to $M = \operatorname{coker}(\varphi)$.*

Using Proposition 2.1.8, we can translate this result into the following

Corollary 2.5.13. *Let $f \in R$ be an irreducible homogeneous polynomial defining a hypersurface $X \subset \mathbb{P}^n$. Then, there is a one-to-one correspondence between two-factor matrix factorizations (φ, ψ) of f and aCM sheaves on X given by*

$$(\varphi, \psi) \longrightarrow \widetilde{(\operatorname{coker} \varphi)} = \mathcal{E}_{\varphi}.$$

Furthermore, this gives a one-to-one correspondence between reduced two-factor matrix factorizations of f and aCM sheaves on X without free summands.

Remark 2.5.14. Keeping the above notations, Eisenbud proved that if $\det(\varphi) = u \cdot f^r$ with u an unit of R and $r \in \mathbb{Z}_{\geq 1}$, then the rank of $\mathcal{E}_{\varphi}$ equals to r (see [80, Proposition 5.6]).

Example 2.5.15. Under the correspondence given in Corollary 2.5.13, the graded two-factor matrix factorization of Example 2.5.11 corresponds to one of the spinor bundles on $Q_4 \subset \mathbb{P}^5$. In fact, by Corollary 2.5.13 it corresponds with an aCM bundle on $Q_4 \subset \mathbb{P}^5$ without free summands and thus, by Theorem 2.3.4, it is one of the spinor bundles.

To determine the numerical conditions under which any aCM bundle on a general hypersurface splits, we will use the following technical result about homogenous polynomials on R.

Proposition 2.5.16. *Let $n \geq 3$, $d \geq 2$, and $\alpha \geq 1$ be integers. If n, d, and α satisfy the inequality*

$$\alpha\left(n + 1 + \binom{n+d-1}{d-1}\right) + 2 \leq \binom{n+d}{d},$$

then general homogenous polynomials in $R = \mathbf{k}[x_0, \ldots, x_n]$ of degree d cannot be expressed as a sum of at most α products of two homogenous polynomials of positive degree.

Proof. [190, Lemma 2.2]. □

Now we can state and prove that under certain numerical conditions on r, d, and n, any rank r aCM bundle on a general hypersurface of degree d in $\mathbb{P}^n$ splits as a direct sum of line bundles.

Theorem 2.5.17. *Let n, d, r be integers such that $n \geq 3$, $d \geq 2$, and $r \geq 1$. If n, d, and r satisfy the inequality*

$$rd\left(n + 1 + \binom{n+d-1}{d-1}\right) + 2 \leq \binom{n+d}{d}, \tag{2.21}$$

then any rank r aCM bundle on a general hypersurface of degree d in $\mathbb{P}^n$ splits as a direct sum of line bundles.

Proof. Assume that n, d, and r satisfy the inequality (2.21). By Proposition 2.5.16, there is a nonempty open set $U \subset |\mathcal{O}_{\mathbb{P}^n}(d)|$ such that any homogeneous polynomial in U cannot be expressed as a sum of at most rd products of two homogenous polynomials of positive degree.

By Bertini's theorem, we may assume that all hypersurfaces in U are nonsingular. We will prove that any rank r aCM bundle on hypersurfaces in U splits as a direct sum of line bundles.

Take $f \in U$ and denote by X the corresponding hypersurface. Let $\mathcal{E}$ be a rank r indecomposable aCM bundle on X. In particular, $\mathcal{E}$ has no free summand (unless $\mathcal{E} \cong \mathcal{O}_X(a)$ for some $a \in \mathbb{Z}$). Let (φ, ψ) be the reduced two-factor matrix factorization corresponding to $\mathcal{E}$ (see Corollary 2.5.13). Since all entries of φ are nonunits,

$$\operatorname{size}(\varphi) \le \deg(\det \varphi).$$

On the other hand, by Remark 2.5.14 we have

$$\deg(\det \varphi) = \deg(u \cdot f^r) = rd.$$

Hence, $l = \operatorname{size}(\varphi) \le rd$. Looking at the $(1, 1)$ entry of $\varphi\psi = f \cdot I_l$, we have the equality

$$\sum_{k=1}^{l} \varphi_{1k}\psi_{k1} = f.$$

Since (φ, ψ) is reduced, this means that f can be expressed as a sum of at most rd products of two homogeneous polynomials of positive degree and this contradicts Proposition 2.5.16. Hence, $\mathcal{E} \cong \mathcal{O}_X(a)$, and we are done. □

Remark 2.5.18. (1) There are integers verifying the inequality (2.21). For instance, $r = 2$, $d = 3$, and $n = 17$. Nevertheless, the inequality (2.21) implies $n \ge rd^2 - d$. Therefore, since $r \ge 2$ and $d \ge 3$, we necessarily have $n \ge 16$.

(2) Remark 2.5.8 (1) shows that there are cases where the inequality does not hold and the assertion of the above theorem is not true.

The above result can also be interpreted as a contribution in the direction of the conjectural scenario in which one expects that any aCM bundle of a fixed rank, over a sufficiently high dimensional hypersurface (no matter its degree) splits as a direct sum of line bundles. The precise conjecture is

Conjecture 2.5.19. *Let $X \subset \mathbb{P}^n$ be a hypersurface. Let $\mathcal{E}$ be an aCM bundle on X. If $\operatorname{rank}(\mathcal{E}) < 2^e$, where $e = \lfloor \frac{n-2}{2} \rfloor$, then $\mathcal{E}$ splits as a direct sum of line bundles.*

The above conjecture was formulated in [38] by Buchweitz, Greuel, and Schreyer, and it is still open in full generality. The motivation for the above conjecture comes from the classification of aCM bundles on quadric hypersurfaces $Q \subset \mathbb{P}^n$ where indeed the bound is reached by the spinor bundles (see Theorem 2.3.4). We have already seen that the cases of rank two and three are understood (Theorems 2.5.7 and 2.5.9) and that we have some bounds for arbitrary rank depending on the degree and the dimension (Theorem 2.5.17) but, in general, for the rank greater or equal to four few results are known.

In this last section we have focused our attention on the existence of low rank aCM bundles on smooth hypersurfaces $X \subset \mathbb{P}^n$ and we have seen that there are few examples. Nevertheless, on any hypersurface $X \subset \mathbb{P}^n$ of degree $d \geq 3$, there exist families of aCM bundles of arbitrary high rank and dimension, i. e., they are of wild representation type. Even more, by [93, Theorem 6.1], we know that any aCM smooth variety $X \subset \mathbb{P}^n$ of dimension $d \geq 2$ is of wild representation type unless it is a linear space, a quadric hypersurface, a smooth rational surface scroll of degree $3 \leq d \leq 4$, or the Veronese surface in $\mathbb{P}^5$.

2.6 Final comments and additional reading

The algebraic counterpart of the theory developed in this chapter, namely the theory of Cohen–Macaulay rings and modules, is a cornerstone of modern commutative algebra. The reader is advised to consult [36] for a thorough account. In the same way, the study of the representation type of projective varieties, as well as their classification as in terms of finite, tame, or wild type, is inspired by predating research in the representation type of algebras and quivers. Reference [205] can be used as an introduction to these topics. The bridge between the algebraic and geometric side offered by Proposition 2.1.8 can be found in [39, Proposition 2.1]. A previous link between both areas is [128, Proposition 2.2.4], based on Grothendieck's results. Grothendieck's Theorem 2.2.1 has a very long history that can be tracked down to Hilbert in 1905. The reader can find a short summary of this history in [169]. A modern proof was given in Grauert and Remmert [101]. The result was proved by Dedekind and Weber in [75] in an algebraic setting. Horrocks' Theorem 2.2.2 was first stated in [117]. The proof presented here is based on [19]. For a refinement of Horrocks' result for vector bundles of low rank the reader can see [91]. Particular families of aCM varieties and related problems are treated in [149] and [179].

The first proof of Knörrer's Theorem 2.3.4 was given in [132] using the theory of matrix factorizations. Later on, Ancona and Ottaviani [4] provided an alternative proof based on the description of the derived category of coherent sheaves on Q_n given by Kapranov in [123]. We have followed the approach of Ancona and Ottaviani. The results from Section 2.4 are based on [69].

The proof of Theorem 2.5.7 is due to Kumar, Rao, and Ravindra, and we have followed their paper [134], from where we have extracted the idea of Proposition 2.5.3. It should be mention that the first part of Theorem 2.5.7 was already known by Kleppe (see [129]) and proved in a much more general context. In [26], Biswas and Ravindra extended Theorem 2.5.7 to complete intersection subvarieties by proving that any aCM bundle of rank two on a general, smooth, complete intersection subvariety of sufficiently high multi-degree and dimension at least four splits.

The proof of Theorem 2.5.17 is due to Sawada and can be found in [190].

As mentioned in the text, most of the contributions to Conjecture 2.5.19 are in the case of low rank, as they can be found in [199] and [181]. Earlier contributions to this conjecture are due to Chiantini and Madonna. Their work is essentially focused on the case of low degree hypersurfaces in $\mathbb{P}^4$ and $\mathbb{P}^5$ (see, for instance, [52, 53], and [54]). For a forey into the theory of aCM sheaves on non-reduced schemes the reader can check [18].

3 Ulrich bundles

This chapter is entirely devoted to the history of Ulrich bundles and provides an introduction to a series of equivalent definitions, first examples, and the basic properties of Ulrich bundles used in the sequel. The existence of Ulrich bundles on a projective variety is a challenging problem with a long and interesting history behind and few known examples; in the remaining chapters of the book we will classify Ulrich bundles on different varieties.

The notion of Ulrich sheaf has its origins in its algebraic counterpart. Ulrich sheaves made their first appearance in the 1980s in commutative algebra, being associated to MCM graded modules with maximal number of generators. In fact, Ulrich in [200] proved that the minimal number $\mu(M)$ of generators of an MCM R-module M of positive rank is bounded as follows: $\mu(M) \leq e(R)\,\mathrm{rank}(M)$ where $e(R)$ denotes the multiplicity of R. Modules attaining the bound have received various names in the literature: they were called linear MCM modules, or maximally generated MCM modules, or Ulrich modules. In the algebraic setup their existence has been proved, for instance, on hypersurface rings [35] and on complete intersection rings [113]. They entered the realm of algebraic geometry 20 years later with the beautiful papers by Eisenbud and Schreyer [88] and Beauville [22]. Since then big effort has been exerted to answer the fundamental and intriguing question, which goes back to [200] (see also [88]): Does any smooth projective variety support an Ulrich bundle?

In algebraic geometry, their importance was underlined in [22] and [88], and relies, among other things, on the following relevant developments: (1) after the Boij–Söderberg theory had been developed, it has been proved that the existence of an Ulrich bundle on a smooth projective variety X of dimension d implies that the cone of cohomology tables of vector bundles on X is the same as that of vector bundles on $\mathbb{P}^d$ [87, Theorem 4.2]; and (2) let $X \subset \mathbb{P}^n$ be a projective variety of dimension d and $D_X \subset \mathrm{Gr}(n-d-1,n)$ the *Cayley–Chow divisor* defined by the $(n-d-1)$-planes that intersect X. If X supports an Ulrich bundle then the defining equation of D_X, the so-called *Cayley–Chow form* of X, is given by the determinant of a square matrix with linear entries in the Plücker coordinates (see Section 4.6 for more details).

We have structured this chapter as follows. In Section 3.1, we offer a brief history of the development of the theory of Ulrich sheaves from its roots in the field of commutative algebra to its irruption in algebraic geometry. In Section 3.2, we establish the equivalence between the different definitions of Ulrich bundles scattered through the literature and give the first examples of Ulrich bundles. In Section 3.3, we collect the basic properties of Ulrich bundles needed in the sequel. The classification of Ulrich bundles is usually arranged according the dimension of the underlying variety. In Section 3.4, we classify Ulrich bundles on a curve $C \subset \mathbb{P}^n$ of degree $\deg(C)$ and genus g and postpone the study of Ulrich bundles on varieties of dimension greater than 1 to the next chapters.

https://doi.org/10.1515/9783110647686-003

3.1 History of Ulrich bundles

In this section, we propose a short introduction to the history of the developments of Ulrich bundles. The origins of this notion have theirs roots in the field of commutative algebra. We are going to present the main landmarks without given further details of the definitions involved. To complete this introduction, the reader can consult any standard book on commutative algebra, for example, [36].

Given a local Cohen–Macaulay ring $(S, \mathfrak{m})$, it has been of high interest to give criteria for S to be a Gorenstein ring. A possible approach would be to understand which kind of module with some sort of extremal behavior supports a local Gorenstein ring S. For instance, one could be interested in S-modules M with a large minimal number of generators. For an S-module M, it was proved that M is Cohen–Macaulay if and only if $\dim(M/\underline{x}M) = e(S)\operatorname{rank}(M)$ where $\underline{x}$ is a maximal regular sequence and $e(S)$ is the multiplicity of the ring. Therefore, it turns out that there is a natural bound on the cardinality of a minimal set of generators of a Cohen–Macaulay module M:

$$\mu(M) = \dim(M/\mathfrak{m}M) \leq \dim(M/\underline{x}M) = e(S)\operatorname{rank}(M).$$

It was proven in [200, Theorem 3.1] that a local ring $(S, \mathfrak{m})$ supporting a finitely generated module M of positive rank such that

(i) $\mu(M) \geq 1/2e(S)\operatorname{rank}(M)$ and
(ii) $\operatorname{Ext}^i_S(M, S) = 0$ for $1 \leq i \leq \dim S$

is a Gorenstein ring. The modules with a large minimal number of generators turned out to have a great importance, and Ulrich raised the following question (see [200, p. 26]):

Question 3.1.1. *Let S be a local Cohen–Macaulay ring with positive dimension and infinite residue class field. Does S support a Cohen–Macaulay module M with positive rank and maximal permitted cardinality of a minimal set of generators (namely, $\mu(M) = e(S)\operatorname{rank}(M)$)?*

In [35], where such modules were called Maximally Generated Maximal Cohen–Macaulay modules (MGMCM), their existence was proven on certain classes of rings. For instance, when $\dim S = 1$, for rings of minimal multiplicity, on the class of dimension 2 homogeneous Cohen–Macaulay domains with infinite residue class field, etc. Moreover, a certain number of questions were raised. The first appearance of the term "Ulrich modules" for MGMCM modules was in [16], where, moreover, their existence was proved for hypersurface rings.

When we move to the realm of homogenous rings $S = \mathbf{k}[x_0, \ldots, x_n]/I$ and graded modules, we have seen in Proposition 2.1.8 that there exists a bijection between MCM S-modules and aCM sheaves on the projective variety $\operatorname{Proj}(S)$. It was in this setting that Ulrich sheaves (without using this terminology, yet) on projective hypersurfaces made their first appearance (see [22]). In this paper, Beauville made clear the link between the existence of rank-one (respectively rank-two) Ulrich sheaves on hypersurfaces and their representation as the determinant (respectively the pfaffian) of a square matrix of linear forms. With this modern costume from algebraic geometry, classical results on the determinantal and pfaffian representation of curves and surfaces were reobtained and new results on threefolds and fourfolds were proved (see Chapter 4).

The definite irruption of Ulrich bundles in algebraic geometry was caused by the paper [88]. Its main motivation was to study the Cayley–Chow form of a projective variety. Given a projective variety $X \subset \mathbb{P}^n$ of dimension d, the *Cayley–Chow divisor* D_X in the Grassmann variety $\mathrm{Gr}(n-d-1, n)$ is composed of the $(n-d-1)$-dimensional planes that intersect X. Ever since its definition, it has been considered a challenging problem to give an explicit presentation of the defining equation of D_X, *the Cayley–Chow form.* In [88], the authors realized that this presentation turns out to be particularly easy when X supports a special kind of vector bundles, the so-called *weakly Ulrich* bundles.

In this context, Eisenbud and Schreyer in [88] raised the following questions:

Question 3.1.2.
(1) *Does any variety $X \subset \mathbb{P}^n$ support an Ulrich sheaf?*
(2) *If so, which is the lowest possible rank of such a sheaf?*

These two questions have been generating a lot of research during the recent years and, despite some important achievements that we are going to explain in the following chapters, they are far from being definitively settled.

3.2 Definition, first examples, and characterization

In this section we introduce several equivalent definitions of Ulrich bundles and illustrate these definitions with many examples.

Definition 3.2.1. Given a projective variety $X \subset \mathbb{P}^n$ embedded by the very ample line bundle $\mathcal{O}_X(1)$ and a coherent sheaf $\mathcal{E}$ on X, we say that $\mathcal{E}$ is an *Ulrich sheaf* if $\mathcal{E}$ is an initialized aCM sheaf and $\mathrm{h}^0(\mathcal{E}) = \deg(X)\,\mathrm{rank}(\mathcal{E})$.

Remark 3.2.2. (1) In Proposition 2.1.8 we have seen that there exists a bijection between aCM sheaves $\mathcal{E}$ on a variety $X \subset \mathbb{P}^n$ and MCM R_X-modules. Under this bijection, MGMCM R_X-modules (also called linear MCM modules or Ulrich modules) correspond to Ulrich sheaves on X.

(2) The definition of Ulrich sheaf depends on the very ample line bundle $\mathcal{O}_X(1)$ of the underlying variety X and the change of the very ample line bundle can affect the existence of Ulrich sheaves. As a first example of this subtle question, the reader can

see Example 3.2.3 (3), and this dependence is also strongly reflected, for instance, in Theorem 5.2.7.

The existence of Ulrich bundles on a given variety is, in general, a nontrivial problem. Let us start with examples of Ulrich line bundles.

Example 3.2.3. (1) $\mathcal{O}_{\mathbb{P}^n}$ is an Ulrich bundle on $\mathbb{P}^n$ and, by Theorem 2.2.2, it is the only indecomposable Ulrich bundle on $\mathbb{P}^n$, up to a twist.

(2) Projective varieties $X \subset \mathbb{P}^n$ with $\operatorname{Pic}(X) \cong \mathbb{Z}$ do not carry Ulrich line bundles except when X is a linear space. The result is no longer true for varieties $X \subset \mathbb{P}^n$ with $\operatorname{Pic}(X) \cong \mathbb{Z}^\rho$, $\rho > 1$. For instance, the quadric $X \cong \mathbb{P}^1 \times \mathbb{P}^1 \subset \mathbb{P}^3$ supports two Ulrich line bundles, $\mathcal{O}_{\mathbb{P}^1\times\mathbb{P}^1}(1,0)$ and $\mathcal{O}_{\mathbb{P}^1\times\mathbb{P}^1}(0,1)$ (see Example 2.1.3). According to Definition 3.3.4, these two Ulrich line bundles are Ulrich dual. It follows from Section 2.5 that a smooth cubic surface $S \subset \mathbb{P}^3$ carries a finite number of Ulrich line bundles, namely $\mathcal{O}_S(D)$ with $D \subset S$ being a rational normal cubic curve (see also Proposition 5.3.21).

(3) We consider the Segre variety $\mathbb{P}^n \times \mathbb{P}^m$ embedded in $\mathbb{P}^{nm+n+m}$ by the very ample line bundle $\mathcal{O}_{\mathbb{P}^n\times\mathbb{P}^m}(1,1)$. Using the *Künneth formula* (see (2.2)), we can prove that a line bundle $\mathcal{O}_{\mathbb{P}^n\times\mathbb{P}^m}(a,b)$ on a Segre variety $\mathbb{P}^n \times \mathbb{P}^m \subset \mathbb{P}^{nm+n+m}$ is an Ulrich line bundle if and only if $(a,b) \in \{(m,0),(0,n)\}$. If we change the very ample line bundle and, for example, fix as very ample line bundle $\mathcal{O}_{\mathbb{P}^n\times\mathbb{P}^m}(2,3)$, we check, using again Künneth formula, that there is no Ulrich line bundle on $\mathbb{P}^n \times \mathbb{P}^m$ with respect to the very ample line bundle $\mathcal{O}_{\mathbb{P}^n\times\mathbb{P}^m}(2,3)$.

(4) A codimension c variety $X \subset \mathbb{P}^n$ is said to be a *determinantal variety* if its homogeneous saturated ideal $I(X) \subset R$ is generated by the maximal minors of a homogeneous $t \times (t+c-1)$ matrix A. The Picard group $\operatorname{Pic}(X)$ of a smooth codimension c determinantal variety $X \subset \mathbb{P}^n$, $n - c \geq 2$, is isomorphic to $\mathbb{Z}^2$ and generated by its hyperplane section H and the codimension $c+1$ determinantal variety Y defined by the maximal minors of the $(t-1) \times (t+c-1)$ matrix B obtained deleting the last row of A (see, for instance, [131, Theorem 4.1]). A line bundle $\mathcal{O}_X(aY + bH)$ is aCM if and only if $-1 \leq a \leq c$. In addition, if the entries of A are general linear forms, we say that X is a linear determinantal variety and in this case the line bundle $\mathcal{O}_X(aY + bH)$ is Ulrich if and only if $a = -1$ or $a = c$ (see [131, Proposition 4.2] and [81, Theorem A2.10]). As examples of linear determinantal varieties we have the rational normal curves and the cubic scroll that will be treated in Example 4.6.10.

In general, computing $h^0(X, \mathcal{E})$ for a vector bundle $\mathcal{E}$ on a variety $X \subset \mathbb{P}^n$ is not always easy. So we would like to have alternative definitions of Ulrich bundles. To state them, we need some preliminary results.

Lemma 3.2.4. *Let $\mathcal{E}$ be an initialized rank r vector bundle on a smooth curve $X \subset \mathbb{P}^n$ with genus g. It holds:*
(1) $h^0(\mathcal{E}) \leq \deg(X)r$,
(2) $\deg(\mathcal{E}) \leq r(\deg(X) + g - 1)$, *and*
(3) $\chi(\mathcal{E}(t)) \leq \deg(X)r(t+1)$ *for all* $t \in \mathbb{Z}$.

Moreover, the equality in any of (1), (2), or (3) implies the equality in the other two inequalities and it is equivalent to $\mathcal{E}$ being Ulrich.

Proof. (1) Any vector bundle $\mathcal{E}$ on a smooth curve is aCM. Let $M = \mathrm{H}^0_*(\mathcal{E})$ be its associated MCM R_X-module. Since $\mathcal{E}$ is initialized, we have $\mathrm{H}^0(\mathcal{E}(-1)) = 0$, and hence $\mathrm{h}^0(\mathcal{E}) \leq \mu(M)$. Ulrich [200] proved that $\mu(M) \leq e(R_X)\,\mathrm{rank}(M)$ where $e(R_X)$ denotes the multiplicity of the coordinate ring of X. Since $e(R_X) = \deg(X)$ and $\mathrm{rank}(\mathcal{E}) = \mathrm{rank}(M)$, we get $\mathrm{h}^0(\mathcal{E}) \leq \deg(X)r$.

(2) We apply Riemann–Roch theorem to $\mathcal{E}(-1)$ (see Theorem 1.4.3) and use again the fact that $\mathrm{H}^0(\mathcal{E}(-1)) = 0$ to obtain

$$\begin{aligned}\deg(\mathcal{E}) - r\deg(X) + r(1-g) &= \chi(\mathcal{E}(-1)) \\ &= \mathrm{h}^0(\mathcal{E}(-1)) - \mathrm{h}^1(\mathcal{E}(-1)) \\ &= -\mathrm{h}^1(\mathcal{E}(-1)) \leq 0.\end{aligned}$$

So, we conclude that $\deg(\mathcal{E}) \leq r(\deg(X) + g - 1)$.

(3) Applying once more Riemann–Roch theorem and using (2), we get

$$\begin{aligned}\chi(\mathcal{E}(t)) &= \deg(\mathcal{E}) + r\deg(X)t + r(1-g) \\ &\leq r(\deg(X) + g - 1) + r\deg(X)t + r(1-g) \\ &= \deg(X)r(t+1).\end{aligned}$$

Observe that the equality in (2) is equivalent to the equality in (3). Let us prove that the equality in (2) implies the equality in (1). First we observe that $\deg(\mathcal{E}) = r(\deg(X) + g - 1)$ if and only if $\mathrm{H}^1(\mathcal{E}(-1)) = 0$. Since a general plane section H of X consists of $\deg(X)$ points, the exact cohomology sequence associated to

$$0 \longrightarrow \mathcal{E}(-1) \longrightarrow \mathcal{E} \longrightarrow \mathcal{E}_{|H} \longrightarrow 0 \tag{3.1}$$

gives us $\mathrm{h}^0(\mathcal{E}) = \mathrm{h}^0(\mathcal{E}_{|H}) = \deg(X)r$. Finally, let us show that the equality in (1) forces the equality in (2). The equality in (1) implies the injectivity $\mathrm{H}^1(\mathcal{E}(-1)) \hookrightarrow \mathrm{H}^1(\mathcal{E})$. Since the map $\mathrm{H}^0(\mathcal{E}(t)) \longrightarrow \mathrm{H}^0(\mathcal{E}_{|H}(t))$ is surjective for all $t \geq 0$, we deduce that the morphism $\mathrm{H}^1(\mathcal{E}(t-1)) \longrightarrow \mathrm{H}^1(\mathcal{E}(t))$ is injective for all $t \geq 0$. Therefore, we have

$$\mathrm{h}^1(\mathcal{E}(-1)) \leq \mathrm{h}^1(\mathcal{E}) \leq \mathrm{h}^1(\mathcal{E}(1)) \leq \cdots \leq \mathrm{h}^1(\mathcal{E}(\ell)) = 0$$

for $\ell \gg 0$, where the last vanishing follows from Serre's theorem (see Theorem 1.3.1). So, $\chi(\mathcal{E}(-1)) = \mathrm{h}^0(\mathcal{E}(-1)) - \mathrm{h}^1(\mathcal{E}(-1)) = 0$ and $\deg(\mathcal{E}) = r(\deg(X) + g - 1)$, which proves what we wanted. □

From Proposition 3.2.4 we can see that the degree of an Ulrich bundle on a projective variety is fixed.

Proposition 3.2.5. *Let $\mathcal{E}$ be a rank r Ulrich bundle on a projective variety $(X, \mathcal{O}_X(H))$ of dimension d. Then, the degree of $\mathcal{E}$ is*

$$\deg(\mathcal{E}) = \frac{r((d+1)H + K_X)H^{d-1}}{2}.$$

Proof. By Proposition 3.2.4, we know that

$$\deg(\mathcal{E}) = \deg(\mathcal{E}_{|C}) = r(\deg(C) + g(C) - 1),$$

where C is a curve defined as the intersection of $d-1$ generic hyperplane sections. In particular, $\deg(C) = H^d$. Now the result follows immediately from the adjunction formula

$$g(C) - 1 = \frac{(K_X + (d-1)H)H^{d-1}}{2}.$$

□

Example 3.2.6. Let $(S, \mathcal{O}_S(H))$ be a smooth surface in $\mathbb{P}^3$ of degree $d \geq 4$ and with $\mathrm{Pic}(S) \cong \mathbb{Z}$. Let $\mathcal{E}$ be a rank r Ulrich bundle on S. The first Chern class of $\mathcal{E}$ is $c_1(\mathcal{E}) = \lambda H$ for certain $\lambda \in \mathbb{Z}$. Therefore, we have $\deg(\mathcal{E}) = c_1(\mathcal{E})H = \lambda H^2 = \lambda d = r\binom{d}{2}$. Hence, for any rank r vector bundle $\mathcal{E}$ on a very general surface S of degree d in $\mathbb{P}^3$, $(d-1)r$ has to be even and this implies that there are no Ulrich bundles of odd rank on a very general surface of even degree.

We are now ready to state a useful criterion for determining whether a vector bundle on a smooth projective variety is Ulrich. Indeed, we have

Proposition 3.2.7. *Let $(X, \mathcal{O}_X(1))$ be a smooth variety of dimension d and let $\mathcal{E}$ be a coherent sheaf on X. Then, $\mathcal{E}$ is an Ulrich bundle if and only if*

$$\mathrm{H}^i(X, \mathcal{E}(-t)) = 0 \quad \text{for any } i \geq 0 \text{ and } 1 \leq t \leq d.$$

Proof. We prove the proposition by induction on d and using the short exact sequence

$$0 \longrightarrow \mathcal{E}(-1) \longrightarrow \mathcal{E} \longrightarrow \mathcal{E}_{|H} \longrightarrow 0 \tag{3.2}$$

where H is a general hyperplane section of X. For $d = 1$, the results follows from Lemma 3.2.4. Assume $d > 1$. Suppose that $\mathrm{H}^i(X, \mathcal{E}(-t)) = 0$ for any $i \geq 0$ and $1 \leq t \leq d$. From the fact that $\mathrm{H}^i(\mathcal{E}(-i)) = 0$ for $i \geq 1$, we deduce that $\mathcal{E}$ is 0-regular. Hence, $\mathrm{H}^i(\mathcal{E}(t-i)) = 0$ for $i \geq 1$ and $t \geq 0$. By Serre's duality $\mathcal{E}^\vee \otimes \omega_X$ is d-regular. Therefore, $\mathrm{H}^j((\mathcal{E}^\vee \otimes \omega_X)(d+t-j)) = 0$ for $t \geq 0$ and $j \geq 1$, and we conclude that $\mathrm{H}^i_*(\mathcal{E}) = 0$ for $1 \leq i \leq d-1$, i. e., $\mathcal{E}$ is an aCM bundle. It remains to see that $\mathrm{h}^0(\mathcal{E}) = \deg(X)\,\mathrm{rank}(\mathcal{E})$. Using the exact sequence (3.2), we get $\mathrm{H}^i(\mathcal{E}_{|H}(-t)) = 0$ for any $i \geq 0$ and $1 \leq t \leq d-1$. Hence, by induction $\mathcal{E}_{|H}$ is an Ulrich bundle. In particular, $\mathrm{h}^0(\mathcal{E}_H) = \deg(X)\,\mathrm{rank}(\mathcal{E})$, and, using once more (3.2), we conclude that $\mathrm{h}^0(\mathcal{E}) = \deg(X)\,\mathrm{rank}(\mathcal{E})$.

Vice versa, if $\mathcal{E}$ is an initialized Ulrich bundle on X, we have $\mathrm{H}^i(\mathcal{E}(-t)) = 0$ for $0 \le i \le d-1$ and $1 \le t \le d$. It remains to check that $\mathrm{H}^d(\mathcal{E}(-t)) = 0$ for $1 \le t \le d$. The exact sequence (3.2) gives us that $\mathcal{E}_{|H}$ is an aCM bundle and $\mathrm{h}^0(\mathcal{E}_{|H}) = \deg(X)\,\mathrm{rank}(\mathcal{E}_{|H})$. By induction, $\mathrm{H}^{d-1}(\mathcal{E}_{|H}(-t)) = 0$ for $1 \le t \le d-1$. Even more, $\mathrm{H}^{d-1}(\mathcal{E}_{|H}(\ell)) = 0$ for $\ell \ge -d+1$. This last vanishing implies the injectivity of $\mathrm{H}^d(\mathcal{E}(\ell-1)) \longrightarrow \mathrm{H}^d(\mathcal{E}(\ell))$ for all $\ell \ge -d+1$ which, together with Serre's theorem asserting that $\mathrm{H}^d(\mathcal{E}(\ell)) = 0$ for $\ell \gg 0$, gives us $\mathrm{H}^d(\mathcal{E}(-t)) = 0$ for $1 \le t \le d$. □

As first examples of Ulrich bundles of higher rank we have the following:

Example 3.2.8. (1) The spinor bundles $\mathcal{S}$ on a quadric $Q_n \subset \mathbb{P}^{n+1}$ are Ulrich bundles of rank $2^{\lfloor \frac{n-1}{2} \rfloor}$ by Theorem 2.3.2. Moreover, by Theorem 2.3.4, they are the only Ulrich bundles on Q_n, up to a twist.

(2) In Example 3.2.3 (2) we have seen that a smooth cubic surface $X \subset \mathbb{P}^3$ supports Ulrich line bundles $\mathcal{O}_X(C)$ corresponding to the rational normal cubic curves C contained in X. Let us now construct an example of rank 2 indecomposable Ulrich bundle on X. We choose a set $Z = \{p_1, \dots, p_5\} \subset X$ of 5 general points. Then Z is an aG variety with canonical module K_Z generated in degree -1 by a unique element ε which provides an extension:

$$\varepsilon : \quad 0 \longrightarrow R_X \longrightarrow M \longrightarrow I_{Z/X}(2) \longrightarrow 0 \tag{3.3}$$

via the isomorphism $K_Z = \mathrm{Ext}^1(I_{Z/X}(1), R_X(-1))$. Applying $\mathrm{Hom}(-, R_X(-1))$ to the exact sequence (3.3), we get that $\mathrm{Ext}^1(M, K_X) = 0$, i. e., M is an MCM R_X-module. Therefore, by Proposition 2.1.8, its sheafification $\mathcal{E} = \widetilde{M}$ is an aCM bundle which fits into an exact sequence

$$0 \longrightarrow \mathcal{O}_X \longrightarrow \mathcal{E} \longrightarrow \mathcal{I}_{Z/X}(2) \longrightarrow 0.$$

Moreover, we have $\mathrm{h}^0(\mathcal{E}) = \mathrm{h}^0(\mathcal{O}_X) + \mathrm{h}^0(\mathcal{I}_{Z/X}(2)) = 6 = 2\deg(X)$ and conclude that $\mathcal{E}$ is a rank 2 Ulrich bundle on X. Now, if $\mathcal{E}$ was decomposable, by Proposition 3.3.1, $\mathcal{E}$ would be the sum of two Ulrich line bundles, namely, it would be of the form $\mathcal{O}_X(C) \oplus \mathcal{O}_X(2H - C)$ for $C \subset X$ a rational normal cubic curve and H the hyperplane section. But since $\mathrm{h}^0(\mathcal{O}_X(C)) = \mathrm{h}^0(\mathcal{O}_X(2H - C)) = 3$, we see that the map

$$\mathbb{P}^2 \times \mathbb{P}^2 \cong \mathbb{P}(\mathrm{H}^0(\mathcal{O}_X(C))) \times \mathbb{P}(\mathrm{H}^0(\mathcal{O}_X(2H - C))) \longrightarrow \mathrm{Hilb}^5(X)$$

that sends a couple of sections to the intersection of their zero sets cannot be surjective. Therefore, for a general set Z of five points on X, we get a rank-two indecomposable Ulrich bundle (see Proposition 5.3.16 for a more general result).

(3) We consider the Grassmann variety $\mathrm{Gr}(3,7) \subset \mathbb{P}^{69}$ which has dimension 16 and the irreducible homogeneous bundle $\mathcal{E} := \Sigma^{(9,8,7,6)}\mathcal{Q} \otimes \Sigma^{(6,6)}\mathcal{S}^\vee$. Observe that $\mathrm{rank}(\mathcal{E}) = 16 \cdot 3 \cdot 4 \cdot 8!$ and that $\deg(\mathrm{Gr}(3,7)) = 14 \cdot 13 \cdot 11 \cdot 8 \cdot 3$ (see [68, (2.2)]). By

Theorem 2.4.9, $\mathcal{E}$ is an aCM bundle on $\mathrm{Gr}(3,7)$ and, applying Borel–Bott–Weil theorem (see Theorem 2.4.4), we get $\mathrm{H}^0(\mathrm{Gr}(3,7),\mathcal{E}(-1)) = 0$ (i. e., $\mathcal{E}$ is initialized) and $\mathrm{h}^0(\mathrm{Gr}(3,7),\mathcal{E}) = \mathrm{rank}(\mathcal{E})\deg(\mathrm{Gr}(3,7))$. Therefore, $\mathcal{E}$ is an Ulrich bundle on $\mathrm{Gr}(3,7)$. It could also be checked that $\mathcal{E}$ is an Ulrich on $\mathrm{Gr}(3,7)$ using the criterion given in Proposition 3.2.7. Indeed, $\mathcal{E}$ is an Ulrich bundle if and only if $\mathrm{H}^i(\mathrm{Gr}(3,7),\mathcal{E}(-t)) = 0$ for $i \geq 0$ and $1 \leq t \leq 16 = \dim \mathrm{Gr}(3,7)$; these conditions can be verified using Theorem 2.4.4.

As we already said, Ulrich bundles are our main objects of study, and they can be characterized in many different ways.

Theorem 3.2.9. *Let $(X,\mathcal{O}_X(1))$ be a d-dimensional smooth variety and let $\mathcal{E}$ be an initialized vector bundle on X. The following conditions are equivalent:*
(1) *$\mathcal{E}$ is Ulrich.*
(2) *$\mathrm{H}^i(\mathcal{E}(-t)) = 0$ for $i \geq 0$ and $1 \leq t \leq d$.*
(3) *$\mathrm{H}^i(\mathcal{E}(-i)) = 0$ for $i > 0$ and $\mathrm{H}^i(\mathcal{E}(-i-1)) = 0$ for $i < d$.*
(4) *For some (respectively all) finite linear projections $\pi : X \longrightarrow \mathbb{P}^d$, the vector bundle $\pi_*\mathcal{E}$ is the trivial vector bundle $\mathcal{O}_{\mathbb{P}^d}^t$ for some t.*
(5) *$\mathcal{E}$ admits a linear $\mathcal{O}_{\mathbb{P}^n}$-resolution of the form*

$$0 \longrightarrow \mathcal{O}_{\mathbb{P}^n}(-n+d)^{a_{n-d}} \longrightarrow \cdots \longrightarrow \mathcal{O}_{\mathbb{P}^n}(-1)^{a_1} \longrightarrow \mathcal{O}_{\mathbb{P}^n}^{a_0} \longrightarrow \mathcal{E} \longrightarrow 0. \tag{3.4}$$

Proof. Lemma 3.2.7 gives equivalence between (1) and (2). Moreover, (3) is a particular case of (2).

Let us prove that (3) implies (4). We choose a finite projection $\pi : X \longrightarrow \mathbb{P}^d$. Since π is a finite morphism, $\mathrm{H}^i(X,\mathcal{E}(\ell)) = \mathrm{H}^i(\mathbb{P}^d,\pi_*(\mathcal{E}(\ell)))$ for all i and ℓ. Therefore, $\pi_*(\mathcal{E})$ satisfies the same vanishing conditions as $\mathcal{E}$. Let us quickly see that a coherent sheaf $\mathcal{F}$ on $\mathbb{P}^d$ with the vanishing conditions (3) should be of the form $\mathcal{O}_{\mathbb{P}^d}^t$ for some $t \in \mathbb{Z}$. First of all, $\mathcal{F}$ is 0-regular and therefore $\mathrm{H}^i(\mathcal{F}(t_i)) = 0$ for $i > 0$ and $t_i \geq -i$. To conclude, we are going to see by induction on d that $\mathcal{F}$ is an aCM bundle and, therefore, by Horrocks' theorem (Theorem 2.2.2), of the desired form. The case $d = 1$ is immediate by Grothendieck's theorem. So let us suppose that it is true for $d-1$ and consider the short exact sequence

$$0 \longrightarrow \mathcal{F}(-1) \longrightarrow \mathcal{F} \longrightarrow \mathcal{F}_{|H} \longrightarrow 0$$

for a generic hyperplane section $H \cong \mathbb{P}^{d-1}$. Considering the long exact sequence associated to it, we see that $\mathcal{F}_{|H}$ satisfies the same vanishing conditions of the statement for $d-1$. Therefore, by induction hypothesis, $\mathcal{F}_{|H}$ splits as a direct sum of trivial line bundles $\mathcal{O}_{\mathbb{P}^{d-1}}$. In particular, we have injections $\mathrm{H}^i(\mathcal{F}(t-1)) \hookrightarrow \mathrm{H}^i(\mathcal{F}(t))$ for $1 \leq i \leq d-1$ and $t \leq -1$. Therefore $\mathcal{F}$ is an aCM bundle, and we are done.

Let us see that (4) implies (5). The hypothesis $\pi_*\mathcal{E} \cong \mathcal{O}_{\mathbb{P}^d}^t$ for some t implies that $M = \mathrm{H}^0_*(\mathcal{E})$ is a free module over $\mathbf{k}[x_0,\dots,x_d] = \mathrm{H}^0_*(\mathcal{O}_{\mathbb{P}^d})$ generated in degree 0. This means that M is an MCM of dimension $d+1$ over R with a linear resolution

$$0 \longrightarrow R(d-n)^{a_{n-d}} \longrightarrow \cdots \longrightarrow R(-1)^{a_1} \longrightarrow R^{a_0} \longrightarrow M \longrightarrow 0.$$

Sheafifying this last resolution, we get (3.4).

Finally, we will prove that (5) implies (2). Cutting the linear resolution (3.4) into short exact sequences, taking cohomology, and using the fact that $\mathrm{H}^i(\mathbb{P}^d, \mathcal{O}_{\mathbb{P}^d}(-t)) = 0$ for all $i \geq 0$ and $1 \leq t \leq d$, we get $\mathrm{H}^i(\mathcal{E}(-t)) = 0$ for $i \geq 0$ and $1 \leq t \leq d$, which proves what we wanted. □

From the resolution (3.4) and its linearity, we can deduce the following properties of Ulrich bundles.

Corollary 3.2.10. *Let $(X, \mathcal{O}_X(1))$ be a d-dimensional smooth variety and let $\mathcal{E}$ be an Ulrich bundle on X. It holds:*
(1) *$\mathcal{E}$ is 0-regular and globally generated.*
(2) *The ranks of the free modules in the resolution (3.4) are given by*

$$a_0 = \deg(X)\operatorname{rank}(\mathcal{E}) \quad \textit{and} \quad a_i = \binom{n-d}{i} a_0, \quad \textit{for } 1 \leq i \leq n-d.$$

In particular, $a_{n-d-i} = a_i$.
(3) $\chi(\mathcal{E}(t)) = \mathrm{h}^0(\mathcal{E})\binom{t+n}{n}$.

Constructing Ulrich bundles is, in general, a difficult task. So as a first approximation one often tries to produce vector bundles which are not far from being Ulrich, the so-called weakly Ulrich bundles.

Definition 3.2.11. Let $(X, \mathcal{O}_X(1))$ be a d-dimensional smooth variety and let $\mathcal{E}$ be an initialized vector bundle on X. We say that $\mathcal{E}$ is *weakly Ulrich* if the following conditions hold:
(1) $\mathrm{H}^j(\mathcal{E}(-m)) = 0$ for $1 \leq j \leq d$ and $m \leq j-1 \leq d-1$, and
(2) $\mathrm{H}^j(\mathcal{E}(-m)) = 0$ for $0 \leq j \leq d-1$ and $m \geq j+2$.

The importance of this notion stems from the fact that if a projective variety $X \subset \mathbb{P}^n$ of dimension d supports a weakly Ulrich bundle, then its Cayley–Chow form has a presentation as the determinant of a map between two vector bundles on the Grassmann variety $\mathrm{Gr}(n-d-1, n)$ (see also Section 4.6). By definition, Ulrich bundles are weakly Ulrich but, as can be seen in the next example, the converse is not true.

Example 3.2.12. The vector bundle $\Omega^1_{\mathbb{P}^2}(2)$ is weakly Ulrich but it is not Ulrich since it is not an aCM bundle on $\mathbb{P}^2$ (see Theorem 2.2.2).

3.3 Properties of Ulrich bundles

The goal of this section is to introduce the main properties of Ulrich bundles needed in the sequel. Among other interesting properties, Ulrich bundles of any rank on smooth projective varieties of arbitrary dimension will be shown to be μ-semistable. Nevertheless, the first thing we will see is that Ulrich property behaves well in short exact sequences. This fact, will allow us to deduce other important properties and to construct new examples of Ulrich bundles.

Proposition 3.3.1. *Consider the following short exact sequence of coherent sheaves on a smooth variety* $(X, \mathcal{O}_X(1))$ *of dimension* d*:*

$$0 \longrightarrow \mathcal{F} \longrightarrow \mathcal{E} \longrightarrow \mathcal{G} \longrightarrow 0.$$

If two of $\mathcal{F}$, $\mathcal{E}$, *and* $\mathcal{G}$ *are Ulrich bundles, then so is the third. In particular,* $\bigoplus_{i=1}^{s} \mathcal{E}_i$ *is an Ulrich bundle if and only if* $\mathcal{E}_i$ *is an Ulrich bundle for any* $1 \leq i \leq s$.

Proof. Let f, e, and $e - f$ be the respective ranks of $\mathcal{F}$, $\mathcal{E}$, and $\mathcal{G}$ and let $\pi : X \longrightarrow \mathbb{P}^d$ be a finite linear projection. Denote by $k = \deg(X)$. Since π is a finite morphism, we have the following short exact sequence of sheaves on $\mathbb{P}^d$:

$$0 \longrightarrow \pi_*\mathcal{F} \longrightarrow \pi_*\mathcal{E} \longrightarrow \pi_*\mathcal{G} \longrightarrow 0. \tag{3.5}$$

If $\mathcal{F}$ and $\mathcal{G}$ are Ulrich bundles, by Theorem 3.2.9 (4), $\pi_*\mathcal{F}$ and $\pi_*\mathcal{G}$ are trivial vector bundles on $\mathbb{P}^d$. Therefore, $\pi_*\mathcal{E}$, being an extension of trivial bundles on $\mathbb{P}^d$, is also trivial. Thus, again by the characterization given in Theorem 3.2.9, $\mathcal{E}$ is Ulrich.

If $\mathcal{E}$ and $\mathcal{G}$ are Ulrich bundles, then $\mathcal{F}$ is locally free. By Theorem 3.2.9, $\pi_*\mathcal{E}$ and $\pi_*\mathcal{G}$ are trivial. So, dualizing (3.5), we get the short exact sequence

$$0 \longrightarrow \mathcal{O}_{\mathbb{P}^d}^{k(e-f)} \longrightarrow \mathcal{O}_{\mathbb{P}^d}^{ke} \longrightarrow (\pi_*\mathcal{F})^\vee \longrightarrow 0.$$

Taking cohomology, we see that $(\pi_*\mathcal{F})^\vee$ is a globally generated vector bundle on $\mathbb{P}^d$ of rank kf with exactly kf global sections. Hence, $\pi_*\mathcal{F} = \mathcal{O}_{\mathbb{P}^d}^{kf}$ and thus, by Theorem 3.2.9, $\mathcal{F}$ is Ulrich.

Finally, assume that $\mathcal{F}$ and $\mathcal{E}$ are Ulrich bundles. Applying the functor $\mathcal{H}om_{\mathcal{O}_{\mathbb{P}^d}}(-, \mathcal{O}_{\mathbb{P}^d})$ to (3.5), we show that $\pi_*\mathcal{G}$ is a torsion-free sheaf of rank $k(e - f)$ with exactly $k(e - f)$ global sections. Hence, $\pi_*\mathcal{G}$ is trivial. Since π is a finite flat surjective morphism and $\pi_*\mathcal{G}$ is locally free, the same is true for $\mathcal{G}$. Hence, we conclude as above that $\mathcal{G}$ is an Ulrich bundle. □

Since we are interested in indecomposable Ulrich bundles, we need a criterion which guarantees that the iterated extension of indecomposable vector bundles is again an indecomposable vector bundle.

Proposition 3.3.2. *Let $X \subset \mathbb{P}^n$ be a smooth variety and $\mathcal{E}_i$, $i = 1, \dots, r$ simple bundles on X. Set $U = \mathbb{P}(\mathrm{Ext}^1(\mathcal{E}_r, \mathcal{E}_1)) \times \cdots \times \mathbb{P}(\mathrm{Ext}^1(\mathcal{E}_r, \mathcal{E}_{r-1}))$ and assume that $\mathrm{Hom}(\mathcal{E}_i, \mathcal{E}_j) = 0$ for all $1 \le i \ne j \le r$. It holds:*

(1) Any vector bundle $\mathcal{E}$ coming from an extension $e \in U$,

$$e: \quad 0 \longrightarrow \bigoplus_{i=1}^{r-1} \mathcal{E}_i \longrightarrow \mathcal{E} \longrightarrow \mathcal{E}_r \longrightarrow 0 \tag{3.6}$$

is simple and, in particular, indecomposable. Moreover, for two vector bundles $\mathcal{E}$ and $\mathcal{E}'$ coming from extensions in U, $\mathrm{Hom}(\mathcal{E}, \mathcal{E}') \ne 0$ if and only if they correspond to the same extension $e \in U$ (and therefore $\mathcal{E} \cong \mathcal{E}'$).

(2) Fix integers $a, b \ge 0$. If, moreover, $\mathcal{E}_1$ is stable then a vector bundle $\mathcal{E}$ coming from a general extension of $\mathcal{E}_1^a$ by $\mathcal{E}_1^b$, i. e.,

$$0 \longrightarrow \mathcal{E}_1^b \longrightarrow \mathcal{E} \longrightarrow \mathcal{E}_1^a \longrightarrow 0,$$

is indecomposable.

Proof. (1) We apply the functor $\mathrm{Hom}(-, \bigoplus_{i=1}^{r-1} \mathcal{E}_i)$ to the exact sequence (3.6). Taking into account that $\mathrm{Hom}(\mathcal{E}_i, \mathcal{E}_j) = 0$ for all $1 \le i \ne j \le r$, we get

$$0 \longrightarrow \mathrm{Hom}\left(\mathcal{E}, \bigoplus_{i=1}^{r-1} \mathcal{E}_i\right) \longrightarrow \mathrm{Hom}\left(\bigoplus_{i=1}^{r-1} \mathcal{E}_i, \bigoplus_{i=1}^{r-1} \mathcal{E}_i\right) \xrightarrow{\delta} \mathrm{Ext}^1\left(\mathcal{E}_r, \bigoplus_{i=1}^{r-1} \mathcal{E}_i\right).$$

Since δ is injective, we deduce that $\mathrm{Hom}(\mathcal{E}, \bigoplus_{i=1}^{r-1} \mathcal{E}_i) = 0$. Now, we apply $\mathrm{Hom}(-, \mathcal{E}_r)$ to the exact sequence (3.6) and obtain the exact sequence

$$0 \longrightarrow \mathrm{Hom}(\mathcal{E}_r, \mathcal{E}_r) \longrightarrow \mathrm{Hom}(\mathcal{E}, \mathcal{E}_r) \longrightarrow 0,$$

and hence $\mathrm{Hom}(\mathcal{E}, \mathcal{E}_r) = \mathbf{k}$. Finally, applying the functor $\mathrm{Hom}(\mathcal{E}, -)$ to (3.6), we get

$$0 \longrightarrow \mathrm{Hom}\left(\mathcal{E}, \bigoplus_{i=1}^{r-1} \mathcal{E}_i\right) \longrightarrow \mathrm{Hom}(\mathcal{E}, \mathcal{E}) \longrightarrow \mathrm{Hom}(\mathcal{E}, \mathcal{E}_r) \cong \mathbf{k},$$

which allows us to conclude that $\mathrm{Hom}(\mathcal{E}, \mathcal{E}) \cong \mathbf{k}$, i. e., $\mathcal{E}$ is simple and hence indecomposable.

(2) See [93, Theorem A, i)]. □

Now we will see that Ulrich bundles also behave well under restrictions to general hyperplanes and dualizing.

Proposition 3.3.3. *Let $\mathcal{E}$ be a rank r Ulrich bundle on a smooth variety $(X, \mathcal{O}_X(1))$ of dimension d and $\deg(X) = k$. Denote by ω_X the canonical line bundle on X. Then:*

1. *The restriction $\mathcal{E}_{|H}$ of $\mathcal{E}$ to a general hyperplane section is an Ulrich bundle.*
2. *$\mathcal{E}^\vee(d+1) \otimes \omega_X$ is an Ulrich bundle on X.*
3. *If $\mathcal{E} \cong \mathcal{E}^\vee(d+1) \otimes \omega_X$, then $\mathcal{E}$ is an Ulrich bundle if and only if $\mathcal{E}$ is 0-regular.*

Proof. (1) We consider the exact sequence

$$0 \longrightarrow \mathcal{E}(-1) \longrightarrow \mathcal{E} \longrightarrow \mathcal{E}_{|H} \longrightarrow 0$$

for a general hyperplane section H, which, by Bertini's theorem, we may assume to be smooth. By Proposition 2.1.13, $\mathcal{E}_{|H}$ is an aCM bundle. Furthermore, from the above exact sequence we see that $\mathrm{H}^0(\mathcal{E}_{|H}(-1)) = 0$ and $\mathrm{h}^0(\mathcal{E}_{|H}) = kr$. Hence, by definition $\mathcal{E}_{|H}$ is an Ulrich bundle.

(2) Since $\mathcal{E}^\vee(d+1) \otimes \omega_X$ is an initialized vector bundle, by Proposition 3.2.7, we have that $\mathcal{E}^\vee(d+1) \otimes \omega_X$ is an Ulrich bundle if and only if for any $1 \leq t \leq d$ and any $i \geq 0$,

$$\mathrm{H}^i(\mathcal{E}^\vee(d+1-t) \otimes \omega_X) = 0.$$

By Serre's duality, this is equivalent to the fact that $\mathcal{E}$ is an Ulrich bundle.

(3) If $\mathcal{E}$ is an Ulrich bundle, then by Corollary 3.2.10, it is 0-regular. Conversely, 0-regularity implies that $\mathrm{H}^i(\mathcal{E}(j)) = 0$ for $j > i$. The rest of the vanishing follows from Serre duality and the fact that $\mathcal{E} \cong \mathcal{E}^\vee(d+1) \otimes \omega_X$. □

In some sense, Proposition 3.3.3 tells us that Ulrich bundles come in pairs and this suggests introducing the following definition.

Definition 3.3.4. Let $\mathcal{E}$ be a rank r Ulrich bundle on a smooth variety $(X, \mathcal{O}_X(1))$ of dimension d and denote by ω_X the canonical line bundle on X. We define the *Ulrich dual* of $\mathcal{E}$ by $\mathcal{E}^\vee(d+1) \otimes \omega_X$.

While dealing with rank 2 vector bundles $\mathcal{E}$ on a projective variety X, we have $\mathcal{E}^\vee \cong \mathcal{E} \otimes \mathcal{O}_X(-c_1(\mathcal{E}))$. This, together with Proposition 3.3.3, brings us to the next concept.

Definition 3.3.5. A rank-two Ulrich bundle $\mathcal{E}$ on a smooth variety $(X, \mathcal{O}_X(1))$ of dimension d is said to be a *special Ulrich bundle* if $\mathcal{O}_X(c_1(\mathcal{E})) \cong \mathcal{O}_X(d+1) \otimes \omega_X$, i. e., if $\mathcal{E}$ is isomorphic to its Ulrich dual.

By means of the next example, we will illustrate how, using the properties that we have recently pointed out, we can construct rank 2 special Ulrich bundles.

Example 3.3.6. Let $\pi : X \to C$ be a geometrically ruled surface of invariant $e > 0$ over a curve C of genus $g \geq 0$. We fix $p \in C$ and denote by f the fiber of π over p and by C_0 the section of self-intersection $-e$. Any line bundle $\mathcal{L}$ on X is of the form $\mathcal{L} \cong \mathcal{O}_X(aC_0 + \pi^*(\mathfrak{b})f)$, with $\mathfrak{b} \in \mathrm{Pic}(C)$ of degree b. If there is no confusion, we also write $\mathcal{O}_X(aC_0 + bf) \otimes \pi^*\mathcal{B}$ where $\mathcal{B} \in \mathrm{Pic}^0(C)$ instead of $\mathcal{O}_X(aC_0 + \pi^*(\mathfrak{b})f)$ when $\mathfrak{b}$ has degree b. In particular, we have (see also Section 5.2)

$$\omega_X \cong \mathcal{O}_X(K_X) = \mathcal{O}_X(-2C_0 + (2g-2-e)f) \otimes \pi^*\mathcal{K}$$

with $\mathcal{K} \in \mathrm{Pic}^0(C)$. Consider the following ample divisor $H = C_0 + (e+1)f$ on X. We are going to construct a special rank 2 Ulrich bundle on $(X, \mathcal{O}_X(H))$. First of all, let us see that

$$\mathcal{O}_X(L_1) := \mathcal{O}_X((g+e+1)f) \otimes \pi^* \mathcal{L}_1$$

with $\mathcal{L}_1 \in \mathrm{Pic}^0(C)$ general is an Ulrich line bundle on X with respect to $\mathcal{O}_X(H)$. To this end, observe that

$$HL_1 = \frac{H(K_X + 3H)}{2}$$

and, by Riemann–Roch theorem (Theorem 1.4.3),

$$\chi(\mathcal{O}_X(L_1 - H)) = (1-g) - \frac{1}{2}(L_1 - H)K_X + \frac{1}{2}(L_1 - H)^2 = 0.$$

These two facts also imply that $\chi(\mathcal{O}_X(L_1 - 2H)) = 0$. On the other hand, according to Lemma 3.2.7, $\mathcal{O}_X(L_1)$ is an Ulrich line bundle on $(X, \mathcal{O}_X(H))$ if and only if for any $i \geq 0$ and any $1 \leq t \leq 2$, we have $\mathrm{H}^i(\mathcal{O}_X(L_1 - tH)) = 0$. Notice that the coefficient of C_0 in $L_1 - H$ equals -1. Hence

$$\mathrm{h}^0(\mathcal{O}_X(L_1 - H)) = 0 \quad \text{and} \quad \mathrm{h}^0(\mathcal{O}_X(L_1 - 2H)) = 0.$$

The coefficient of C_0 in $H - L_1 + K_X$ also equals -1. Thus,

$$\mathrm{h}^2(\mathcal{O}_X(L_1 - H)) = \mathrm{h}^0(\mathcal{O}_X(H - L_1 + K_X)) = 0.$$

These vanishings, together with the fact that $\chi(\mathcal{O}_X(L_1 - H)) = 0$, give us $\mathrm{h}^1(\mathcal{O}_X(L_1 - H)) = 0$. Notice that $\mathcal{O}_X(2H - L_1 + K_X) = \mathcal{O}_X((g-1)f) \otimes \pi^* \mathcal{L}_2$ with $\mathcal{L}_2 \in \mathrm{Pic}^0(C)$ general. Therefore,

$$\mathrm{h}^2(\mathcal{O}_X(L_1 - 2H)) = \mathrm{h}^0(\mathcal{O}_X(2H - L_1 + K_X)) = \mathrm{h}^0(\mathcal{O}_C((g-1)p) \otimes \mathcal{L}_2) = 0.$$

Since $\chi(\mathcal{O}_X(L_1 - 2H)) = 0$, we also have $\mathrm{h}^1(\mathcal{O}_X(L_1 - 2H)) = 0$. Putting everything altogether, we get that $\mathcal{O}_X(L_1)$ is an Ulrich line bundle. By Proposition 3.3.3 (3),

$$\mathcal{O}_X(L_2) := \mathcal{O}_X(K_X + 3H) \otimes \mathcal{O}_X(L_1)^\vee$$

is also an Ulrich line bundle on X. In addition, since $\mathrm{Ext}^1(\mathcal{O}_X(L_1), \mathcal{O}_X(L_2)) \neq 0$, any nontrivial extension

$$0 \longrightarrow \mathcal{O}_X(L_2) \longrightarrow \mathcal{E} \longrightarrow \mathcal{O}_X(L_1) \longrightarrow 0$$

gives us, by Proposition 3.3.1, a rank-two Ulrich bundle on X. Finally, since $\mathcal{O}_X(c_1(\mathcal{E})) \cong \mathcal{O}_X(L_1 + L_2) \cong \mathcal{O}_X(3H + K_X)$, $\mathcal{E}$ is a special Ulrich bundle on $(X, \mathcal{O}_X(H))$.

Special Ulrich bundles will play an important role in Chapter 5 when the underlying variety is a surface. Meanwhile, we will see another interesting feature of Ulrich bundles.

Proposition 3.3.7. *Let $(X, \mathcal{O}_X(H))$ be a smooth variety of dimension d and $\sigma : \tilde{X} \longrightarrow X$ the blow-up of X at a point $p \in X$. Denote by E the exceptional divisor. Assume that $\tilde{H} := \sigma^* H - E$ is very ample. If $\mathcal{E}$ is an Ulrich bundle on $(X, \mathcal{O}_X(H))$, then $\tilde{\mathcal{E}} := \sigma^*\mathcal{E} \otimes \mathcal{O}_{\tilde{X}}(-E)$ is an Ulrich bundle on $(\tilde{X}, \mathcal{O}_{\tilde{X}}(\tilde{H}))$.*

Proof. According to Proposition 3.2.7, it is enough to check that

$$\mathrm{H}^i(\tilde{X}, \tilde{\mathcal{E}}(-j\tilde{H})) = 0 \quad \text{for } 0 \leq i \leq d, \quad 1 \leq j \leq d.$$

We first deduce from the exact sequence

$$0 \longrightarrow \mathcal{O}_{\tilde{X}}(-E) \longrightarrow \mathcal{O}_{\tilde{X}} \longrightarrow \mathcal{O}_E \cong \mathcal{O}_{\mathbb{P}^{d-1}} \longrightarrow 0$$

that

$$\sigma_*(\mathcal{O}_{\tilde{X}}(jE)) \cong \mathcal{O}_X \quad \text{for } 0 \leq j \leq d-1$$

and $R^i\sigma_*\mathcal{O}_{\tilde{X}}(jE) = 0$ for $i > 0$ and $0 \leq j \leq d-1$. Applying the projection formula, we have that for $i > 0$ and $1 \leq j \leq d$,

$$R^i\sigma_*(\tilde{\mathcal{E}}(-j\tilde{H})) = \mathcal{E}(-jH) \otimes R^i\sigma_*\mathcal{O}_{\tilde{X}}((j-1)E).$$

Hence, Leray spectral sequence implies that for $0 \leq i \leq d$ and $1 \leq j \leq d$,

$$\begin{aligned}
\mathrm{H}^i(\tilde{X}, \tilde{\mathcal{E}}(-j\tilde{H})) &= \mathrm{H}^i(X, \sigma_*(\tilde{\mathcal{E}}(-j\tilde{H}))) \\
&\cong \mathrm{H}^i(X, \mathcal{E}(-jH) \otimes \sigma_*\mathcal{O}_{\tilde{X}}((j-1)E)) \\
&\cong \mathrm{H}^i(X, \mathcal{E}(-jH)) = 0
\end{aligned}$$

where the last equality follows from the fact that $\mathcal{E}$ is an Ulrich bundle on $(X, \mathcal{O}_X(H))$. □

Example 3.3.8. We consider the projective plane $\mathbb{P}^2$ and the cotangent bundle $\Omega^1_{\mathbb{P}^2}$. Consider the blow-up $\sigma : \widetilde{\mathbb{P}}^2 \longrightarrow \mathbb{P}^2$ of $\mathbb{P}^2$ at a point $p \in \mathbb{P}^2$ and denote by $E = \sigma^{-1}(p)$. Since $\Omega^1_{\mathbb{P}^2}(3)$ is a rank 2 Ulrich bundle on $(\mathbb{P}^2, \mathcal{O}_{\mathbb{P}^2}(2))$ and $\mathcal{O}_{\widetilde{\mathbb{P}}^2}(\widetilde{H}) := \sigma^*(\mathcal{O}_{\mathbb{P}^2}(2)) \otimes \mathcal{O}_{\widetilde{\mathbb{P}}^2}(-E)$ is very ample, by Proposition 3.3.7, we conclude that $\sigma^*(\Omega^1_{\mathbb{P}^2}(3)) \otimes \mathcal{O}_{\widetilde{\mathbb{P}}^2}(-E)$ is a rank 2 Ulrich bundle on $(\widetilde{\mathbb{P}}^2, \mathcal{O}_{\widetilde{\mathbb{P}}^2}(\widetilde{H}))$.

Ulrichness is also resilient under pushing forward vector bundles for the blow-up morphism.

Theorem 3.3.9. *Let $(X, \mathcal{O}_X(H))$ be a smooth variety of dimension $d \geq 2$ and $\sigma : \tilde{X} \longrightarrow X$ the blow-up of X at a point $p \in X$. Denote by E the exceptional divisor. Assume that $\tilde{H} := \sigma^* H - E$ is very ample. If $\mathcal{E}$ is a rank r Ulrich bundle on $(\tilde{X}, \mathcal{O}_{\tilde{X}}(\tilde{H}))$ and $\mathcal{E}_E \cong \mathcal{O}_E(1)^r$ then $\sigma_*\mathcal{E}(E)$ is an Ulrich bundle on $(X, \mathcal{O}_X(H))$.*

Proof. By hypothesis, we have $\mathcal{E}_E(sE) \cong (\mathcal{O}_E(1)^r) \otimes \mathcal{O}_E(-s) \cong \mathcal{O}_E(1-s)^r$ for all $s \in \mathbb{Z}$. In particular, $\mathcal{E}_E(E) \cong \mathcal{O}_E^r$ and therefore, by [45, Corollary 2.3], $\mathcal{F} := \sigma_*\mathcal{E}(E)$ is a vector bundle such that $\mathcal{E} \cong \sigma^*\mathcal{F}(-E)$. Moreover, by [45, Theorem 2.1], $R^i\sigma_*\mathcal{E}(E) = 0$ for all $i \geq 1$. Hence,

$$\mathrm{H}^i(X, \sigma_*\mathcal{E}(E) \otimes \mathcal{O}_X(-tH)) = \mathrm{H}^i(\tilde{X}, \mathcal{E}(-t\tilde{H} - (t-1)E)).$$

So it will be enough to show that the cohomology groups on the right-hand side of the equation are zero for $i \geq 0$ and $t = 1, \ldots, d$. The case $t = 1$ is trivial by hypothesis. For the remaining values of t, let us consider the structural exact sequence

$$0 \longrightarrow \mathcal{O}_{\tilde{X}}(-E) \longrightarrow \mathcal{O}_{\tilde{X}} \longrightarrow \mathcal{O}_E \longrightarrow 0. \tag{3.7}$$

Now, observe that $\mathcal{E}(-a\tilde{H} - bE)_E \cong \mathcal{O}_E(1-(a-b))^r$, so all of its cohomology groups vanish for any a, b such that $-d-1 \leq b-a \leq -2$. Then, tensoring the short exact sequence (3.7) by $\mathcal{E}(-a\tilde{H} - bE)$, we can see that $\mathrm{H}^i(\mathcal{E}(-a\tilde{H} - bE)) = 0$ for all $a = 1, \ldots, d$ and $0 \leq b \leq a-1$. In particular, we obtain the needed vanishings. □

The following lemma describes Ulrich bundles on the product of two smooth projective varieties.

Lemma 3.3.10. *Let $\mathcal{E}$ and $\mathcal{F}$ be Ulrich bundles on smooth projective varieties $(X, \mathcal{O}_X(1))$ and $(Y, \mathcal{O}_Y(1))$, respectively. Denote by $l = \dim X$ and $m = \dim Y$. Then*

$$\mathcal{E} \boxtimes \mathcal{F}(l) \quad \textit{and} \quad \mathcal{E}(m) \boxtimes \mathcal{F}$$

are both Ulrich bundles on $(X \times Y, \mathcal{O}_{X\times Y}(1,1)) \cong (X \times Y, \mathcal{O}_X(1) \boxtimes \mathcal{O}_Y(1))$.

Proof. We will prove that $\mathcal{E} \boxtimes \mathcal{F}(l)$ is an Ulrich bundle (the other case can be seen in a similar way). Since $\dim(X \times Y) = l + m$, according to Proposition 3.2.7, it is enough to see that for any $1 \leq t \leq l+m$ and $i \geq 0$,

$$\mathrm{H}^i(X \times Y, \mathcal{E} \boxtimes \mathcal{F}(l) \otimes (\mathcal{O}_X(-t) \boxtimes \mathcal{O}_Y(-t))) = 0. \tag{3.8}$$

By the Künneth formula, for $1 \leq i \leq l+m$,

$$\mathrm{H}^i(X \times Y, \mathcal{E} \boxtimes \mathcal{F}(l) \otimes (\mathcal{O}_X(-t) \boxtimes \mathcal{O}_Y(-t))) = \bigoplus_{p+q=i} \mathrm{H}^p(X, \mathcal{E}(-t)) \otimes \mathrm{H}^q(Y, \mathcal{F}(l-t)).$$

Since $\mathcal{E}$ is an Ulrich bundle on $(X, \mathcal{O}_X(1))$, for any $1 \leq t \leq l$ and $i \geq 0$, $\mathrm{H}^i(X, \mathcal{E}(-t)) = 0$ and, since $\mathcal{F}$ is an Ulrich bundle on $(Y, \mathcal{O}_Y(1))$, for any $l+1 \leq t \leq l+m$ and $i \geq 0$, $\mathrm{H}^i(X, \mathcal{F}(l-t)) = 0$. Hence (3.8) holds and $\mathcal{E} \boxtimes \mathcal{F}(l)$ is an Ulrich bundle. □

Example 3.3.11. (1) From Lemma 3.3.10, we recover Example 3.2.3 (3), namely that $\mathcal{O}_{\mathbb{P}^n\times\mathbb{P}^m}(m,0)$ and $\mathcal{O}_{\mathbb{P}^n\times\mathbb{P}^m}(0,n)$ are Ulrich line bundles on $\mathbb{P}^n \times \mathbb{P}^m$ with respect to $\mathcal{O}_{\mathbb{P}^n\times\mathbb{P}^m}(1,1)$.

(2) Using inductively Lemma 3.3.10 as well as Example 3.2.3, we find Ulrich line bundles on $X = \mathbb{P}^{n_1} \times \cdots \times \mathbb{P}^{n_s}$ with respect to $\mathcal{O}_X(1, \dots, 1)$. Indeed, any line bundle of the form $\mathcal{O}_X(a_1, \dots, a_s)$, where if we order the coefficients $0 = a_{i_1} \leq \cdots \leq a_{i_k} \leq \cdots \leq a_{i_s}$ then $a_{i_k} = \sum_{1 \leq j < k} n_{i_j}$, is an Ulrich line bundle. We are going to study deeply Ulrich bundles on Segre varieties in Chapter 7, and we will see that the last list is exhaustive (see in particular Proposition 7.1.3).

Lemma 3.3.12. *Let $\pi : X \longrightarrow Y$ be a finite surjective morphism of smooth projective varieties and let $\mathcal{E}$ be a vector bundle on X. Then $\mathcal{E}$ is an Ulrich bundle on $(X, \pi^*\mathcal{O}_Y(1))$ if and only if $\pi_*\mathcal{E}$ is an Ulrich bundle on $(Y, \mathcal{O}_Y(1))$.*

Proof. The claim follows from the isomorphism given by the projection formula

$$\mathrm{H}^i(Y, \pi_*\mathcal{E} \otimes \mathcal{O}_Y(t)) \cong \mathrm{H}^i(X, \mathcal{E} \otimes \pi^*\mathcal{O}_Y(t))$$

for any $t \in \mathbb{Z}$ and $i \geq 0$. □

As a consequence of the above two results, we can prove the existence of Ulrich bundles on $(\mathbb{P}^n, \mathcal{O}_{\mathbb{P}^n}(m))$ for $m \geq 1$. In fact, we have

Corollary 3.3.13. *Let $(X, \mathcal{O}_X(1))$ be a smooth variety of dimension d carrying a rank r Ulrich bundle. Then, $(X, \mathcal{O}_X(m))$ for $m \geq 1$ carries a rank $rd!$ Ulrich bundle. In particular, $(\mathbb{P}^n, \mathcal{O}_{\mathbb{P}^n}(m))$ carries a rank $n!$ Ulrich bundle.*

Proof. Let us start proving that $(\mathbb{P}^n, \mathcal{O}_{\mathbb{P}^n}(m))$ carries a rank $n!$ Ulrich bundle. First of all, we consider the quotient map

$$\pi : \mathbb{P}^1 \times \cdots \times \mathbb{P}^1 \longrightarrow \mathrm{Sym}^n(\mathbb{P}^1) \cong \mathbb{P}^n.$$

The pullback $\pi^*\mathcal{O}_{\mathbb{P}^n}(1)$ is $\mathcal{O}_{\mathbb{P}^1}(1) \boxtimes \cdots \boxtimes \mathcal{O}_{\mathbb{P}^1}(1)$ and since $\mathcal{O}_{\mathbb{P}^1}(m-1)$ is Ulrich on $(\mathbb{P}^1, \mathcal{O}_{\mathbb{P}^1}(m))$, by Lemma 3.3.10, the line bundle $\mathcal{L} = \mathcal{O}_{\mathbb{P}^1}(m-1) \boxtimes \cdots \boxtimes \mathcal{O}_{\mathbb{P}^1}(nm-1)$ is an Ulrich bundle for $(\mathbb{P}^1 \times \cdots \times \mathbb{P}^1, \pi^*\mathcal{O}_{\mathbb{P}^n}(m))$. Therefore, by Lemma 3.3.12, $\pi_*\mathcal{L}$ is a rank $n!$ Ulrich bundle on $(\mathbb{P}^n, \mathcal{O}_{\mathbb{P}^n}(m))$.

Now, let $\pi : X \longrightarrow \mathbb{P}^d$ be a finite projection, $\mathcal{E}$ a rank r Ulrich bundle on $(X, \mathcal{O}_X(1))$ and $\mathcal{F}$ a rank $d!$ Ulrich bundle on $(\mathbb{P}^d, \mathcal{O}_{\mathbb{P}^d}(m))$. By the projection formula, $\mathcal{E} \otimes \pi^*\mathcal{F}$ is a rank $rd!$ Ulrich bundle on $(X, \mathcal{O}_X(m))$. □

We will end this section revealing one of the most important properties of Ulrich bundles when dealing with classification problems: their stability and μ-stability (see Section 1.5).

Proposition 3.3.14. *Let $(X, \mathcal{O}_X(1))$ be a d-dimensional smooth variety of degree k and let $\mathcal{E}$ be a rank r Ulrich bundle on X. Then $\mathcal{E}$ is semistable and, in particular, μ-semistable.*

Proof. Let $\mathcal{F}$ be a rank s torsion-free coherent subsheaf of $\mathcal{E}$. For any finite linear projection $\pi : X \longrightarrow \mathbb{P}^d$, since $\mathcal{E}$ is Ulrich, by Theorem 3.2.9, $\pi_*\mathcal{F}$ is a rank ks torsion-free coherent subsheaf of $\pi_*\mathcal{E} = \mathcal{O}_{\mathbb{P}^d}^{kr}$. Since $\mathcal{O}_{\mathbb{P}^d}^{kr}$ is semistable, we have that $P_{\pi_*\mathcal{F}}(m) \leq P_{\pi_*\mathcal{E}}(m)$ for all $m \gg 0$ and, since π is finite, $dP_{\pi_*\mathcal{F}}(m) = P_{\mathcal{F}}(m)$ and $dP_{\pi_*\mathcal{E}}(m) = P_{\mathcal{E}}(m)$. Hence, $P_{\mathcal{F}}(m) \leq P_{\mathcal{E}}(m)$ for all $m \gg 0$ and thus $\mathcal{E}$ is semistable. □

Remark 3.3.15. Thanks to Proposition 3.3.14, it makes sense to denote by $M^U_{X,H}(r; c_1, \ldots, c_{\min(d,n)}) \subset M_{X,H}(r; c_1, \ldots, c_{\min(d,n)})$ the subspace (perhaps empty) inside the moduli space parameterizing those rank r semistable vector bundles with fixed Chern classes c_i that are Ulrich bundles. Analogously, $M^{U,s}_{X,H}(r; c_1, \ldots, c_{\min(d,n)}) \subset M^s_{X,H}(r; c_1, \ldots, c_{\min(d,n)})$ will denote the subspace parameterizing the stable ones.

In the following result we will see that if $\mathcal{E}$ is an Ulrich bundle then it admits a Jordan–Hölder filtration by Ulrich factors.

Proposition 3.3.16. *Let $\mathcal{E}$ be a rank r Ulrich bundle on a d-dimensional smooth variety $X \subset \mathbb{P}^n$ of degree k. Then, for any Jordan–Hölder filtration of $\mathcal{E}$,*

$$0 = \mathcal{E}_0 \subset \mathcal{E}_1 \subset \cdots \subset \mathcal{E}_l = \mathcal{E},$$

the $\mathcal{E}_i$ are also Ulrich bundles for $1 \leq i \leq l$. In particular, if $\mathcal{E}$ is a strictly semistable Ulrich bundle of rank $r \geq 2$, then there exists a subbundle $\mathcal{F}$ of $\mathcal{E}$ of rank $s < r$ which is Ulrich.

Proof. Since $\mathcal{E}$ is an Ulrich bundle, take a finite linear projection $\pi : X \longrightarrow \mathbb{P}^d$ so that $\pi_*\mathcal{E} \cong \mathcal{O}_{\mathbb{P}^d}^{kr}$ and consider a Jordan–Hölder filtration of $\mathcal{E}$ as in the statement. By definition, for any $1 \leq i \leq l$, $\mathcal{E}_i$ is torsion-free and the quotient $\mathcal{E}_i/\mathcal{E}_{i-1}$ is torsion-free and stable with $P_{\mathcal{E}_i/\mathcal{E}_{i-1}}(m) = P_{\mathcal{E}}(m) = k\binom{m+n}{n}$ where this last equality follows from Corollary 3.2.10. Assume that $\mathcal{E}_1$ is an Ulrich bundle. Then, Proposition 3.3.1 implies that $\mathcal{E}/\mathcal{E}_1$ is an Ulrich bundle, and we conclude by induction on the rank of $\mathcal{E}$. Hence, it only remains to prove that $\mathcal{E}_1$ is an Ulrich bundle.

First of all, notice that since

$$P_{\pi_*\mathcal{E}_1}(m) = \frac{P_{\mathcal{E}_1}(m)}{k} = \frac{P_{\mathcal{E}}(m)}{k} = P_{\pi_*\mathcal{E}}(m)$$

we have that $\pi_*\mathcal{E}_1$ is a destabilizing subsheaf of $\pi_*\mathcal{E} \cong \mathcal{O}_{\mathbb{P}^d}^{kr}$. It follows from the semistability of $\mathcal{O}_{\mathbb{P}^d}^{kr}$ that $\pi_*\mathcal{E}_1$ is a semistable sheaf on $\mathbb{P}^d$, and hence it has a Jordan–Hölder filtration

$$0 = \mathcal{H}_0 \subset \mathcal{H}_1 \subset \cdots \subset \mathcal{H}_{s-1} \subset \mathcal{H}_s = \pi_*\mathcal{E}_1. \tag{3.9}$$

Consider now the short exact sequence

$$0 \longrightarrow \mathcal{E}_1 \longrightarrow \mathcal{E} \longrightarrow \mathcal{E}/\mathcal{E}_1 \longrightarrow 0$$

and its pushforward into $\mathbb{P}^d$

$$0 \longrightarrow \pi_*\mathcal{E}_1 \longrightarrow \pi_*\mathcal{E} \longrightarrow \pi_*\mathcal{E}/\pi_*\mathcal{E}_1 \longrightarrow 0.$$

Since $\pi_*\mathcal{E}_1$ and $\pi_*\mathcal{E}$ have the same Hilbert polynomial, so does $\pi_*\mathcal{E}/\pi_*\mathcal{E}_1$, and since $\pi_*\mathcal{E}$ is semistable we deduce that $\pi_*\mathcal{E}/\pi_*\mathcal{E}_1$ is also semistable. Thus, we may concatenate (3.9) with the Jordan–Hölder filtration of $\pi_*\mathcal{E}/\pi_*\mathcal{E}_1$ to obtain a Jordan–Hölder filtration of $\pi_*\mathcal{E} \cong \mathcal{O}_{\mathbb{P}^d}^{kr}$. But the filtration

$$0 \subset \mathcal{O}_{\mathbb{P}^d} \subset \mathcal{O}_{\mathbb{P}^d}^2 \subset \cdots \subset \mathcal{O}_{\mathbb{P}^d}^{kr}$$

is also a Jordan–Hölder filtration of $\mathcal{O}_{\mathbb{P}^d}^{kr}$. Hence, for $1 \leq j \leq s$, the successive quotients $\mathcal{H}_j/\mathcal{H}_{j-1}$ are isomorphic to $\mathcal{O}_{\mathbb{P}^d}$. In particular, $\mathcal{H}_1 \cong \mathcal{O}_{\mathbb{P}^d}$ and $\mathcal{H}_2/\mathcal{H}_1 \cong \mathcal{O}_{\mathbb{P}^d}$. Therefore, $\mathcal{H}_2$ is an extension of $\mathcal{O}_{\mathbb{P}^d}$ by $\mathcal{O}_{\mathbb{P}^d}$, and since $\operatorname{Ext}^1(\mathcal{O}_{\mathbb{P}^d}, \mathcal{O}_{\mathbb{P}^d}) = 0$ this means that the extension splits and $\mathcal{H}_2 \cong \mathcal{O}_{\mathbb{P}^d}^2$. Continuing in this way, we see that $\mathcal{H}_s = \pi_*\mathcal{E}_1$ is an extension of trivial line bundles on $\mathbb{P}^d$ and thus $\pi_*\mathcal{E}_1$ is also trivial. Since π is a finite flat morphism and $\pi_*\mathcal{E}_1$ is locally free, the same is true for $\mathcal{E}_1$. Hence, by Theorem 3.2.9, $\mathcal{E}_1$ is an Ulrich bundle. □

Corollary 3.3.17. *Let $\mathcal{E}$ be an Ulrich bundle of rank $r \geq 2$ on a d-dimensional smooth variety $X \subset \mathbb{P}^n$ of degree k. Then,*
(1) *If $\mathcal{E}$ is not stable, it is an extension of Ulrich bundles of smaller rank.*
(2) *If $\mathcal{E}$ is stable, then it is also μ-stable.*

Proof. (1) By Proposition 3.3.16, if $\mathcal{E}$ is strictly semistable, then there exists a subbundle $\mathcal{F}$ of $\mathcal{E}$ of rank $s < r$ which is Ulrich. Thus, by Proposition 3.3.1, $\mathcal{E}/\mathcal{F}$ is Ulrich and $\mathcal{E}$ is an extension of Ulrich bundles.

(2) Assume that $\mathcal{E}$ is stable. We want to see that $\mathcal{E}$ is μ-stable, i. e., that for any $\mathcal{F} \subset \mathcal{E}$, with $\mathcal{E}/\mathcal{F}$ torsion-free, $\mu(\mathcal{F}) < \mu(\mathcal{E})$. Since $\mathcal{E}$ is μ-semistable, by Proposition 3.3.14, we have $\mu(\mathcal{F}) \leq \mu(\mathcal{E})$. If the equality holds, using the short exact sequence

$$0 \longrightarrow \mathcal{F} \longrightarrow \mathcal{E} \longrightarrow \mathcal{E}/\mathcal{F} \longrightarrow 0$$

and its restriction to a general hyperplane section, by induction we can prove that $\mathcal{F}$ is also an Ulrich bundle. Therefore, by Corollary 3.2.10, $P_{\mathcal{F}}(m) = P_{\mathcal{E}}(m)$ which contradicts the stability of $\mathcal{E}$. Hence, $\mathcal{E}$ is μ-stable. □

3.4 On the existence of Ulrich bundles: first results towards Eisenbud–Schreyer conjecture

As it has been mention in Section 3.1, since the entrance of Ulrich sheaves in the realm of algebraic geometry, a leading question has been the existence of Ulrich bundles on a given projective variety. This problem, as it will be seen throughout the rest of this book, is far from being solved. However, in the particular case of curves, the existence of Ulrich bundles can be settled. Indeed, in the seminal work [88], the existence of Ulrich line bundles on curves of arbitrary genus was already stated. Building upon this result, in this section we are going to prove that any curve of genus $g \geq 1$ supports stable (and, therefore, indecomposable) Ulrich sheaves of arbitrary rank.

Definition 3.4.1. Let $C \subset \mathbb{P}^n$ be a curve. A coherent sheaf $\mathcal{E}$ on C has *no cohomology* if $\mathrm{h}^0(\mathcal{E}) = \mathrm{h}^1(\mathcal{E}) = 0$.

By Lemma 3.2.7, we immediately obtain the following:

Lemma 3.4.2. *Let $(C, \mathcal{O}_C(1))$ be a smooth curve of degree k and genus g and let $\mathcal{G}$ be a rank r vector bundle on C. Then, the following are equivalent:*
(1) *$\mathcal{G}$ is Ulrich.*
(2) *$\mathcal{G}(-1)$ is a vector bundle with no cohomology.*
(3) *$\deg(\mathcal{G}) = r(k + g - 1)$ and $\mathrm{h}^0(C, \mathcal{G}(-1)) = 0$.*

Proof. To see the equivalence between (1) and (2), we argue as follows. If $\mathcal{G}$ is an Ulrich bundle, we know, by Lemma 3.2.7, that $\mathcal{G}(-1)$ has no cohomology. Conversely, assume that $\mathcal{G}(-1)$ has no cohomology. In particular, it is 0-regular and, therefore, globally generated. Hence, $\mathcal{G}$ is an Ulrich bundle.

Finally, to prove the equivalence between (2) and (3), notice that, by Riemann–Roch theorem, $\chi(\mathcal{G}(-1)) = 0$ if and only if $\deg(\mathcal{G}) = r(k + g - 1)$. □

Example 3.4.3. (1) The line bundle $\mathcal{O}_{\mathbb{P}^1}(-1)$ has no cohomology. Hence, by Lemma 3.4.2, $\mathcal{O}_{\mathbb{P}^1}(d-1)$ is an Ulrich bundle on $(\mathbb{P}^1, \mathcal{O}(d))$ for any $d \geq 1$.

(2) Let C be a smooth elliptic curve and $\mathcal{L} \in \mathrm{Pic}^0(C)$ a nontrivial line bundle of degree 0. Since $\mathcal{L}$ has no cohomology, by Lemma 3.4.2, $\mathcal{L}(1)$ is an Ulrich line bundle on $(C, \mathcal{O}_C(1))$ for any very ample line bundle $\mathcal{O}_C(1)$.

Therefore, the main question for curves is to prove the existence of vector bundles with no cohomology. As a main tool to find them, we will use the geometrical version of the Riemann–Roch theorem (see [11, p. 12]). To this end, let us introduce some notation.

Let C be a smooth curve and $\phi : C \longrightarrow \mathbb{P}^r$ be a map. For any effective divisor D on C, $\overline{\phi(D)}$ denotes the intersection of all the hyperplanes $H \subset \mathbb{P}^r$ such that either $\phi(C) \subset H$ or $\phi^*(H) - D$ is still an effective divisor.

Obviously, in the particular case that ϕ is an embedding and $D = \sum_{i=1}^d p_i$ consists of different points, $\overline{\phi(D)}$ is the usual linear span of $\phi(D)$ in $\mathbb{P}^r$. On the other hand, associated to a line bundle $\mathcal{L}$ on C we have the map

$$\phi_{\mathcal{L}} : C \longrightarrow \mathbb{P}(\mathrm{H}^0(\mathcal{L})^\vee)$$

which is a closed embedding if and only if $\mathcal{L}$ is very ample. This is the case for the canonical line bundle $\mathcal{O}_C(K_C)$ whenever the genus $g(C) > 1$. To simplify, ϕ_K will denote $\phi_{\mathcal{O}_C(K_C)}$. Taking into account these observations, we can state the geometric version of the Riemann–Roch theorem.

Theorem 3.4.4. *For an effective divisor D of degree k on a smooth curve of genus g > 1,*

$$h^0(\mathcal{O}_C(D)) = k - \dim \overline{\phi_K(D)}.$$

Proof. See [11, p. 12]. □

Theorem 3.4.5. *Let $(C, \mathcal{O}_C(1))$ be a smooth curve. Then C supports an Ulrich line bundle.*

Proof. By Lemma 3.4.2, it is enough to show that C supports a line bundle $\mathcal{L}$ with no cohomology. Notice that a line bundle $\mathcal{L}$ on C has no cohomology if and only if $h^0(\mathcal{L}) = 0$ and $\deg(\mathcal{L}) = g(C) - 1$. Let us study first the cases of genus $g = g(C) \leq 1$. In Example 3.2.3 we have seen that $\mathcal{O}_{\mathbb{P}^1}$ is an Ulrich line bundle on $\mathbb{P}^1$. Assume now that $g = 1$. In this case, C is an elliptic curve and, taking two distinct points $p, q \in C$, $\mathcal{O}_C(p - q)$ will work. So let us assume that $g(C) > 1$. Then the span of g general points $p_1, \dots, p_g \in C \subset \mathbb{P}(H^0(\mathcal{O}_C(K_C))^\vee) \cong \mathbb{P}^{g-1}$ will have dimension $g - 1$ and therefore, by Theorem 3.4.4, $h^0(\mathcal{O}_C(\sum_{i=1}^g p_i)) = 1$. Therefore, taking another general point $q \in C$, the line bundle $\mathcal{O}_C(\sum_{i=1}^g p_i - q)$ will have no cohomology. □

Regarding indecomposable Ulrich bundles of higher rank, by Grothendieck's theorem, they cannot exist on rational curves. Vector bundles on elliptic curves where classified by Atiyah in [14]. Using this classification, we can state the following result:

Theorem 3.4.6. *Let $(C, \mathcal{O}_C(1))$ be a smooth elliptic curve. For any $r \geq 1$, there exists a one-dimensional family of indecomposable rank r Ulrich bundles on C. They are strictly semistable.*

Proof. By Lemma 3.4.2, we are looking for vector bundles $\mathcal{G}$ of degree 0 and no cohomology. Atiyah proved (see [14, Theorem 7]) that there exists a unique vector bundle $\mathcal{E}_r$ of rank r and degree 0 with nonzero global sections. Therefore, the vector bundles of the form $\mathcal{G} \cong \mathcal{E}_r \otimes \mathcal{L}$ for $\mathcal{L} \in \operatorname{Pic}^0(C)$, $\mathcal{L} \not\cong \mathcal{O}_C$ are the required ones. We know that Ulrich bundles are semistable. On the other hand, the existence of an exact sequence of the form (see [14])

$$0 \longrightarrow \mathcal{O}_X \longrightarrow \mathcal{E}_r \longrightarrow \mathcal{E}_{r-1} \longrightarrow 0$$

shows that they are not stable. □

Let us focus now our attention on smooth curves $C \subset \mathbb{P}^n$ of genus $g > 1$.

Theorem 3.4.7. *Let $(C, \mathcal{O}_C(1))$ be a smooth curve of genus $g > 1$ and degree k and take $H \in |\mathcal{O}_C(1)|$. Then C supports stable rank r Ulrich bundles and they form an open subset of the moduli space $M_{C,H}(r; r(k + g - 1))$ of stable vector bundles of rank r and degree $r(k + g - 1)$.*

Proof. For $r = 1$, we already obtained this result in Theorem 3.4.5. So let us assume that $r > 1$. We will take r distinct Ulrich line bundles $\mathcal{L}_i$, $i = 1, \dots, r$. To prove their existence, we fix g general points $p_1, \dots, p_g \in C$. By Theorem 3.4.5, any general point $q_i \in C$ gives rise to a different Ulrich line bundle $\mathcal{L}_i = \mathcal{O}_C(\sum_{j=1}^{g} p_j - q_i)$, $1 \le i \le r$.

Since $\mathrm{ext}^1(\mathcal{L}_i, \mathcal{L}_j) = \mathrm{h}^1(\mathcal{L}_i^\vee \otimes \mathcal{L}_j) = -\chi(\mathcal{L}_i^\vee \otimes \mathcal{L}_j) = g - 1 > 0$, by Proposition 3.3.2 (1), a general extension of the form

$$0 \longrightarrow \bigoplus_{i=1}^{r-1} \mathcal{L}_i \longrightarrow \mathcal{E} \longrightarrow \mathcal{L}_r \longrightarrow 0$$

will be a rank r simple Ulrich bundle $\mathcal{E}$ on C. The dimension of the moduli space of vector bundles near $[\mathcal{E}]$ is given by

$$\mathrm{h}^1(\mathcal{E}^\vee \otimes \mathcal{E}) = 1 - \chi(\mathcal{E}^\vee \otimes \mathcal{E}) = r^2(g-1) + 1.$$

On the other hand, as it is explained in [41, Remark 4.6], if there were no stable bundles among them, the general one would be given as an extension of a stable rank $r - 1$ Ulrich bundle and an Ulrich line bundle. The dimension of the family of such extensions is given by the formula

$$(r-1)^2(g-1) + 1 + g + (r-1)(g-1) + 1$$

which is smaller than $r^2(g-1)+1$. Hence, we can conclude that there exist stable rank r Ulrich bundles. Since being Ulrich is an open condition, we deduce that there exists a nonempty open subset $M^U_{C,H}(r, r(k+g-1))$ of the moduli space $M_{C,H}(r, r(k+g-1))$ of stable rank r bundles of degree $r(k+g-1)$ whose geometric points are parameterizing stable Ulrich bundles. □

Although, as we have just seen, the existence of Ulrich bundles on arbitrary curves is known, things become much more involved when we move forward to higher dimensions. Indeed, already in the case of smooth projective surfaces, the existence of Ulrich bundles is only known in some cases, as they can be surfaces $S \subset \mathbb{P}^3$, surfaces of minimal degree, del Pezzo surfaces, etc. However, for instance, in the case of surfaces of general type, very sparse results have been obtained by now. Chapter 5 will be devoted to the study of the existence of Ulrich bundles on projective surfaces. In the rest of the chapters, we will deal with the existence of Ulrich bundles and related issues on varieties of higher dimension.

All these results support and can be seen as a partial answer to the following conjecture due to Eisenbud and Schreyer (see also Question 3.1.2).

Conjecture 3.4.8. *Any smooth projective variety $(X, \mathcal{O}_X(1))$ supports an Ulrich bundle.*

3.5 Final comments and additional reading

Section 3.1 intends to motivate historically the study of Ulrich bundles in the algebraic geometry setting. The riveted reader can complete this lecture with further readings from, for example, [24] and [59]. Ulrich modules' advent can be fixed, up to our knowledge, in [200], where Ulrich gave criteria for a local Cohen–Macaulay ring to be Gorenstein. Of course, in [200] Ulrich modules had not been christened, yet. The first mention of an "Ulrich module" can be tracked down to [16]. The main references for Sections 3.2 and 3.3 are the papers [24, 41, 65], and [88] where the reader can extend his/her acquaintance with Ulrich bundles. In [22], Beauville made a systematic study of the relation between Ulrich bundles and determinantal representation of hypersurfaces. In algebraic geometry their importance was underlined in [88] and [39]. In particular, the questions raised in [88] spurred many algebraic geometers to tackle the problem of determining which varieties support an Ulrich bundle. As the reader will perceive, this question will become a central leitmotif for the remainder of this book.

For sake of readability, most of the results in this chapter were stated for smooth projective varieties. However, it is important to point out that for some of them the hypothesis can be relaxed. For instance, thanks to [109] and [110], Lemma 3.2.4 holds true for integral locally Gorenstein curves (using the arithmetic genus of the curve instead of the geometric one). The existence of Ulrich line bundles on curves from Theorem 3.4.5 is also true for purely 1-dimensional scheme, as explained in Section 3.4 from [88]. Also in Proposition 3.2.7 and Theorem 3.2.9 the smoothness condition can be dropped (see [88, Proposition 2.1]).

4 Ulrich bundles on complete intersections

The roots of Ulrich bundles date back more than a century ago when algebraists and geometers sought for determinantal representations of hypersurfaces. More precisely, given a homogeneous polynomial f of degree d, they looked for a square matrix $M = (\ell_{ij})$ of size d with linear entries such that $f = \det(M)$. We can reinterpret this problem geometrically and coordinate-free as follows: the hypersurface in $\mathbb{P}^n$ of equation $f = 0$ supports a rank-one Ulrich bundle (see Proposition 4.2.18). In this chapter we develop this thread of ideas, studying systematically linear matrix factorizations of homogeneous forms f and working out their relationship with the existence of Ulrich bundles supported on the hypersurface $X \subset \mathbb{P}^n$ associated to f.

We will start the chapter giving in Section 4.1 a brief historical introduction to the results concerning the representation of (powers of) homogenous polynomials as the determinant of matrices with linear forms. Then, in Section 4.2, we will systematically study the linear matrix factorizations associated to the expression of a homogeneous polynomial f written as the sum of the products of linear forms. This allows us to see that any homogeneous form has such a linear matrix factorization and to bound its size. In particular, we get an integer r and an expression of f^r as the determinant of a matrix with linear entries. Then we relate linear matrix factorizations with the construction of Ulrich bundles and also get an upper bound for their rank. In Section 4.3, we deal with the problem of the existence of Ulrich bundles on complete intersections and again get an upper bound for the minimum rank of an Ulrich bundle on a general complete intersection $X \subset \mathbb{P}^n$ in terms of the degrees of the generators of $I(X)$. In Section 4.4 we focus on the case of surfaces $S \subset \mathbb{P}^3$. We apply the results obtained in Section 4.2 to study the existence of Ulrich bundles of rank as low as possible on a smooth surface of degree d with $\mathrm{Pic}(S) \cong \mathbb{Z}$, as well as constructing families of surfaces supporting Ulrich bundles of reasonably low rank. In Section 4.5 we introduce a rather new approach of constructing Ulrich bundles based on representation theory. Finally, we conclude this chapter in Section 4.6, where we present an introduction to the theory of Cayley–Chow forms and its relations with the existence of Ulrich bundles on a given variety.

4.1 Determinantal representation of homogeneous forms

In this section, we discuss whether a homogeneous polynomial $f \in \mathbf{k}[x_0, \dots, x_n]$ (or a power of f) can be written as the determinant of a matrix with linear entries. This problem is closely related to the existence of Ulrich bundles on the hypersurface $X \subset \mathbb{P}^n$ defined by f. Indeed, the interest in looking for Ulrich bundles on hypersurfaces stems partly on the fact that a rank r Ulrich bundle on a hypersurface $X \subset \mathbb{P}^n$ defined by a homogeneous form f of degree d exists if and only if f^r can be written as the deter-

https://doi.org/10.1515/9783110647686-004

minant of a square matrix of size dr with linear entries (see, for instance, [77, 192, 58, 100]).

The interest in determinantal representations of homogeneous polynomials rose during the nineteenth century. Up to our knowledge, the problem of representing a homogenous form as the determinant of a matrix of linear forms was first clearly stated by Hesse in [114]. In addition, in [115] he was able to prove that a smooth quartic plane curve has a symmetric determinantal representation. Moreover, he showed that these representations are in bijection with the 36 families of contact cubics. Hesse also proved that for smooth plane cubic curves the representation follows from the fact that any smooth plane cubic curve can be written in three ways as the Hessian curve. The cases of cubic and quartic surfaces were already solved by Grassmann in [100] and Cremona in [74], respectively. Indeed, Grassmann's result could be rephrased as follows: a smooth cubic surface S can be defined by an equation $\det(M) = 0$, where M is a linear matrix of size 3. There are 72 such representations (up to the action of $GL(3) \times GL(3)$ by left and right multiplication); They correspond in a one-to-one way to the linear systems of rational normal cubic curves on S.

Later on, Cayley approached the symmetric representational case for these degrees in [51]. It was Coble [57] who first understood when a quartic surface S_4 can be represented as a determinant of a matrix of linear forms. It turns out that S_4 can be defined by the determinant of a matrix with linear entries if and only if it contains an aCM curve of degree 6 and genus 3. By Noether–Lefschetz theorem, a very general surface $S \subset \mathbb{P}^3$ of degree $d \geq 4$ has Picard group generated by the hyperplane section and therefore a very general quartic surface does not contain such curves. Other examples of curves and surfaces were treated by Schur in [192]. Already in [77], it was realized that only a general plane curve of arbitrary degree, quadric surfaces, and cubic surfaces can be represented by a determinant of linear forms. Their existence for smooth plane curves of arbitrary degree was given in [78]. For determinantal representations of singular hypersurfaces, the reader can look at [125]. Therefore, people realized that it would be worth finding determinantal representation for powers f^r, $r \geq 2$, of the form f defining an integral surface in $\mathbb{P}^3$, or more generally, an integral hypersurface in $\mathbb{P}^n$, $n \geq 3$. The first case would be $r = 2$ and, adding the condition that the matrix M is skew-symmetric ($M^t = -M$), $\det(M)$ is a perfect square of an element known as the pfaffian $\mathrm{Pf}(M)$ of M and $f = \mathrm{Pf}(M)$. This situation has been less studied in the literature. In [1] it is proved that a generic cubic threefold in $\mathbb{P}^4$ can be written as a linear pfaffian. Beauville (see [22]) systematically studied which hypersurfaces have a pfaffian representation. In any case, by the results in this chapter, we will see that for any hypersurface X defined by a homogeneous form f, there exists an integer $r \geq 1$ such that f^r has a linear determinantal representation (Corollary 4.2.11).

4.2 Ulrich bundles on hypersurfaces in $\mathbb{P}^n$, $n \geq 3$

This section has two aims. First, we effectively construct Ulrich bundles on a smooth hypersurface $X \subset \mathbb{P}^n$ using linear matrix factorizations. Then we provide an upper bound for the Ulrich complexity of a homogeneous polynomial (see Definition 4.2.17), as well as an upper bound for

$$r(d,n) := \min\{r \mid \text{a general hypersurface } X \subset \mathbb{P}^n \text{ of degree } d \text{ has an Ulrich bundle of rank } r\}.$$

We will see in Corollary 4.2.16 that $r(d,n) < \infty$.

Definition 4.2.1. Let $0 \neq f \in R = \mathbf{k}[x_0, \ldots, x_n]$ be a homogeneous polynomial of degree $d \geq 2$. A *linear matrix factorization* of f of size m is an equation

$$fI_m = \alpha_0\alpha_1 \cdots \alpha_{d-1}$$

where the α_i are square matrices of size m whose entries are linear forms in R and I_m stands for the identity matrix of size m.

Given a linear matrix factorization $fI_m = \alpha_0\alpha_1 \cdots \alpha_{d-1}$ and $j \in \mathbb{Z}$, we set $\alpha_j := \alpha_i$ where $0 \leq i \leq d-1$ and $j \equiv i \bmod d$. Notice that if $fI_m = \alpha_0\alpha_1 \cdots \alpha_{d-1}$, then $fI_m\alpha_0 = \alpha_0\alpha_1 \cdots \alpha_{d-1}\alpha_0$. Since $fI_m\alpha_0 = \alpha_0 fI_m$, this implies that also $fI_m = \alpha_1\alpha_2 \cdots \alpha_{d-1}\alpha_0$ is a linear matrix factorization of f and, in general, $fI_m = \alpha_i\alpha_{i+1} \cdots \alpha_{i+d-1}$ is also a linear matrix factorization for all $i \in \mathbb{Z}$.

Two linear matrix factorizations $fI_m = \alpha_0\alpha_1 \cdots \alpha_{d-1}$ and $fI_m = \beta_0\beta_1 \cdots \beta_{d-1}$ of the same size m are equivalent if there exist matrices $S_j \in \mathrm{GL}(m)$ such that $\beta_j = S_j\alpha_j S_{j+1}^{-1}$ for all j.

Remark 4.2.2. Notice that any linear matrix factorization $fI_m = \alpha_0\alpha_1 \cdots \alpha_{d-1}$ gives rise to a two-factor matrix factorization $fI_m = \alpha_0\beta$ with $\beta = \alpha_1 \cdots \alpha_{d-1}$ but, in general, the converse is not true (see Definition 2.5.10).

Definition 4.2.3. The *sum* of two linear matrix factorizations $fI_{m_1} = \alpha_0\alpha_1 \cdots \alpha_{d-1}$ and $fI_{m_2} = \beta_0\beta_1 \cdots \beta_{d-1}$ of size m_1 and m_2, respectively, is the linear matrix factorization $fI_{m_1+m_2} = \gamma_0\gamma_1 \cdots \gamma_{d-1}$ of size $m_1 + m_2$ where for all i

$$\gamma_i = \begin{pmatrix} \alpha_i & 0 \\ 0 & \beta_i \end{pmatrix}.$$

The linear matrix factorization $fI_m = \alpha_0\alpha_1 \cdots \alpha_{d-1}$ is called *indecomposable* if it is not a sum of linear matrix factorizations of f of smaller size.

Example 4.2.4. Let $C \subset \mathbb{P}^2$ be a smooth plane cubic curve. It is well known that after changing coordinates C is the zero locus of the polynomial $f \in \mathbf{k}[x,y,z]$,

$$f = y^2z - x(x-z)(x-\lambda z)$$

where $\lambda \in \mathbf{k}$. It is called the *Weierstrass equation* of the plane cubic curve. Let ξ be a primitive root of order 3 of the unit and consider the following matrices:

$$\alpha_0 = \begin{pmatrix} x - \lambda z & y & 0 \\ 0 & x - z & \xi y \\ -\xi^2 z & 0 & x \end{pmatrix}, \quad \alpha_1 = \begin{pmatrix} x & \xi y & 0 \\ 0 & x - \lambda z & \xi^2 y \\ -z & 0 & x - z \end{pmatrix},$$

$$\alpha_2 = \begin{pmatrix} x - z & \xi^2 y & 0 \\ 0 & x & y \\ -\xi z & 0 & x - \lambda z \end{pmatrix}.$$

Then $(y^2 z - x(x-z)(x-\lambda z))I_3 = \alpha_0\alpha_1\alpha_2$ is a linear matrix factorization of f.

Next result will allow us to determine a linear matrix factorization of any homogenous polynomial f of degree d.

Theorem 4.2.5. *Let $0 \neq f, g \in \mathbf{k}[x_0, \ldots, x_n]$ be homogeneous polynomials of degree d. Assume that f has a linear matrix factorization of size m and that g has a linear matrix factorization of size l. Then, $h := f + g$ has a linear matrix factorization, not necessarily indecomposable, of size dlm.*

Proof. Let $fI_m = \alpha_0\alpha_1 \cdots \alpha_{d-1}$ and $gI_l = \beta_0\beta_1 \cdots \beta_{d-1}$ be linear matrix factorizations of size m and l of f and g, respectively. For any $0 \leq i \leq d-1$, we write l copies of α_i into the main diagonal and define the following square matrices of size ml with linear entries:

$$\Delta_i = \begin{pmatrix} \alpha_i & 0 & \cdots & \cdots & 0 \\ 0 & \alpha_i & 0 & \cdots & 0 \\ 0 & 0 & \ddots & & 0 \\ \vdots & & & \ddots & 0 \\ 0 & & & & \alpha_i \end{pmatrix}.$$

Similarly, for any $0 \leq i \leq d-1$, we write m copies of β_i into the main diagonal and define the following square matrices of size ml with linear entries:

$$\Omega_i = \begin{pmatrix} \beta_i & 0 & \cdots & \cdots & 0 \\ 0 & \beta_i & 0 & \cdots & 0 \\ 0 & 0 & \ddots & & 0 \\ \vdots & & & \ddots & 0 \\ 0 & & & & \beta_i \end{pmatrix}.$$

Finally, denote by ξ a dth root of the unit and for any $0 \leq j \leq d-1$ define the following square matrices of size dlm:

$$\gamma_j = \begin{pmatrix} \Delta_{j-1} & \xi^{j-1}\Omega_0 & 0 & \cdots & \cdots & 0 \\ 0 & \Delta_{j-2} & \xi^{j-2}\Omega_1 & 0 & \cdots & 0 \\ 0 & 0 & \Delta_{j-3} & \xi^{j-3}\Omega_2 & 0 & 0 \\ \vdots & \vdots & \cdots & \ddots & \ddots & \vdots \\ 0 & 0 & \cdots & \cdots & \Delta_{j-d+1} & \xi^{j-d+1}\Omega_{d-2} \\ \xi^{j-d}\Omega_{d-1} & 0 & \cdots & \cdots & 0 & \Delta_{j-d} \end{pmatrix},$$

where the Δ_i, Ω_i are regarded as blocks and $\Delta_i := \Delta_j$ for $i \equiv j \bmod d$. Using the facts that ξ is a primitive dth root of the unity, $\Delta_i\Omega_j = \Omega_j\Delta_i$, and, for any j,

$$\Delta_j\Delta_{j+1}\cdots\Delta_{j+d-1} = \Delta_0\cdots\Delta_{d-1},$$
$$\Omega_j\Omega_{j+1}\cdots\Omega_{j+d-1} = \Omega_0\cdots\Omega_{d-1},$$

we can see that

$$\gamma_0\gamma_1\cdots\gamma_{d-1} = fI_{dml} + gI_{dml} = hI_{dml}$$

is a linear matrix factorization of size dml of h, as desired. □

In the next example we illustrate how we can apply the above result to explicitly construct a linear matrix factorization of any homogeneous polynomial f.

Example 4.2.6. Fix $n \geq 2$ and let $Q_n \subset \mathbb{P}^{n+1}$ be a smooth quadric defined by a homogeneous polynomial f of degree 2.

First of all, assume that n is odd and write $n+1 = 2m$. After a change of coordinates, we can assume that f is given by

$$f = x_0^2 + x_1x_2 + \cdots + x_{2m-1}x_{2m}.$$

Let us construct inductively, using Theorem 4.2.5, a linear matrix factorization of f of size 2^m. We start with

$$x_0^2 I_1 = (x_0)(x_0)$$

a linear matrix factorization of x_0^2 of size 1 and

$$x_1x_2I_1 = (x_1)(x_2)$$

a linear matrix factorization of x_1x_2 of size 1. Then, by Theorem 4.2.5,

$$(x_0^2 + x_1x_2)I_2 = \alpha_1\beta_1$$

with

$$\alpha_1 = \begin{pmatrix} x_0 & x_1 \\ -x_2 & x_0 \end{pmatrix} \quad \text{and} \quad \beta_1 = \begin{pmatrix} x_0 & -x_1 \\ x_2 & x_0 \end{pmatrix}$$

is a linear matrix factorization of size 2 of $x_0^2 + x_1x_2$. Assume inductively that

$$(x_0^2 + x_1x_2 + \cdots + x_{2m-3}x_{2m-2})I_{2^{m-1}} = \alpha_{m-1}\beta_{m-1}$$

is a linear matrix factorization of $x_0^2 + x_1x_2 + \cdots + x_{2m-3}x_{2m-2}$ of size 2^{m-1} and take $x_{2m-1}x_{2m}I_1 = (x_{2m-1})(x_{2m})$ a linear matrix factorization of $x_{2m-1}x_{2m}$ of size 1. Then, by Theorem 4.2.5,

$$fI_{2^m} = (x_0^2 + x_1x_2 + \cdots + x_{2m-1}x_{2m})I_{2^m} = \alpha_m\beta_m$$

with the block matrices

$$\alpha_m = \begin{pmatrix} \alpha_{m-1} & x_{2m-1}I_{2^{m-1}} \\ -x_{2m}I_{2^{m-1}} & \beta_{m-1} \end{pmatrix} \quad \text{and} \quad \beta_m = \begin{pmatrix} \beta_{m-1} & -x_{2m-1}I_{2^{m-1}} \\ x_{2m}I_{2^{m-1}} & \alpha_{m-1} \end{pmatrix}$$

is a linear matrix factorization of size 2^m of f defining $Q_n \subset \mathbb{P}^{n+1}$.

If n is even, we write $n = 2m$ and, after a change of coordinates, can assume that f is given by

$$f = x_0x_1 + x_2x_3 + \cdots + x_{2m}x_{2m+1}.$$

Let us see that f has a linear matrix factorization of size 2^m. We start with $x_0x_1I_1 = (x_0)(x_1)$ and $x_2x_3I_1 = (x_2)(x_3)$ linear matrix factorization of size 1 of x_0x_1 and x_2x_3 respectively. Then, by Theorem 4.2.5,

$$(x_0x_1 + x_2x_3)I_2 = \alpha_1\beta_1$$

with

$$\alpha_1 = \begin{pmatrix} x_0 & x_2 \\ -x_3 & x_1 \end{pmatrix} \quad \text{and} \quad \beta_1 = \begin{pmatrix} x_1 & -x_2 \\ x_3 & x_0 \end{pmatrix}$$

is a linear matrix factorization of size 2 of $x_0x_1 + x_2x_3$. Proceeding as in the odd case, inductively we construct an explicit linear matrix factorization of size 2^m of f (for more details, see [139]).

Corollary 4.2.7. *Let $0 \neq f \in \mathbf{k}[x_0, \ldots, x_n]$ be a homogeneous polynomial of degree d with s summands of products of linear forms. Then, f has a linear matrix factorization, not necessarily indecomposable, of size $m = d^{s-1}$.*

Proof. We will proceed by induction on the number of summands. If $s = 1, f = l_1 l_2 \cdots l_d$ with l_i linear forms. Trivially, $l_1 l_2 \cdots l_d I_1 = (l_1)(l_2) \cdots (l_d)$ is a linear matrix factorization of f of size 1. Let f be a homogeneous polynomial of degree d with $s \geq 2$ summands. We can write f as

$$f = g + l_1' l_2' \cdots l_d'$$

where g is a homogeneous polynomial of degree d and $s - 1$ summands and the l_i' are linear forms. By induction hypothesis, g has a linear matrix factorization of size d^{s-2} and $l_1' l_2' \cdots l_d'$ has a linear matrix factorization of size 1. Then, by Theorem 4.2.5, f has a linear matrix factorization of size d^{s-1}. □

According to Corollary 4.2.7, the size of a linear matrix factorization of f depends on the number of summands of f. This bring us to recalling the following definitions.

Definition 4.2.8. Let $0 \neq f \in \mathbf{k}[x_0, \ldots, x_n]$ be a homogeneous polynomial of degree d. The *Waring rank* of f, which will be denoted by $w(f)$, is defined as the minimum number of terms in a expression of f as a linear combination of powers of linear forms:

$$f = \sum_{i=1}^{w(f)} l_i^d$$

where the $l_i \in \mathbf{k}[x_0, \ldots, x_n]$ are linear forms. The *Chow rank* of f, which will be denoted by $ch(f)$, is defined as the minimum number of terms in an expression of f as a linear combination of products of linear forms as follows:

$$f = \sum_{i=1}^{ch(f)} l_{i1} l_{i2} \cdots l_{id}$$

with the $l_{ij} \in \mathbf{k}[x_0, \ldots, x_n]$ being linear forms.

Remark 4.2.9. Alexander and Hirschowitz observed that for a general homogeneous polynomial f of degree d, it should be true that $w(f)(n+1) \geq \binom{n+d}{d}$ and, for a general f, they proved that $w(f)$ is as small as possible, namely

$$w(f) = \left\lceil \frac{1}{n+1} \binom{n+d}{d} \right\rceil \tag{4.1}$$

with a short list of exceptions (see [2, Theorem 2]):
- $d = 2$ and any n where $w(f) = n + 1$
- $d = 3$ and $n = 4$ where $w(f) = 8$
- $d = 4$ and $n = 2$ where $w(f) = 6$
- $d = 4$ and $n = 3$ where $w(f) = 10$, and
- $d = 4$ and $n = 4$ where $w(f) = 15$

Notice than for all these cases, except for the quadrics, it holds that $w(f) = \lceil \frac{1}{n+1} \binom{n+d}{d} \rceil + 1$.

On the other hand, it is conjectured that for a general homogeneous polynomial $f \in \mathbf{k}[x_0, \dots, x_n]$ of degree $d \neq 2$,

$$ch(f) = \left\lceil \frac{1}{dn+1}\binom{n+d}{d} \right\rceil. \tag{4.2}$$

In general, it is a difficult problem to compute $ch(f)$ and few cases are known. For instance, the answer is known for general homogeneous forms $f \in \mathbf{k}[x_0, x_1, x_2, x_3]$ of degree $d \geq 2$ and for general cubics $f \in \mathbf{k}[x_0, \dots, x_n]$ (see [197, Corollary 1.5] and [198, Corollary 1.3]).

On the other hand, it is easy to see that for a general homogenous polynomial $f \in \mathbf{k}[x_0, x_1, \dots, x_n]$,

$$ch(f) \leq \left\lceil \frac{w(f)}{2} \right\rceil. \tag{4.3}$$

In fact, if f can be written as $f = \sum_{i=1}^{w(f)} l_i^d$ with l_i being linear forms, we can suppose, without loss of generality, that $l_i = x_i$ for $i = 1, 2$ and therefore $(x_0^d + x_1^d) = \prod_{j=1}^{d}(x_0 - \zeta^j x_1)$ for ζ a primitive dth root of -1. Performing this trick for each pair of linear forms l_i, l_{i+1}, for odd $i < w(f)$, we get the bound.

Example 4.2.10. (1) Consider $R = \mathbf{k}[x, y, z]$ and f a general form of degree 3. We have

$$w(f) = \left\lceil \frac{1}{3}\binom{5}{3} \right\rceil = 4.$$

Hence, there are 4 linear forms $l_1, \dots, l_4 \in R$ such that

$$f = l_1^3 + l_2^3 + l_3^3 + l_4^3.$$

On the other hand, as we pointed out in Example 4.2.4, we know that the equation of a smooth plane cubic can be written in the Weierstrass form as

$$y^2 z = x(x - z)(x - \lambda z)$$

where $\lambda \in \mathbf{k}$. Therefore, $ch(f) = 2$.

(2) Let $f = x_0^2 + x_1x_2 + \dots + x_{2m-1}x_{2m}$ be a homogeneous polynomial of degree 2 defining a smooth quadric $Q_{n_1} \subset \mathbb{P}^{n_1+1}$ of odd dimension $n_1 = 2m - 1$ and let $g = x_0x_1 + x_2x_3 + \dots + x_{2l}x_{2l+1}$ be a homogeneous polynomial of degree 2 defining a smooth quadric $Q_{n_2} \subset \mathbb{P}^{n_2+1}$ of even dimension $n_2 = 2l$. Then, $ch(f) = m + 1$ and $ch(g) = l + 1$.

(3) Let $X \subset \mathbb{P}^n$ be the *Fermat hypersurface* defined as the zero locus of the degree d form $f = x_0^d + x_1^d + \dots + x_n^d$. It is clear that $w(f) = n + 1$. On the other hand, if we denote by ξ a dth root of the unit, we can write the Fermat form as follows:

$$f = \begin{cases} \sum_{j=0}^{\frac{n-1}{2}} \prod_{i=1}^{d}(x_{2j} + \xi^i x_{2j+1}) & \text{if } n \text{ is odd,} \\ \sum_{j=0}^{\frac{n-2}{2}} \prod_{i=1}^{d}(x_{2j} + \xi^i x_{2j+1}) + x_n^d & \text{if } n \text{ is even.} \end{cases}$$

In particular, $ch(f) = \frac{n+1}{2}$ if n is odd and $ch(f) = \frac{n+2}{2}$ if n is even. Notice that this value is far from the conjectured value in (4.2) for the general form of degree d.

Corollary 4.2.11. *Any homogenous polynomial f of degree d has a linear matrix factorization of size $d^{w(f)-1}$ and a linear matrix factorization of size $d^{ch(f)-1}$.*

Proof. The claim follows from Corollary 4.2.7. □

In Chapter 2, we have seen the relation between two-factor matrix factorizations of a homogeneous polynomial $f \in \mathbf{k}[x_0, \ldots, x_n]$ and the existence of aCM bundles on a smooth hypersurface on $\mathbb{P}^n$ defined by f. Let us now see the relation between linear matrix factorizations and Ulrich bundles.

Proposition 4.2.12. *Let $X \subset \mathbb{P}^n$ be an integral hypersurface of degree d defined by a homogeneous polynomial f with a linear matrix factorization of size m. Then, X supports an Ulrich bundle, not necessarily indecomposable, of rank $\frac{m}{d}$.*

Proof. By assumption, f has a linear matrix factorization $fI_m = \alpha_0\alpha_1 \cdots \alpha_{d-1}$ of size m. Denote by $\psi = \alpha_0\alpha_1 \cdots \alpha_{d-2}$ and $\varphi = \alpha_{d-1}$. Then

$$fI_m = \psi\varphi$$

is a two-factor matrix factorization of f. Hence, by Corollary 2.5.13, there is an aCM bundle $\mathcal{E} = \operatorname{coker}(\varphi)$ on X given by the exact sequence

$$0 \longrightarrow \mathcal{O}^m_{\mathbb{P}^n}(-1) \xrightarrow{\varphi} \mathcal{O}^m_{\mathbb{P}^n} \longrightarrow \mathcal{E} \longrightarrow 0.$$

In particular,

$$\mathrm{H}^0(X, \mathcal{E}(-1)) = 0 \quad \text{and} \quad \mathrm{H}^{n-1}(X, \mathcal{E}(1-n)) = 0.$$

Thus, by Proposition 3.2.7, $\mathcal{E}$ is an Ulrich bundle on X. Finally, it follows from the above exact sequence that

$$m = \mathrm{h}^0(\mathcal{E}) = \deg(X)\operatorname{rank}(\mathcal{E}),$$

from which we deduce that $\operatorname{rank}(\mathcal{E}) = \frac{m}{d}$. □

Remark 4.2.13. (1) The above proposition shows that Conjecture 3.4.8 is true for smooth hypersurfaces.

(2) It follows from the above result that the size of any linear matrix factorization of a homogeneous polynomial of degree d defining an integral hypersurface is a multiple of d, which a priori is not obvious from the definition.

(3) Let $X \subset \mathbb{P}^n$ be an integral hypersurface of degree d defined by a homogeneous polynomial f and assume that $fI_m = \alpha_0\alpha_1 \cdots \alpha_{d-1}$ is a linear matrix factorization of f of size m. According to Proposition 4.2.12, to any linear matrix α_i we can associate an Ulrich bundle which we denote by $\mathcal{F}_i$. By [17, Theorem 4.1], the bundles $\mathcal{F}_i(-i)$, $0 \leq i \leq d-1$, give us a filtration of $\mathcal{O}^m_X$ by twist of Ulrich bundles.

As an immediate consequence, we have the following result.

Theorem 4.2.14. *Let $X \subset \mathbb{P}^n$ be an integral hypersurface of degree d defined by a homogeneous polynomial f. Then:*

(1) *If $f = \sum_{i=1}^{s} m_i$ is a monomial presentation of f, then X supports an Ulrich bundle, not necessarily indecomposable, of rank d^{s-2}.*
(2) *If $w(f)$ denotes the Waring rank of f, then X supports an Ulrich bundle, not necessarily indecomposable, of rank $d^{w(f)-2}$.*
(3) *If $ch(f)$ denotes the Chow rank of f, then X supports an Ulrich bundle, not necessarily indecomposable, of rank $d^{ch(f)-2}$.*

Proof. The result follows from Corollary 4.2.7 and Proposition 4.2.12. □

Remark 4.2.15. It is not surprising that the rank of the Ulrich bundle $\mathcal{E}$ and the degree of the hypersurface X have the same parity. In fact, it follows from Corollary 3.2.5 that $\deg(\mathcal{E}) = \frac{r(d-1)}{2} \in \mathbb{Z}$. Thus, even degree hypersurfaces do not support Ulrich bundles of odd rank.

Corollary 4.2.16. *Fix integers $d, n \geq 3$. The minimum rank $r(d, n)$ such that any general hypersurface $X \subset \mathbb{P}^n$ of degree d has an Ulrich bundle of rank $r(d, n)$ is bounded by*

$$r(d,n) \leq d^{f(d,n)-2} \quad \textit{where } f(d,n) := \left\lceil \frac{1}{2(n+1)} \binom{n+d}{d} \right\rceil,$$

unless $f(4,4) = 8$. In particular, $r(d, n) < \infty$.

Proof. There exists a dense open subset U of the space of homogeneous polynomials of degree d such that, for any $f \in U$,

$$w(f) = \begin{cases} \lceil \frac{1}{(n+1)} \binom{n+d}{d} \rceil & \text{if } (d,n) \notin \{(3,4),(4,3),(4,4)\}, \\ 8 & \text{if } (d,n) = (3,4), \\ 10 & \text{if } (d,n) = (4,3), \\ 15 & \text{if } (d,n) = (4,4). \end{cases}$$

On the other hand, according to (4.3), for any $f \in U$, $ch(f) \leq \lceil \frac{w(f)}{2} \rceil$. Finally, the result follows from Theorem 4.2.14. □

Definition 4.2.17. The *Ulrich complexity* $uc(f)$ of a homogeneous polynomial $f \in \mathbf{k}[x_0, \dots, x_n]$ of degree d is defined as the smallest number r such that there exists a square matrix M of homogeneous linear forms with

$$\det(M) = f^r$$

and such that there is a matrix N with $MN = fI_{dr}$.

As was explained in Section 4.1, the notion of Ulrich complexity goes back to the nineteenth century when mathematicians sought for determinantal representations of hypersurfaces, and in the setup of vector bundles this notion was introduced by Bläser, Eisenbud, and Schreyer [27]. We devote the remaining part of this section to getting upper bounds for the Ulrich complexity.

Proposition 4.2.18. *Let $0 \neq f \in R = \mathbf{k}[x_0, \ldots, x_n]$ be a homogeneous polynomial of degree d defining an integral hypersurface $X \subset \mathbb{P}^n$. Then, the Ulrich complexity $uc(f)$ of f is the minimum rank of an Ulrich bundle $\mathcal{E}$ on X. In addition, if f admits a linear matrix factorization of size m then*

$$uc(f) \leq \frac{m}{d}$$

and, in particular, $uc(f) \leq d^{ch(f)-2}$.

Proof. Let $\mathcal{E}$ be an Ulrich bundle on X. By Theorem 3.2.9, to say that $\mathcal{E}$ is Ulrich means that its minimal free resolution has the following shape:

$$0 \longrightarrow \mathcal{O}_{\mathbb{P}^n}^m(-1) \xrightarrow{M} \mathcal{O}_{\mathbb{P}^n}^m \longrightarrow \mathcal{E} \longrightarrow 0,$$

where M is a square matrix of linear forms. Since $\mathcal{E}$ is annihilated by f, the multiplication by f on $\mathcal{O}_{\mathbb{P}^n}^m$ factors through M. So, we can find a matrix N of degree $d-1$ forms such that $MN = fI_m$. In particular, $\det(M)\det(N) = f^m$. Thus, if f is irreducible, the determinant of M is a power f^r of f, as in the definition of Ulrich complexity. By Remark 2.5.14, r is the rank of $\mathcal{E}$. The last assertion follows from the fact that, by Corollary 4.2.12, if f admits a linear matrix factorization of size m then it supports an Ulrich bundle of rank $\frac{m}{d}$. Hence, $uc(f) \leq \frac{m}{d}$ and, by Theorem 4.2.14, $uc(f) \leq d^{ch(f)-2}$. □

Example 4.2.19. In Example 4.2.10 we have seen that the Fermat hypersurface $X \subset \mathbb{P}^n$ of degree d is represented by a form $f = x_0^d + x_1^d + \cdots + x_n^d$ with $ch(f) = \frac{n+1}{2}$ if n is odd and $ch(f) = \frac{n+2}{2}$ if n is even. Therefore, X is the support of an Ulrich bundle of rank $d^{\frac{n-3}{2}}$ if n is odd and of rank $d^{\frac{n-2}{2}}$ if n is even. In particular, $uc(f) \leq d^{\frac{n-3}{2}}$ if n is odd and $uc(f) \leq d^{\frac{n-2}{2}}$ if n is even.

In general, the bound $uc(f) \leq d^{ch(f)-2}$ is not sharp, and one expects to have Ulrich bundles of smaller rank. Nevertheless, in some cases the linear matrix factorization of f gives us Ulrich bundles of the smallest possible rank on the hypersurface defined by f as it is shown in the following examples.

Example 4.2.20. (1) Let $Q_n \subset \mathbb{P}^{n+1}$ be a smooth quadric defined by a homogeneous polynomial f. According to Example 4.2.10 and by Theorem 4.2.14, Q_n supports an Ulrich bundle of rank $2^{\frac{n-1}{2}}$ if n is odd and of rank $2^{\frac{n-2}{2}}$ if n is even. By Example 3.2.8, they correspond to the spinor bundles. Moreover, notice that the linear matrix factorization explicitly described in Example 4.2.6 gives us a constructive presentation of the spinor bundles.

(2) According to Cayley [48] and Salmon [189], a general cubic surface $S \subset \mathbb{P}^3$ can be represented by an equation

$$\ell_1\ell_2\ell_3 + \ell_4\ell_5\ell_6 = 0$$

where ℓ_i are linear forms. It is the so-called *Cayley–Salmon form* of the cubic surface S. Therefore, the Chow rank $ch(f)$ of a general form $f \in \mathbf{k}[x_0, x_1, x_2, x_3]$ of degree 3 is $ch(f) = 2$ and the Ulrich complexity of f is $uc(f) = 1$, since

$$f = \ell_1\ell_2\ell_3 + \ell_4\ell_5\ell_6 = \det\begin{pmatrix} 0 & \ell_1 & \ell_4 \\ \ell_5 & 0 & \ell_2 \\ \ell_3 & \ell_6 & 0 \end{pmatrix}$$

and S supports an Ulrich bundle $\mathcal{E}$ of rank 1, as it was already pointed out in Example 3.2.3 (2).

(3) Let $X \subset \mathbb{P}^4$ be a smooth cubic hypersurface defined by f. It follows from [197, Theorem 1.4] that $ch(f) = 3$. Therefore, by Theorem 4.2.14, X has an Ulrich bundle $\mathcal{E}$ of rank 3. Since $\mathrm{Pic}(X) \cong \mathbb{Z}$, X cannot support an Ulrich line bundle. Hence $\mathcal{E}$ is an Ulrich bundle of the smallest odd rank. Moreover, Theorem 4.2.5 gives us an explicit presentation of $\mathcal{E}$.

4.3 Ulrich bundles on complete intersections

In this short section we will deal with general complete intersection varieties $X \subset \mathbb{P}^n$ of codimension c and type $(d_1, \dots, d_c)$, namely projective varieties of codimension c whose homogeneous ideal $I(X)$ is generated by a regular sequence $f_1, \dots, f_c \in R = \mathbf{k}[x_0, \dots, x_n]$ where $\deg(f_i) = d_i$. The existence of Ulrich bundles $\mathcal{E}$ on X was stated by Herzog, Ulrich, and Backelin [113]. Nevertheless, in their construction the rank of $\mathcal{E}$ was out of control. As application of the results obtained in this section, we will get an upper bound for

$$r(n; d_1, \dots, d_c) = \min\{r \mid \text{a general complete intersection } X \subset \mathbb{P}^n \text{ of type } d_1, \dots, d_c \text{ supports an Ulrich bundle of rank } r\}.$$

Like for the products of varieties, if we know the existence of Ulrich bundles on varieties X and Y, we can build an Ulrich bundle on their intersection $X \cap Y$ provided they meet properly. This result was already proved in [113] (see also [44]), and it will be extremely useful while proving the existence of Ulrich bundles on complete intersection varieties (Theorem 4.3.2).

Proposition 4.3.1. *Let $X, Y \subset \mathbb{P}^n$ be smooth varieties endowed with Ulrich bundles $\mathcal{E}$ and $\mathcal{F}$ of rank e and f, respectively. Assume that X and Y intersect properly at Z, that is,* $\dim Z = \dim(X \cap Y) = \dim X + \dim Y - n$. *Assume also that* $\dim Z \geq 1$. *Then, $\mathcal{E} \otimes_{\mathcal{O}_{\mathbb{P}^n}} \mathcal{F}$ is an Ulrich bundle of rank ef on $Z = X \cap Y$.*

Proof. Denote by $l = \dim X$ and $m = \dim Y$. Since $\mathcal{E}$ is a rank e Ulrich bundle on X, by Theorem 3.2.9, $\mathcal{E}$ admits a linear $\mathcal{O}_{\mathbb{P}^n}$-resolution of the form

$$0 \longrightarrow \mathcal{O}_{\mathbb{P}^n}(-n+l)^{a_{n-l}} \longrightarrow \cdots \longrightarrow \mathcal{O}_{\mathbb{P}^n}(-1)^{a_1} \longrightarrow \mathcal{O}_{\mathbb{P}^n}^{a_0} \longrightarrow \mathcal{E} \longrightarrow 0 \tag{4.4}$$

for some integers $a_i \in \mathbb{Z}_{>0}$. Since X and Y meet properly, the above sequence tensored by $\mathcal{O}_Y$ is exact along Y. Thus, since we have the isomorphism

$$\mathcal{E} \otimes_{\mathcal{O}_{\mathbb{P}^n}} \mathcal{O}_Y \otimes_{\mathcal{O}_Y} \mathcal{F} \cong \mathcal{E} \otimes_{\mathcal{O}_{\mathbb{P}^n}} \mathcal{F},$$

tensoring by $\mathcal{F}$, we get the following complex on Y:

$$0 \longrightarrow \mathcal{F}(-n+l)^{a_{n-l}} \xrightarrow{\phi_{n-l}} \cdots \xrightarrow{\phi_2} \mathcal{F}(-1)^{a_1} \xrightarrow{\phi_1} \mathcal{F}^{a_0} \longrightarrow \mathcal{E} \otimes_{\mathcal{O}_{\mathbb{P}^n}} \mathcal{F} \longrightarrow 0, \tag{4.5}$$

which is exact on $X \cap Y$. According to Proposition 3.2.7, $\mathcal{E} \otimes_{\mathcal{O}_{\mathbb{P}^n}} \mathcal{F}$ is an Ulrich bundle on $Z = X \cap Y$ if and only if for any $i \geq 0$,

$$\mathrm{H}^i(Z, \mathcal{E} \otimes_{\mathcal{O}_{\mathbb{P}^n}} \mathcal{F}(-t)) = 0 \quad \text{for } 1 \leq t \leq l+m-n. \tag{4.6}$$

For any $1 \leq p \leq n-l$, denote by $C_p := \mathrm{im}(\phi_p)$. Notice that $C_0 \cong \mathcal{E} \otimes_{\mathcal{O}_{\mathbb{P}^n}} \mathcal{F}$, $C_{n-l} \cong \mathcal{F}(-n+l)$, and for any $1 \leq p \leq n-l$, we have the short exact sequence

$$0 \longrightarrow C_p \longrightarrow \mathcal{F}(-p+1) \longrightarrow C_{p-1} \longrightarrow 0.$$

Twisting them by $\mathcal{O}_Y(-t)$ for any $1 \leq t \leq l+m-n$ and using the fact that, since $\mathcal{F}$ is an Ulrich bundle on Y, for any $i \geq 0$ we have $\mathrm{H}^i(\mathcal{F}(-t)) = 0$ for $1 \leq t \leq m$, we get that for any $1 \leq t \leq l+m-n$ and $i \geq 0$,

$$\mathrm{H}^i(C_{n-l}(-t)) = \mathrm{H}^i(C_{n-l-1}(-t)) = \cdots = \mathrm{H}^i(C_0(-t)) = 0.$$

In particular, from (4.6) we get that $\mathcal{E} \otimes_{\mathcal{O}_{\mathbb{P}^n}} \mathcal{F}$ is an Ulrich bundle on $X \cap Y$ of rank ef. □

As an application of the above result, we have:

Theorem 4.3.2. *Let $X \subset \mathbb{P}^n$ be a general complete intersection variety of codimension c and type $(d_1, \ldots, d_c)$. Let $f_1, \ldots, f_c \in R$ be the generators of $I(X)$ of degree d_i and call n_i the Chow rank of f_i. Then, X supports an Ulrich bundle (possibly decomposable) of rank $d_1^{n_1-2} \cdots d_c^{n_c-2}$*

Proof. By Theorem 4.2.14, there is an Ulrich bundle of rank $d_i^{n_i-2}$ on the hypersurface $X_i \subset \mathbb{P}^n$ defined by f_i. We now apply Proposition 4.3.1 and get what we wanted. □

Corollary 4.3.3. *The minimum rank $r(n; d_1, \ldots, d_c)$ such that any general complete intersection $X \subset \mathbb{P}^n$ of type $d_1, \ldots, d_c$ supports an Ulrich bundle of rank $r(n; d_1, \ldots, d_c)$ is bounded by*

$$r(n; d_1, \ldots, d_c) \leq \prod_{i=1}^{c} d_i^{f(d_i,n)-2} \quad \text{where } f(d_i, n) := \left\lceil \frac{1}{2(n+1)} \binom{n+d_i}{d_i} \right\rceil.$$

In particular, $r(n; d_1, \ldots, d_c) < \infty$.

Remark 4.3.4. (1) Theorem 4.3.2 shows that Conjecture 3.4.8 holds for a general complete intersection variety.

(2) In [56, Remark 4.19], the authors made the following observation: Ulrich bundles of rank 2^{2m-2} exist on a complete intersection of two quadrics Q_1 and Q_2 in $\mathbb{P}^{2m+1}$. Since the Chow rank of each quadric is $m+1$, it follows from the above theorem that the complete intersection of Q_1 and Q_2 supports an Ulrich bundle of rank 2^{2m-2}.

(3) Nevertheless, for small values of m, the bound is far of being sharp. In fact, $r(4;2,2) \leq 4$ while the existence of rank 2 Ulrich bundles on del Pezzo surfaces $X \subset \mathbb{P}^4$ was established by Eisenbud and Schreyer in [88, Corollary 6.5] and by Miró-Roig and Pons-Llopis in [158, Theorem 4.9]. Also, according to our result $r(4;3,2) \leq 6$ and since a complete intersection X of hypersurfaces in $\mathbb{P}^4$ of degree 2 and 3 is a K3 surface, by [92, Theorem 1], X supports a rank 2 Ulrich bundle (see also Theorem 5.4.3).

4.4 Low rank Ulrich bundles on a general surface in $\mathbb{P}^3$

As we saw in Proposition 2.5.1 on a smooth surface $S \subset \mathbb{P}^3$ of degree $d \geq 3$ there always exist indecomposable rank 2 aCM bundles. Much subtler is the existence of indecomposable lower rank Ulrich bundles, and we will discuss it in this section. Recall that in Section 4.2 we defined the number $r(d,3)$ as the minimum r such that a general surface $S \subset \mathbb{P}^3$ of degree d supports an Ulrich bundle of rank r. Moreover, we manage to find an upper bound for $r(d,3)$ in terms of the Chow rank of the homogeneous form f associated to S (see (4.3) and Proposition 4.2.18). However, it would be very interesting to know the precise value of $r(d,3)$ for arbitrary d. Since the existence of an Ulrich bundle of rank 1 (respectively rank 2) on a given surface $S \subset \mathbb{P}^3$ defined by the form f is equivalent to the fact that $f = \det(M)$ (respectively $f = \mathrm{Pf}(M)$) for a size d square matrix M of linear forms (respectively $2d$ and M being skew-symmetric), we are going to see that $r(d,3) \geq 3$ as soon as d is large enough. However, the precise value of $r(d,3)$ is, for the time being, far from being well-understood. In this section we are going to collect the available information about this topic.

As we saw in the proof of Proposition 4.2.18, having a rank r Ulrich bundle on the degree d surface $S \subset \mathbb{P}^3$ defined by f is equivalent to saying that the Ulrich complexity of f is less than or equal to r. As a consequence, we get the following result:

Lemma 4.4.1. *Let $S \subset \mathbb{P}^3$ be a very general surface defined by a form f of degree $d \geq 4$. Then f cannot be represented as the determinant of a matrix of linear forms.*

Proof. By Noether–Lefschetz theorem, it is known that a very general surface of degree $d \geq 4$ has $\mathrm{Pic}(S) \cong \mathbb{Z}$. Since $\mathcal{O}_S$ is not Ulrich, $uc(f) \geq 2$ and, by definition of Ulrich complexity, we conclude the proof. □

Indeed, we can detect exactly which curves $C \subset S$ correspond to Ulrich line bundles $\mathcal{L} \cong \mathcal{O}_S(C)$ on S.

Lemma 4.4.2. *Let $S \subset \mathbb{P}^3$ be a smooth surface defined by a homogeneous form f of degree d. Then, $C \subset S$ is a smooth aCM curve of degree $\frac{1}{2}d(d-1)$ and genus $g(C) = \frac{1}{6}(d-2)(d-3)(2d+1)$ if and only if $\mathcal{O}_S(C)$ is an Ulrich line bundle on S.*

Proof. An aCM curve with these invariants is given by the maximal minors of a $(d-1)\times d$ matrix N with linear entries. Namely, it has a minimal locally free resolution of the following type:

$$0 \longrightarrow \mathcal{O}_{\mathbb{P}^3}(-d)^{d-1} \xrightarrow{N} \mathcal{O}_{\mathbb{P}^3}(-d+1)^d \xrightarrow{\Delta} \mathcal{O}_{\mathbb{P}^3} \longrightarrow \mathcal{O}_C \longrightarrow 0,$$

with Δ given by the maximal minors of N. Since $C \subset S$, $f \in I(C)$, and therefore f is the determinant of a matrix M obtained from N by adding a column of linear forms. Conversely, if $\mathcal{L} \cong \mathcal{O}_S(C)$ is an Ulrich line bundle, by Proposition 3.2.7, we have $\chi(\mathcal{L}(-1)) = \chi(\mathcal{L}(-2)) = 0$. Therefore, from Riemann–Roch formula (see Theorem 1.4.3) we obtain the degree and the genus of C as in the statement. Moreover, from the short exact sequence

$$0 \longrightarrow \mathcal{I}_S \longrightarrow \mathcal{I}_C \longrightarrow \mathcal{I}_{C|S} \cong \mathcal{O}_S(-C) \longrightarrow 0$$

and the fact that $\mathcal{O}_S(-C)\otimes\mathcal{O}_S(d-1)$ is also an Ulrich line bundle (see Proposition 3.3.3), it follows that C is an aCM curve. □

Remark 4.4.3. Lemma 4.4.1 can be obtained using a different approach. Indeed, if f were represented by the determinant of a linear matrix M of size d, the submatrix N of size $(d-1) \times d$ would define an aCM curve $C \subset S$ of degree $\frac{d(d-1)}{2}$ and genus $g(C) = \frac{(d-2)(d-3)(2d+1)}{2}$. Since these are not the invariants of a complete intersection curve of type $(d, \frac{d-1}{2})$, we can conclude that $\mathrm{Pic}(S) \ncong \mathbb{Z}$ and therefore S is not very general.

In the previous lemma, we described Ulrich line bundles on surfaces of low degree. For instance, a smooth quadric surface supports two Ulrich line bundles corresponding to the lines of each of the two families of rulings. We can also obtain a complete list of Ulrich line bundles on a smooth cubic surface.

Lemma 4.4.4. *Let $S \subset \mathbb{P}^3$ be a smooth cubic surface. Then, S supports 72 nonisomorphic Ulrich line bundles corresponding to the linear classes of rational normal cubic curves in S.*

Proof. It is classically known that a smooth cubic surface contains rational normal cubic curves. In (2.12) the list of divisors that correspond to such curves is explicitly given. □

If we focus the attention on rank-two Ulrich bundles, then we look for linear matrices of size $2d$ such that $f^2 = \det(M)$. Moreover, if one takes M skew-symmetric, that is, $M^t = -M$, then f can be written as the pfaffian $\mathrm{Pf}(M)$ of M. This is the content of the next proposition.

Proposition 4.4.5. *There exists a dense subset of $\mathbb{P}(\mathrm{H}^0(\mathcal{O}_{\mathbb{P}^3}(d)))$ parameterizing surfaces $S \subset \mathbb{P}^3$ of degree d with hyperplane section H supporting Ulrich bundles of rank two and first Chern class $c_1 = (d-1)H$ if and only if $d \leq 15$.*

Proof. Let us denote by Σ_d the variety parameterizing linear skew-symmetric matrices M of size $2d$ such that $\mathrm{Pf}(M) = 0$ defines a smooth surface $S \subset \mathbb{P}^3$. We define a map

$$\mathrm{Pf} : \Sigma_d \longrightarrow \mathbb{P}(\mathrm{H}^0(\mathcal{O}_{\mathbb{P}^3}(d)))$$

that sends such a matrix to its pfaffian. Observing that $\mathrm{GL}(2d)$ leaves invariant the fibers of this morphism, we see that

$$\dim(\Sigma_d/\,\mathrm{GL}(2d)) = 4d(2d-1) - 4d^2.$$

It turns out that $\dim(\Sigma_d/\,\mathrm{GL}(2d)) < \binom{3+d}{3}$ for $d \geq 16$, providing the "only if" part of the statement. To prove the other direction, we need to show that Pf is a dominant map for $d \leq 15$. To this end, given a general matrix $M \in \Sigma_d$, denote by M_{ij} the pfaffian of the skew-symmetric matrix obtained from M by deleting the rows and columns of index i and j, respectively. The proof to show that Pf is a dominant map for $d \leq 15$ consists of two independent parts. First, one realizes that dominance can be detected by proving that the differential of Pf is surjective for a general matrix $M \in \Sigma_d$ (as it is explained in [1, Appendix V]). This turns out to be equivalent to the fact that $\mathrm{H}^0(\mathbb{P}^3, \mathcal{O}_{\mathbb{P}^3}(d))$ is spanned by the forms $x_k M_{ij}$. The second part of the proof consists in checking with the computer program Macaulay2 (see [102]) the previous statement. This was done by Schreyer in the appendix of [22]. □

Remark 4.4.6. Notice that the proof of Proposition 4.4.5 is not computer-free. Only in the cases $d = 3, 4, 5$ it is known theoretically that a general surface of degree d supports Ulrich bundles of rank two and first Chern class $c_1 = (d-1)H$. So, they have a pfaffian representation.

However, using the results from the previous sections, for arbitrary degree d, we can construct large families of surfaces of degree d that support Ulrich bundles of low rank.

Proposition 4.4.7. *Let $d \geq 3$ be an integer. Then, there exist*

1. *A family of dimension $2d^2+1$ of smooth surfaces $S \subset \mathbb{P}^3$ of degree d supporting Ulrich line bundles;*
2. *A family of dimension α with $2d^2 + 1 \leq \alpha \leq 4d(d-1)$ of smooth surfaces $S \subset \mathbb{P}^3$ of degree d supporting Ulrich bundles of rank two; and*
3. *A family of dimension 4 of smooth surfaces $S \subset \mathbb{P}^3$ of degree d and hyperplane section H supporting Ulrich bundles of rank d and first Chern class $\frac{d(d-1)}{2}H$.*

Proof. (1) The determinant defines a map

$$\phi : \mathrm{Mat}_d(\mathbf{k}[x_0, x_1, x_2, x_3]_1)/(\mathrm{GL}(d) \times \mathrm{GL}(d)) \longrightarrow \mathbb{P}(\mathrm{H}^0(\mathcal{O}_{\mathbb{P}^3}(d))).$$

The image of ϕ is the constructible set of surfaces supporting Ulrich line bundles. On the other hand, the fibers of this map are finite. Therefore we get the dimension as in the statement.

(2) As in the previous case, the image of the map

$$\mathrm{Pf} : \Sigma_d \longrightarrow \mathbb{P}(\mathrm{H}^0(\mathcal{O}_{\mathbb{P}^3}(d)))$$

that sends a skew-symmetric matrix to its pfaffian is a constructible family of surfaces supporting Ulrich bundles of rank two. The dimension of this family is bounded by $\dim(\Sigma_d/\,\mathrm{GL}(2d)) = 4d(d-1)$. On the other hand, Kleppe and Miró-Roig showed (see [131, Proposition 4.13]) that on linear determinantal surfaces it is possible to construct indecomposable rank-two Ulrich bundles as extensions of Ulrich line bundles. This fact provides the lower bound.

(3) Let $X \subset \mathbb{P}^4$ be the Fermat threefold defined by the equation $x_0^d + \cdots + x_4^d = 0$. By Theorem 4.2.14, we know that X has a linear matrix factorization of size d^2. Therefore, X supports an Ulrich bundle $\mathcal{E}$ of rank d. Now for a general hyperplane $\Lambda \subset \mathbb{P}^4$, $S := X \cap \Lambda$ will be a smooth surface of degree d in $\mathbb{P}^3$. Moreover, by Proposition 3.3.3, $\mathcal{E}_{|S}$ is an Ulrich bundle. Finally, it is showed in [170, Example 9] that the map

$$f\colon \mathrm{Gr}(3,4) \to \mathbb{P}(\mathrm{H}^0(\mathcal{O}_{\mathbb{P}^3}(d))),$$
$$\Lambda \to [X \cap \Lambda]$$

from hyperplanes $\mathbb{P}^3 \subset \mathbb{P}^4$ to surfaces $S := X \cap \Lambda$ of $\mathbb{P}^3$ has maximal variation, so we get a family of dimension 4 of surfaces of degree d supporting Ulrich bundles of rank d. It is proven in [180] that for a smooth hypersurface $X \subset \mathbb{P}^4$ and a general hyperplane $\Lambda \subset \mathbb{P}^4$, $\mathrm{Pic}(X \cap \Lambda) \cong \mathbb{Z}$, and therefore the Ulrich bundles $\mathcal{E}_{|S}$ constructed above will have $c_1(\mathcal{E}_{|S}) = \frac{d(d-1)}{2}H$. □

For a general surface $S \subset \mathbb{P}^3$ of degree d, we can state the following result:

Theorem 4.4.8. *Let $S \subset \mathbb{P}^3$ be a general surface of degree d defined by a form f. Then S supports an Ulrich bundle of rank $d^{ch(f)-2}$ with $ch(f) \leq \lceil \frac{1}{8}\binom{3+d}{d} \rceil$.*

Proof. The result follows from the bound given in (4.3) and the fact that for a general form f defining a surface of degree d on $\mathbb{P}^3$ we know that $ch(f) = \lceil \frac{1}{8}\binom{3+d}{d} \rceil$. □

Remark 4.4.9. (1) On a very general surface $S \subset \mathbb{P}^3$ with hyperplane section H, an Ulrich bundle $\mathcal{E}$ has first Chern class of the form $c_1(\mathcal{E}) = \frac{r(d-1)}{2}H$ (see Remark 4.2.15). Therefore, a very general surface of even degree does not support Ulrich bundles of odd rank. Hence it is natural that on the previous theorem we obtain Ulrich bundles of the same parity as the degree of the underlying surface.

(2) Once we know the minimum rank $r(d,3)$ of an Ulrich bundle on a general surface $S \subset \mathbb{P}^3$ of degree d, it would be interesting to know for which values $r \geq r(d,3)$, S supports indecomposable Ulrich bundles of rank r. This question is difficult to settle,

Table 4.1: Existence of low rank Ulrich bundles on general surfaces of $\mathbb{P}^3$.

degree	$r(d,3)$	**∃ indecomposable Ulrich** r-bundle on the general surface
1	1	$r = 1$
2	1	$r = 1$
3	1	any rank r
4	2	any rank $r = 2a \geq 2$
$5 \leq d \leq 15$	2	any rank $r = 2a \geq 2$
$16 \leq d \leq 22$	$\leq d^{\lceil \frac{1}{3d+1}\binom{3+d}{3}\rceil - 2}$	$r = r(d,3)$
$d \geq 23$	$\leq d^{\lceil \frac{1}{8}\binom{3+d}{d}\rceil - 2}$	$r = r(d,3)$

in general. Let us study a few well-understood cases. Assume $4 \leq d \leq 9$. We saw that a general surface $S \subset \mathbb{P}^3$ of this degree supports an Ulrich bundle $\mathcal{E}$ of rank two. Since they do not support Ulrich line bundles, we apply Proposition 3.3.16 to conclude that $\mathcal{E}$ is stable. Now, applying [41, Proposition 2.12], we get a lower bound for $\operatorname{ext}^1(\mathcal{E},\mathcal{E})$:

$$\operatorname{ext}^1(\mathcal{E},\mathcal{E}) \geq -\chi(\mathcal{E}^\vee \otimes \mathcal{E}) = -4 - \frac{d-1}{3}(d^2 - 11d + 12) \geq 12$$

for d in our range. Therefore, we can apply Theorem 3.3.2 to see that for any $r = 2a$, $a \geq 1$, S supports indecomposable Ulrich bundles of rank r.

(3) Using the explicit computation of the dimension of the Hilbert space for arithmetically Gorenstein set of points in $\mathbb{P}^3$ corresponding to Ulrich bundles of rank 2 (see [130, Theorem 2.6]), we extend the previous construction and conclude that a general surface $S \subset \mathbb{P}^3$ of degree $3 \leq d \leq 15$ supports indecomposable Ulrich bundles of rank $r = 2a$, for arbitrarily $a \geq 1$.

We can summarize the results discussed in this section in Table 4.1.

Remark 4.4.10. (1) From the previous table one can realize that the first open case is the existence of rank-three Ulrich bundles with Chern class $6H$ on a (general) quintic surface $S \subset \mathbb{P}^3$. By Serre's correspondence, the existence of an indecomposable rank 3 Ulrich bundle on a smooth quintic surface $S \subset \mathbb{P}^3$ with $\operatorname{Pic}(S) = \mathbb{Z}$ is equivalent to the existence of a level set $P \subset S \subset \mathbb{P}^3$ of 75 points with h-vector 1 3 6 10 15 20 12 6 2. Recall that a set of points $P \subset \mathbb{P}^n$ is said to be *level* if its canonical module has a set of generators all of the same degree.

(2) There exist smooth quintic surfaces $S \subset \mathbb{P}^3$ which support indecomposable rank 3 Ulrich bundles. Indeed, let $S \subset \mathbb{P}^3$ be a smooth quintic surface defined by the determinant of a 5×5 matrix with entries general linear forms. It follows from [131, Corollary 5.6] that S supports an indecomposable Ulrich bundle of any r for any $r \geq 1$. It is worthwhile to point out that S is smooth and $\operatorname{Pic}(S) = \mathbb{Z}^2$.

4.5 Ulrich bundles on hypersurfaces via representation theory

In this section, we explain a new approach based on representation theory to construct Ulrich bundles on hypersurfaces, particularly of low rank, which, as we have seen, are difficult to find, in general. The main ideas of this approach have been developed by Manivel [150]. The general setting is as follows. We start with an affine algebraic group G that acts on three vector spaces V, A, and B, where A and B have the same dimension. Let us suppose that there exists a generically injective G-equivariant map $\phi : V^\vee \longrightarrow \mathrm{Hom}(A,B)$.

Associated to this data we have a short exact sequence of $\mathcal{O}_{\mathbb{P}(V)}$-modules

$$0 \longrightarrow A \otimes \mathcal{O}_{\mathbb{P}(V)}(-1) \longrightarrow B \otimes \mathcal{O}_{\mathbb{P}(V)} \longrightarrow \overline{\mathcal{E}} \longrightarrow 0.$$

Notice that the reduced support of the coherent sheaf $\overline{\mathcal{E}}$ is a G-invariant hypersurface $X \subset \mathbb{P}(V)$. As we have already seen, in case that X is irreducible and $\overline{\mathcal{E}}$ is the pushforward of a vector bundle $\mathcal{E}$ on X, we can relate the dimension of A to the rank of $\mathcal{E}$ and the degree of X by the formula (Proposition 4.2.12)

$$\dim A = \mathrm{rank}(\mathcal{E}) \deg(X).$$

In particular, if we know a priori that $\dim A = \mathrm{rank}(\overline{\mathcal{E}} \otimes \mathcal{O}_X) \deg(X)$, $\overline{\mathcal{E}}$ should be the pushforward of an Ulrich bundle $\mathcal{E}$ on a nonempty open set of X (see Theorem 3.2.9). This general strategy provides usually singular hypersurfaces, but we can get Ulrich bundles on a smooth hypersurface of lower dimension just cutting X with linear sections of dimension lower than the codimension of the singular locus of X (for this recall that being Ulrich is preserved under taking hyperplane sections, see Proposition 3.3.3).

This approach has been pursued in the case of cubic hypersurfaces, seen as hyperplane sections of the secant variety to the four *Severi varieties*. Recall that Hartshorne conjectured (see [107, Conjecture 4.2]) and Zak proved (see [206, Theorem 2.8]) that the secant variety $S(X)$ to an l-dimensional smooth nondegenerate projective variety $X \subset \mathbb{P}^n = \mathbb{P}(V)$ with $n < 3l/2 + 2$ should satisfy $S(X) = \mathbb{P}^n$. The limit case when the dimension of X satisfies $n = 3l/2 + 2$ and $S(X) \subsetneq \mathbb{P}^n$ has been classified by Zak (see [206, Chapter IV]). There are only four cases, called the Severi varieties, and they are related to the unitary composition algebras $\mathbb{R}$, $\mathbb{C}$, $\mathbb{H}$, and $\mathbb{O}$. A Severi variety has dimension $l = 2a$ for $a = 1, 2, 4, 8$. Its automorphism group is reductive and a finite cover acts linearly on the $3a + 3$ dimensional vector space V. Beside the cone over X, G has another nontrivial orbit closure, the cone over the secant variety $Y = S(X)$, which is a cubic hypersurface. From this presentation it follows that X is the singular locus of Y which has codimension $a + 1$. Therefore a general linear section of Y of dimension less than or equal to a defines a smooth cubic hypersurface.

There is a nice way to understand uniformly the four Severi varieties as "Veronese surfaces over composition algebras over $\mathbb{R}$." Let K_i be the real field $\mathbb{R}$ (respectively, the complex field $\mathbb{C}$, the quaternions $\mathbb{H}$, or the octonions $\mathbb{O}$) for $i = 0$ (respectively, for

Table 4.2: List of Severi varieties.

$a = 2^i$	$1 = 2^0$	$2 = 2^1$	$4 = 2^2$	$8 = 2^3$
G	SL_3	SL_3^2	SL_6	E_6
X	$v_2(\mathbb{P}^2)$	$(\mathbb{P}^2)^2$	$\mathrm{Gr}(1,5)$	$\mathbb{OP}^2$
$V = H_i$	$\mathrm{Sym}^2(\mathbb{C}^3)$	$(\mathbb{C}^3)^{\otimes 2}$	$\wedge^2\mathbb{C}^6$	V_{27}
Hermitian matrices over the composition algebra $\mathbf{k} = K_i$	$\mathbf{k} = \mathbb{R}$	$\mathbf{k} = \mathbb{C}$	$\mathbf{k} = \mathbb{H}$	$\mathbf{k} = \mathbb{O}$

$i = 1, 2, 3$). Let H_i denote the Jordan algebra of Hermitian (3×3)-matrices over K_i (a matrix $A \in H_i$ is *Hermitian* if $\bar{A}^t = A$ where the bar denotes the involution of K_i). Let

$$X_i = \{[A] \in \mathbb{P}(H_i) \mid \mathrm{rank}(A) = 1\} \subset \mathbb{P}(H_i) \cong \mathbb{P}^{2^i 3+2}.$$

Then

$$Y_i := S(X_i) = \{[A] \in \mathbb{P}(H_i) \mid \mathrm{rank}(A) \le 2\} = V(\det(A)) \subset \mathbb{P}(H_i)$$

is a degree 3 hypersurface, and by definition X_i is a Severi variety of dimension $\dim(X_i) = 2^{i+1} = 2\dim(K_i)$ which corresponds to the case $a = 2^i$.

We gather the preceding explanation in Table 4.2 where E_6 stands for the exceptional group of automorphisms of the Cayley plane $\mathbb{OP}^2$ and V_{27} is the minimal representation of E_6.

An homogeneous polynomial $h \in \mathrm{Sym}^3(V)$ giving the cubic hypersurface $Y \subset \mathbb{P}(V)$ defines by polarization a linear map

$$\phi : V^\vee \longrightarrow \mathrm{Sym}^2(V) \subset \mathrm{Hom}(V^\vee, V).$$

This map ϕ is equivariant with respect to the semisimple part of G and generically injective. Therefore, we have a short exact sequence

$$0 \longrightarrow V^\vee \otimes \mathcal{O}_{\mathbb{P}(V)}(-1) \longrightarrow V \otimes \mathcal{O}_{\mathbb{P}(V)} \longrightarrow \mathcal{E} \longrightarrow 0.$$

We get

Lemma 4.5.1. *The sheaf $\mathcal{E}$ is, on the smooth part of the cubic hypersurface $Y \subset \mathbb{P}(V)$, the pushforward of a vector bundle of rank $a + 1$.*

From Lemma 4.5.1 we obtain the following results. In the case $a = 2$, we recover the fact that any smooth cubic surface in $\mathbb{P}^3$ supports Ulrich bundles of rank three. For $a = 4$, we obtain that any smooth pfaffian cubic fourfold supports Ulrich bundles of rank five.

Particularly interesting is the case $a = 8$. The Severi variety in this case is the so-called *Cayley plane*, with the automorphism group being the exceptional group E_6. The secant variety is known as the *Cartan hypersurface*. In this setting, Lemma 4.5.1 is rephrased as follows:

Proposition 4.5.2. *A general linear section of the Cartan cubic hypersurface of dimension less than or equal to eight is a smooth cubic hypersurface supporting Ulrich bundles of rank nine.*

Eight-dimensional cubic hypersurfaces that can be expressed as a linear section of the Cartan hypersurface are not general. However, in dimension seven we have the following result:

Proposition 4.5.3. *A general cubic hypersurface $Y \subset \mathbb{P}^8$ is, up to an isomorphism, a linear section of the Cartan cubic, in a finite number of ways.*

Proof. See [120, Proposition 3.1]. □

Remark 4.5.4. (1) Recall that using the linear matrix factorization associated to cubic forms f, we showed the existence of Ulrich bundles of rank $3^{ch(f)-2}$ where $ch(f)$ is the Chow rank of f (see Theorem 4.2.14). On the other hand, Proposition 4.5.2 shows that we can expect the existence of Ulrich bundles of much lower rank.

(2) It would be very interesting to improve Proposition 4.5.3 by finding out exactly in how many different ways a general cubic hypersurface $Y \subset \mathbb{P}^8$ is, up to an isomorphism, a linear section of the Cartan cubic. Moreover, whether each of these representations gives nonisomorphic line bundles is also of interest.

(3) Due to Proposition 4.5.2, we can focus our attention on the particular case of general cubic fourfolds and wonder whether the Ulrich complexity of such a fourfold is nine or lower. In [22, Proposition 9.2] it was proven that the set of cubic fourfolds supporting a rank-two Ulrich bundle form a hypersurface in the space of all smooth cubic fourfolds. On the other hand, Kim and Schreyer [127, Proposition 2.5] proved, throughout a careful study of the Chern classes of such a bundle, that on a general cubic fourfold the rank of an Ulrich bundle should be divisible by three and greater than or equal to six. Therefore, the Ulrich complexity of such a general cubic fourfold should be either six or nine.

Another analogous construction deals with the case of septic hypersurfaces. If we start with a vector space V of dimension seven, $\mathrm{GL}(V)$ acts on $\bigwedge^3 V$ with finitely many orbits. There is a unique semiinvariant hypersurface $X \subset \mathbb{P}^{34} = \mathbb{P}(\bigwedge^3 V)$ of degree seven. Three-forms $\omega \in \bigwedge^3 V$ define a symmetric map

$$\phi(\omega) : \bigwedge^2 V \longrightarrow \bigwedge^5 V.$$

It was seen in [150, Lemma 2.4] that $\phi(\omega)$ defines an isomorphism for any $\omega \notin X$. Moreover, for a generic $\omega \in X$, $\phi(\omega)$ has corank three. Gathering this information, we see that

$$0 \longrightarrow \bigwedge^2 V \otimes \mathcal{O}_{\mathbb{P}(\bigwedge^3 V)}(-1) \longrightarrow \bigwedge^5 V \otimes \mathcal{O}_{\mathbb{P}(\bigwedge^3 V)} \longrightarrow \mathcal{E} \longrightarrow 0$$

defines an Ulrich sheaf of rank three. Since the complement of the open orbit of the $\mathrm{GL}(V)$-action has codimension three, a general linear section of X of dimension two will be a smooth heptic surface in $\mathbb{P}^3$. Therefore, we have

Proposition 4.5.5. *There exists a family of dimension 76 of degree 7 smooth surfaces $S \subset \mathbb{P}^3$ that support indecomposable Ulrich bundles of rank 3.*

Finally, we introduce a new construction due to Eisenbud–Shamash (see [80, Section 6]) to obtain Ulrich bundles on hypersurfaces. Let $X \subset \mathbb{P}^n$ be a hypersurface defined by a homogeneous polynomial f of degree d and let us consider an arbitrary module M over the ring $S = R/(f)$, $R = \mathbf{k}[x_0, \dots, x_n]$. Let us assume that M, as an R-module, has a minimal graded free resolution

$$0 \longrightarrow F_c \xrightarrow{\psi_c} \cdots \longrightarrow F_1 \xrightarrow{\psi_1} F_0 \longrightarrow M \longrightarrow 0. \tag{4.7}$$

Tensoring it by S, we obtain a graded free resolution over S given by

$$\cdots \to \begin{matrix} \overline{F}_5 \\ \oplus \\ \overline{F}_3(-d) \\ \oplus \\ \overline{F}_1(-2d) \end{matrix} \to \begin{matrix} \overline{F}_4 \\ \oplus \\ \overline{F}_2(-d) \\ \oplus \\ \overline{F}_0(-2d) \end{matrix} \to \begin{matrix} \overline{F}_3 \\ \oplus \\ \overline{F}_1(-d) \end{matrix} \to \begin{matrix} \overline{F}_2 \\ \oplus \\ \overline{F}_0(-d) \end{matrix} \to \overline{F}_1 \to \overline{F_0} \to N \to 0 \tag{4.8}$$

with $N := M \otimes_R S$ and $\overline{F}_i := F_i \otimes_R S$. This (not necessarily minimal) S-resolution becomes 2-periodic after $c-1$ steps. Moreover, one has an explicit understanding of its maps. Indeed, the map at the ith step,

$$\bigoplus_{k\geq 0} \overline{F}_{i-2k}(-kd) \xrightarrow{\phi_i} \bigoplus_{k\geq 0} \overline{F}_{i-1-2k}(-kd), \tag{4.9}$$

is constructed as follows. The component $\overline{F}_{i-2k}(-kd) \longrightarrow \overline{F}_{i-1-2k}(-kd)$ is inherited from (4.7) whereas

$$\overline{F}_i \longrightarrow \bigoplus_{k\geq 1} \overline{F}_{i-1-2k}(-kd)$$

is the zero map. For $i \geq c$, the map ϕ_i from (4.9) gives a two-factor matrix factorization of f and, in particular, an aCM vector bundle on X. If, moreover, we can assure that for even i the graded map ψ_i between free R-modules in the resolution (4.8) is of degree 1, then we have that for c even (respectively c odd), the matrix associated to ϕ_c (respectively ϕ_{c+1}) provides an Ulrich bundle on X.

To apply this approach is, however, not straightforward due to the tight requirements that the resolution (4.8) of N should satisfy. As a first example, let us consider a smooth cubic threefold $X \subset \mathbb{P}^4$. It is known that X contains elliptic normal curves C of degree 5. Any such curve C is arithmetically Gorenstein and thus has an $\mathcal{O}_{\mathbb{P}^4}$-resolution of the form

$$0 \longrightarrow \mathcal{O}_{\mathbb{P}^4}(-5) \longrightarrow \mathcal{O}_{\mathbb{P}^4}(-3)^5 \longrightarrow \mathcal{O}_{\mathbb{P}^4}(-2)^5 \longrightarrow \mathcal{O}_{\mathbb{P}^4} \longrightarrow \mathcal{O}_C \longrightarrow 0.$$

Therefore the aforementioned construction allows us to obtain Ulrich bundles of rank 2 on the cubic hypersurface $X \subset \mathbb{P}^4$.

A more demanding task is to find the appropriate subvariety $Z \subset X \subset \mathbb{P}^8$ on a general cubic hypersurface that allows us to recover the Ulrich bundles of rank nine from Proposition 4.5.2.

A natural candidate to fulfill the requirements would be a threefold $Z \subset X \subset \mathbb{P}^8$ with resolution

$$\begin{aligned} 0 \longrightarrow \mathcal{O}_{\mathbb{P}^8}(-8) \longrightarrow \mathcal{O}_{\mathbb{P}^8}(-6)^{10} \longrightarrow \mathcal{O}_{\mathbb{P}^8}(-5)^{16} \longrightarrow \\ \mathcal{O}_{\mathbb{P}^8}(-3)^{16} \longrightarrow \mathcal{O}_{\mathbb{P}^8}(-2)^{10} \longrightarrow \mathcal{O}_{\mathbb{P}^8} \longrightarrow \mathcal{O}_Z \longrightarrow 0 \end{aligned} \tag{4.10}$$

This corresponds to one of the Mukai's Fano threefolds. More precisely, Z is defined as a linear section of the spinor variety $S_{10} \subset \mathbb{P}^{15}$. Then everything boils down to the following

Proposition 4.5.6. *Any general $\mathbb{P}^{15} \subset \mathbb{P}(V_{27})$ is the linear span of the spinor variety S_{10} contained in the Cartan cubic $Y \subset \mathbb{P}(V_{27})$.*

Proof. See [120, Lemma 5.2]. □

To complete the argument, take $X := Y \cap \Lambda$, $\Lambda \cong \mathbb{P}^8$, a general cubic linear section of the Cartan hypersurface $Y \subset \mathbb{P}(V_{27})$. For a general $\mathbb{P}^{15} \cong \Delta \supset \Lambda$, Proposition 4.5.6 tells us that the Fano threefold $Z := S_{10} \cap \Lambda \subset Y \cap \Lambda = X$ has the resolution (4.10), and we can conclude.

4.6 Ulrich bundles on projective varieties and Cayley–Chow forms

In Section 4.2 we have studied the connection between linear matrix factorizations of a given hypersurface $X \subset \mathbb{P}^n$ and Ulrich bundles supported on $(X, \mathcal{O}_X(1))$. Indeed, this connection can be introduced in full generality for an arbitrary projective variety $(X, \mathcal{O}_X(1))$ in terms of its associated *Cayley–Chow form*. Cayley–Chow forms were introduced for curves in the projective space $\mathbb{P}^3$ in [49] and [50]. Later on, the concept was generalized to arbitrary varieties by Bertini, Chow, and van der Warden (see [203]). The current section is devoted to a concise introduction to this concept.

Let $(X, \mathcal{O}_X(1))$ be a d-dimensional projective variety providing an embedding $X \subset \mathbb{P}^n$. Let us consider the following incidence relation:

$$\begin{array}{ccc} & \mathbb{F} := \{(x, H) \mid x \in H\} \subset \mathbb{P}^n \times \mathrm{Gr}(n-d-1, n) & \\ & \swarrow^{p} \qquad\qquad \searrow^{q} & \\ \mathbb{P}^n & & \mathrm{Gr}(n-d-1, n) \end{array} \tag{4.11}$$

where we have denoted by p and q the corresponding natural projections.

Definition 4.6.1. Let $(X, \mathcal{O}_X(1))$ be a d-dimensional projective variety embedded in $\mathbb{P}^n$. With the above notations, the variety

$$D_X := q \circ p^{-1}(X) = \{H \in \mathrm{Gr}(n-d-1, n) \mid H \cap X \neq \emptyset\} \subset \mathrm{Gr}(n-d-1, n)$$

has codimension 1 in $\mathrm{Gr}(n-d-1, n)$ which means that it is a divisor. Also D_X is called the *Cayley–Chow divisor* of X.

Therefore, D_X is given by a single equation f_X in the Plücker embedding. In fact, $\mathrm{Pic}(\mathrm{Gr}(n-d-1, n)) \cong \mathbb{Z} < \mathcal{O}_{\mathrm{Gr}(n-d-1,n)}(1) >$ and $\mathcal{O}_{\mathrm{Gr}(n-d-1,n)}(1)$ gives the *Plücker embedding* $\mathrm{Gr}(n-d-1, n) \subset \mathbb{P}(V)$, $V = \wedge^{n-d}\mathbf{k}^{n+1}$. Moreover, this is a projectively normal embedding, and therefore $D_X \subset \mathrm{Gr}(n-d-1, n)$ is cut out in $\mathrm{Gr}(n-d-1, n)$ by a single homogeneous polynomial f_X in the *Plücker coordinates.*

Definition 4.6.2. Let $(X, \mathcal{O}_X(1))$ be a d-dimensional projective variety embedded in $\mathbb{P}^n$. With the above notations, the homogeneous form f_X in the Plücker coordinates associated to the variety X is called the *Cayley–Chow form* of X.

Example 4.6.3. (1) Let $X \subset \mathbb{P}^n$ be a hypersurface. Then, obviously $\mathrm{Gr}(n-d-1, n) \cong \mathbb{P}^n$ and D_X is nothing but X itself. Therefore the Cayley–Chow form f_X of X is the homogeneous polynomial defining X in $\mathbb{P}^n$.

(2) Let $C \subset \mathbb{P}^3$ be a rational cubic curve. The ideal of C inside $\mathbb{P}^3$ is generated by the three maximal minors of the matrix

$$\begin{pmatrix} x_0 & x_1 & x_2 \\ x_1 & x_2 & x_3 \end{pmatrix}.$$

The Cayley–Chow form f_C of C in the Plücker coordinates w_{pq} of $\mathrm{Gr}(1, 3) \subset \mathbb{P}(\bigwedge^2 V)$ is the determinant of the matrix

$$f_C = \begin{vmatrix} w_{01} & w_{02} & w_{03} \\ w_{02} & w_{03} + w_{12} & w_{13} \\ w_{03} & w_{13} & w_{23} \end{vmatrix}.$$

We will see in Example 4.6.10 that this presentation of the Cayley–Chow form generalizes to any rational normal curve.

(3) The Cayley–Chow form of $\mathbb{P}^d$ embedded in $\mathbb{P}^{\binom{d+e}{d}-1}$ by the Veronese embedding associated to $\mathcal{O}_{\mathbb{P}^d}(e)$ is the resultant of $d+1$ forms of degree e in $d+1$ variables.

An important feature of Cayley–Chow forms f_X is that one can recover the original variety X from them. Set-theoretically, this result was proved by van der Warden in [203]. Scheme-theoretically, we have the following result:

Theorem 4.6.4. *Let $X \subset \mathbb{P}^n$ be a hypersurface or a smooth variety. Then the Cayley–Chow form f_X of X completely determines the variety X.*

Proof. See [46, Theorem 1.14] and [88, Proposition 1.3]. □

In general, it is an intriguing question to find the Cayley–Chow form f_X of a given variety X. We are going to see that the Cayley–Chow form f_X of a projective variety $(X, \mathcal{O}_X(1))$ has a very pleasant presentation when $(X, \mathcal{O}_X(1))$ supports Ulrich bundles.

Indeed, this theory can be developed in greater generality. For this, let us consider a coherent sheaf $\mathcal{F}$ on $\mathbb{P}^n$. Let us assume that the support $\operatorname{Supp}(\mathcal{F})$ of $\mathcal{F}$ has dimension d and it is represented by the d-cycle

$$\sum_i \operatorname{length}(\mathcal{O}_{\mathbb{P}^n, X_i} \otimes \mathcal{F}) \cdot [X_i],$$

with the sum running over a finite number of subvarieties $X_i \subset \mathbb{P}^n$ of dimension d. Then the *Cayley–Chow divisor* $D_{\mathcal{F}}$ of $\mathcal{F}$ is defined to be the corresponding sum of Cayley–Chow divisors

$$\sum_i \operatorname{length}(\mathcal{O}_{\mathbb{P}^n, X_i} \otimes \mathcal{F}) \cdot D_{X_i}.$$

Analogously, the *Cayley–Chow form* $f_{\mathcal{F}}$ of $\mathcal{F}$ is the homogeneous polynomial on the Plücker coordinates cutting out $D_{\mathcal{F}}$ inside the Grassmann variety $\operatorname{Gr}(n-d-1, n)$.

Remark 4.6.5. (1) The sheaf $\mathcal{G} := q_*(p^*\mathcal{F})$ is supported precisely on the set $q \circ p^{-1}(\operatorname{Supp}(\mathcal{F}))$ and the generic rank of $\mathcal{G}$ on D_{X_i} is $\operatorname{length}(\mathcal{O}_{\mathbb{P}^n, X_i} \otimes \mathcal{F})$. Therefore, the Cayley–Chow divisor of $\mathcal{F}$ is actually the divisor associated to the coherent sheaf $\mathcal{G}$ on $\operatorname{Gr}(n-d-1, n)$.

(2) For $X \subset \mathbb{P}^n$ any d-dimensional variety, the Cayley–Chow divisor of $\mathcal{O}_X$ is the Cayley–Chow divisor of X. In general, for a rank r vector bundle $\mathcal{F}$ on X, the Cayley–Chow divisor of $\mathcal{F}$ is r times the Cayley–Chow divisor of X.

From the previous remark, we see that an approach to determining (a multiple of) the Cayley–Chow divisor (and the Cayley–Chow form) of a variety $X \subset \mathbb{P}^n$ is to find (a low rank) vector bundle $\mathcal{F}$ on X for which we can find its Cayley–Chow divisor. We will see that this is the case for Ulrich bundles. First of all, however, we should understand how to find the Cayley–Chow divisor of $\mathcal{G} := q_*(p^*\mathcal{F})$. This has been obtained, as an extension of Beilinson's results, in [88]. We explain it here.

Let $\mathcal{U} := \mathcal{S}^\vee$ be the be the universal vector subbundle of rank $d+1$ appearing on the universal exact sequence introduced in (2.9)

$$0 \longrightarrow \mathcal{S}^\vee \longrightarrow \mathcal{O}^{n+1}_{\mathrm{Gr}(n-d-1,n)} \longrightarrow \mathcal{Q} \longrightarrow 0.$$

Set $W = V^\vee$ and let us write the exterior algebra $E = \bigwedge W$ with $\mathbb{P}^n \cong \mathbb{P}(V)$. Any element $a \in \bigwedge^p V$ gives rise to a "contraction map" of vector bundles $a : \bigwedge^q \mathcal{U} \longrightarrow \bigwedge^{q-p} \mathcal{U}$. It holds:

(1) The contraction maps make $\bigwedge \mathcal{U}$ into a module over E.
(2) The maps

$$\bigwedge^p V \longrightarrow \mathrm{Hom}\Big(\bigwedge^q \mathcal{U}, \bigwedge^{q-p} \mathcal{U}\Big) \tag{4.12}$$

are isomorphisms for all p, q with $0 \le q-p, q \le n-d-1$.

These two properties allow us to construct the *Tate resolution* $T(\mathcal{F})$ of any coherent sheaf $\mathcal{F}$ on $\mathbb{P}^n$ with support a variety of dimension d. Recall that $T(\mathcal{F})$ is a doubly infinite exact complex of finitely generated free graded E-modules, where the grading on E is such that the elements of V have degree -1.

The eth term of $T(\mathcal{F})$ is

$$T^e(\mathcal{F}) := \bigoplus_j \mathrm{H}^j(\mathcal{F}(e-j)) \otimes E(j-e)$$

where we denote by $E(l)$ the free rank 1 graded E-module with generator in degree $-l$ (so $E(1) = E$). The details for the maps of the complex $T(\mathcal{F})$ can be found in [82].

Now we define the following functor $U(-)$ from the category of graded free E-modules to the category of vector bundles over $\mathrm{Gr}(n-d-1,n)$ by sending $E(p)$ to $U(E(p)) := \bigwedge^p \mathcal{U}$. At the level of maps, the functor $U(-)$ sends an element $a \in \bigwedge^p V$ corresponding to the map $E(q) \longrightarrow E(q-p)$ to the corresponding map $\bigwedge^q \mathcal{U} \to \bigwedge^{q-p} \mathcal{U}$ obtained from the isomorphism (4.12).

Now, for the Tate resolution $T := T(\mathcal{F})$ of a coherent sheaf $\mathcal{F}$ on $\mathbb{P}^n$, it holds that $U(T^e) = 0$ for $|e| \gg 0$, so $U(\mathcal{F}) := U(T)$ is a bounded complex of vector bundle on $\mathrm{Gr}(n-d-1,n)$.

Keeping the above notations, we have the following result:

Theorem 4.6.6. *For any coherent sheaf $\mathcal{F}$ on $\mathbb{P}^n$ with support being a variety X of dimension d, there is a canonical complex of vector bundles $U(\mathcal{F})$ on $G(n-d-1,n)$ with $U(\mathcal{F}) \cong Rq_*(p^*(\mathcal{F}))$ in the derived category of sheaves on the Grassmann variety $\mathrm{Gr}(n-d-1,n)$ such that the lth term of $U(\mathcal{F})$ is*

$$U(\mathcal{F})_l = \bigoplus_j \mathrm{H}^j(\mathcal{F}(l-j)) \otimes \wedge^{j-l}\mathcal{U},$$

which is generically exact and fails to be exact precisely along D_X.

Proof. See [88, Theorem 1.2]. □

The complex $U(\mathcal{F})$ is called the *Cayley–Chow complex*.

Remark 4.6.7. Again, let us suppose that $X \subset \mathbb{P}^n$ is a hypersurface and therefore the Cayley–Chow form f_X is the homogenous polynomial defining X in $\mathbb{P}^n$. Then, for a vector bundle $\mathcal{F}$ on X, the Cayley–Chow complex $U(\mathcal{F})$ on $\mathbb{P}^n$ defined in Theorem 4.6.6 is exactly the Beilinson's complex as described in Theorem A.0.27.

Now, let us suppose that we are working with an Ulrich bundle $\mathcal{F}$ on the variety $X \subset \mathbb{P}^n$. Applying the characterization from Theorem 3.2.9, we see that $U(\mathcal{F})$ has a particularly simple presentation.

Theorem 4.6.8. *Let $(X, \mathcal{O}_X(1))$ be a projective variety of dimension d and let $\mathcal{F}$ be a rank r vector bundle on X. The Cayley–Chow complex $U(\mathcal{F})$ is reduced to a single map*

$$U(\mathcal{F})_{-1} = \mathcal{O}_{\mathrm{Gr}(n-d-1,n)}(-1)^a \xrightarrow{\Psi_{\mathcal{F}}} U(\mathcal{F})_0 = \mathcal{O}^a_{\mathrm{Gr}(n-d-1,n)}$$

with $a = \mathrm{h}^0(\mathcal{F}) = \mathrm{h}^d(\mathcal{F}(-d-1))$ if and only if $\mathcal{F}$ is an Ulrich bundle on $(X, \mathcal{O}_X(1))$. In this case, the Cayley–Chow form f_X of X has the following expression in the Plücker coordinates:

$$f_X^r = \det(\Psi_{\mathcal{F}}).$$

Proof. It comes from the fact that $\mathcal{F}$ is an Ulrich bundle if and only if $U(\mathcal{F})_j = 0$ for $j \neq -1, 0$ (and, in this case, $\mathrm{h}^0(\mathcal{F}) = \mathrm{h}^d(\mathcal{F}(-d-1))$). Hence, the Cayley–Chow complex reduces to

$$\mathrm{H}^d(\mathcal{F}(-d-1)) \otimes \det(\mathcal{U}) \longrightarrow \mathrm{H}^0(\mathcal{F}) \otimes \mathcal{O}_{\mathrm{Gr}(n-d-1,n)}.$$ □

Therefore, in order to give an explicit presentation of the Cayley–Chow form of a variety $X \subset \mathbb{P}^n$, we need to compute $\det(\Psi_{\mathcal{F}})$. This can be done in the following way. Recall that, by Theorem 3.2.9, an Ulrich bundle $\mathcal{F}$ on X has a linear resolution

$$0 \longrightarrow \mathcal{O}_{\mathbb{P}^n}(-n+d)^{a_{n-d}} \xrightarrow{\psi_{n-c}} \cdots \xrightarrow{\psi_2} \mathcal{O}_{\mathbb{P}^n}(-1)^{a_1} \xrightarrow{\psi_1} \mathcal{O}^{a_0}_{\mathbb{P}^n} \longrightarrow \mathcal{F} \longrightarrow 0 \tag{4.13}$$

where the ψ_i are matrices of linear forms or, equivalently, of elements from V. Therefore $\Psi_{\mathcal{F}} := \psi_1 \wedge \psi_2 \cdots \wedge \psi_{n-d}$ can be interpreted as a square matrix (of size dr) with entries on $\bigwedge^{n-d} V$ or, in other words, with entries being linear forms under the Plücker embedding of the Grassmann variety $\mathrm{Gr}(n-d-1, n)$. We summarize the results in the next theorem.

Theorem 4.6.9. *Let $(X, \mathcal{O}_X(1))$ be a variety of dimension d and let $\mathcal{F}$ be an Ulrich bundle on it. Then, $\Psi_{\mathcal{F}}$ is the only nonzero map in the complex $U(\mathcal{F})$. In particular, the Cayley–Chow form f_X of X satisfies*

$$f_X^r = f_{\mathcal{F}} = \det(\Psi_{\mathcal{F}}).$$

Moreover, assuming that $\mathcal{F}$ is a special rank-two Ulrich bundle (i. e., $\bigwedge^2 \mathcal{F} \cong \omega_X(d+1)$), $\Psi_{\mathcal{F}}$ is skew-symmetric and $f_X = \mathrm{Pf}(\Psi_{\mathcal{F}})$.

Proof. See [88, Theorem 0.3]. □

Example 4.6.10. (1) This example is the starting point of the theory of resultants, as developed by Leibniz, Bézout, and Sylvester, among others. We consider the rational normal curve $\mathbb{P}^1 \xrightarrow{\cong} C \subset \mathbb{P}^n$ embedded by the very ample line bundle $\mathcal{O}_{\mathbb{P}^1}(n)$. The ideal of C inside $\mathbb{P}^n$ is generated by the maximal minors of the matrix

$$\begin{pmatrix} x_0 & x_1 & \dots & x_{n-1} \\ x_1 & x_2 & \dots & x_n \end{pmatrix}.$$

We saw in Lemma 3.4.2 that $\mathcal{O}_{\mathbb{P}^1}(n-1)$ is the only Ulrich bundle on $(\mathbb{P}^1, \mathcal{O}_{\mathbb{P}^1}(n))$. Applying Theorem 4.6.8, it is possible to compute the Tate resolution of $\mathcal{O}_{\mathbb{P}^1}(n-1)$, and we obtain that the Cayley–Chow form of $C \subset \mathbb{P}^n$ is the determinant of the $d \times d$ symmetric matrix $M = (m_{ij})$ with

$$m_{ij} = \sum_{\substack{p+q=i+j-1 \\ p<\min(i,j)}} w_{pq}$$

where w_{pq} denotes the pqth Plücker coordinate of $\mathrm{Gr}(n-2, n)$, namely, the coordinate corresponding to the element of the basis $v_0 \wedge \cdots \wedge \hat{v}_p \wedge \cdots \wedge \hat{v}_q \wedge \cdots \wedge v_n$.

(2) Let us consider the cubic scroll $S := S(2,1) \subset \mathbb{P}^4$, a surface of minimal degree 3 in $\mathbb{P}^4$. It is defined by the maximal minors of the matrix

$$\begin{pmatrix} x_0 & x_1 & x_3 \\ x_1 & x_2 & x_4 \end{pmatrix}.$$

We will see in Proposition 5.2.2 that S supports an Ulrich line bundle $\mathcal{L}$ which allows us to write the Cayley–Chow form f_S of S as the determinant of the following 3×3 matrix:

$$\Psi_{\mathcal{L}} = \begin{pmatrix} w_{034} & w_{013} & w_{023} \\ -w_{134} & w_{023} + w_{014} & -w_{123} - w_{024} \\ w_{234} & -w_{024} & w_{124} \end{pmatrix}$$

where w_{ijk} denotes the ijkth Plücker coordinate of $\mathrm{Gr}(1,4) \subset \mathbb{P}^9 = \mathbb{P}(\bigwedge^2 V)$.

4.7 Final comments and additional reading

This chapter tries to highlight the connections between classical geometric problems and the modern theory of Ulrich bundles. We hope to convey the feeling that Ulrich bundles pervade many areas of contemporary Algebraic Geometry. With this leitmotif in mind, we wrote the historical Section 4.1. For the basic notions related to linear matrix factorization, we have followed the works by Backelin and Herzog [16] and

Backelin, Herzog, and Sanders [17]. The notion of Ulrich complexity was introduced by Bläser, Eisenbud, and Schreyer [27]. For additional reading concerning the generalization of the notion of two-factor matrix factorization to complete intersections, we suggest the book [86] by Eisenbud and Paaeva. The reader can find explicit matrix factorizations in the work [127] by Kim and Schreyer where the authors gave explicit constructions of Ulrich bundles found in [150] by Manivel.

Section 4.3 deals with Ulrich bundles on complete intersections. The existence of Ulrich bundles on such varieties was proved in [113] from an algebraic point of view. It is, however, difficult from this paper to figure out an explicit bound for the minimum rank of an Ulrich bundle on a general complete intersection. Combining the results from [113] and [44] with those from Section 4.2, we could write down such a bound in Corollary 4.3.3. However, as it is explained in Remark 4.3.4, this bound is expected to be far from being sharp in many cases.

Part of the content of Section 4.4 can be found in [22]. Maximal variation of hyperplane sections is worked out in [170]. Section 4.5 is based mostly on [150] where the method of using representation theory to construct Ulrich bundles is devised. Some results depend on a previous work from [120] about cubic hypersurfaces of high dimension. Reference [127] also recalls Eisenbud–Shamash's method to construct Ulrich bundles from particular resolutions. The reader can find an introduction to the theory of Severi varieties, as well as their relationship with the four composition algebras over $\mathbb{R}$, in [187] and [206]. The arguments about Cayley–Chow forms and their relation to Ulrich bundles from Section 4.6 are based on the article [88]. An introduction to the theory of Cayley–Chow forms can be found in [47] (see also [136]). For some applications of this theory to computer vision the reader can look at [95].

5 Ulrich bundles on surfaces

In this chapter, we restrict our attention to the existence and construction of rank r Ulrich bundles on smooth projective surfaces $(S, \mathcal{O}_S(H))$. Let us stress that since the Ulrich property depends on the embedding of S in a projective space defined by $|H|$, by writing a smooth projective surface $(S, \mathcal{O}_S(H))$ we always mean that H is a very ample divisor. Low rank vector bundles on smooth surfaces have been intensively studied; indeed, Ulrich bundles of rank 2 (or sometimes rank 1) are known in many cases. The classification of Ulrich bundles on smooth surfaces is usually arranged accordingly to the Enriques–Kodaira classification of the underlying surfaces or, equivalently, according to the Kodaira dimension $\kappa(S)$ of S.

Since the position of the surface in the Enriques–Kodaira classification is of uttermost importance for the study of Ulrich bundles on surfaces, we will organize this chapter according to this classification. In Section 5.1, we state a useful characterization of rank r (special) Ulrich bundles on smooth surfaces and gather together some constructions that will be intensively used in next sections, like Serre's construction based on the Cayley–Bacharach property. In Section 5.2, we focus our attention on the existence of Ulrich bundles on minimal surfaces with Kodaira dimension $-\infty$. We explicitly determine the possible ranks of a d-uple Veronese embedding of $\mathbb{P}^2$ and describe Ulrich bundles of arbitrary rank on a Hirzebruch surface using a sort of Beilinson's spectral sequence. The case of nonminimal surfaces with $\kappa(S) = -\infty$ is treated in Section 5.3 with special emphasis on del Pezzo surfaces, that is, on smooth surfaces with ample anticanonical divisor. Section 5.4 is entirely devoted to surfaces with Kodaira dimension 0, and as one of the main strategies we use elementary transformation of a vector bundle along a suitable effective divisor. We prove that for any minimal surface S with $\kappa(S) = 0$ and any very ample line bundle $\mathcal{O}_S(H)$, the surface $(S, \mathcal{O}_S(H))$ supports a rank 2 simple special Ulrich bundle. We will also see that for any nonminimal surface S with $\kappa(S) = 0$ there exist infinitely many very ample line bundles $\mathcal{O}_S(H)$ such that $(S, \mathcal{O}_S(H))$ has a rank 2 Ulrich bundle. In Section 5.5 we treat smooth surfaces S with positive Kodaira dimension and separately discuss the cases $\kappa(S) = 1$ and $\kappa(S) = 2$. In the later case, little can be said, but at least a few examples are provided. Finally, in Section 5.6, we prove the asymptotic existence of Ulrich bundles on any smooth projective surface. More precisely, we establish that for any smooth projective surface $(S, \mathcal{O}_S(H))$, there exists an integer m_0 such that and for any $m \geq m_0$, $(S, \mathcal{O}_S(mH))$ carries an Ulrich bundle.

The original version of this chapter was revised: the footnote on p. 151, line 1 has been added. An Erratum is available at DOI: https://doi.org/10.1515/9783110647686-012
https://doi.org/10.1515/9783110647686-005

5.1 Constructions and preliminaries

In this section, we introduce the tools that we will use throughout the chapter. Moreover, we are going to give a characterization of Ulrich vector bundles on a smooth surface $(S, \mathcal{O}_S(H))$ that will make much easier checking whether an arbitrary vector bundle holds this property.

As it can be expected from Proposition 3.2.7, the lower the dimension of the underlying variety X, the easier to check that a given vector bundle on X is Ulrich. For instance, we saw in Lemma 3.4.2 that in the case of curves (see Lemma 3.4.2), a vector bundle $\mathcal{E}$ is Ulrich whenever it has the right degree (a numerical condition) and $\mathcal{E}(-1)$ has no sections (a cohomological condition). Now we are going to see that something similar can be expected for surfaces. We recall the Riemann–Roch theorem for a rank r vector bundle $\mathcal{E}$ on a surface S with canonical divisor K_S as it was stated in (1.5):

$$\chi(\mathcal{E}) = \operatorname{rank}(\mathcal{E})\chi(\mathcal{O}_S) + \frac{c_1(\mathcal{E})(c_1(\mathcal{E}) - K_S)}{2} - c_2(\mathcal{E}). \tag{5.1}$$

Proposition 5.1.1. *Let $(S, \mathcal{O}_S(H))$ be a smooth projective surface and let $\mathcal{E}$ be a vector bundle on S. The following two conditions are equivalent:*

(1) *$\mathcal{E}$ is an Ulrich bundle.*

(2) $\mathrm{h}^0(\mathcal{E}(-1)) = \mathrm{h}^2(\mathcal{E}(-2)) = 0$ *and the following two equalities hold:*

$$\begin{aligned} c_1(\mathcal{E})H &= \frac{\operatorname{rank}(\mathcal{E})}{2}(3H^2 + HK_S) \quad \textit{and} \\ c_2(\mathcal{E}) &= \frac{1}{2}(c_1(\mathcal{E})^2 - c_1(\mathcal{E})K_S) - \operatorname{rank}(\mathcal{E})(H^2 - \chi(\mathcal{O}_S)). \end{aligned} \tag{5.2}$$

Proof. Let us see first that condition (1) implies (2). If $\mathcal{E}$ is an Ulrich bundle on S, by Theorem 3.2.9, we have the two cohomological vanishings and $\chi(\mathcal{E}(-1)) = \chi(\mathcal{E}(-2)) = 0$. So, using Riemann–Roch theorem for surfaces, we can set two equations:

$$\begin{aligned} 0 &= \operatorname{rank}(\mathcal{E})\chi(\mathcal{O}_S) + \frac{c_1(\mathcal{E}(-1))(c_1(\mathcal{E}(-1)) - K_S)}{2} - c_2(\mathcal{E}(-1)), \\ 0 &= \operatorname{rank}(\mathcal{E})\chi(\mathcal{O}_S) + \frac{c_1(\mathcal{E}(-2))(c_1(\mathcal{E}(-2)) - K_S)}{2} - c_2(\mathcal{E}(-2)). \end{aligned}$$

Some easy computations using Example 1.4.2 show that they are equivalent to the two conditions (5.2).

Let us suppose now that condition (2) holds. Again, (5.2) implies that $\chi(\mathcal{E}(-1)) = \chi(\mathcal{E}(-2)) = 0$. On the other hand, we have $\mathrm{H}^0(\mathcal{E}(-2)) = 0$. So, this allows us to obtain that $\mathrm{H}^i(\mathcal{E}(-2)) = 0$ for all i. In order to obtain the remaining cohomological vanishings, we take a general hyperplane section H and consider the short exact sequence

$$0 \longrightarrow \mathcal{E}(-1) \longrightarrow \mathcal{E} \longrightarrow \mathcal{E}_{|H} \longrightarrow 0. \tag{5.3}$$

We twist the exact sequence (5.3) by $\mathcal{O}_S(-1)$ and take the long exact sequence of global sections. We obtain $\mathrm{H}^2(\mathcal{E}(-1)) = 0$ and can conclude the proof by Proposition 3.2.7. □

Remark 5.1.2. Notice that, by Proposition 5.1.1, given a smooth surface $(S, \mathcal{O}_S(H))$, a divisor $D \in \mathrm{Pic}(S)$ and $r \geq 2$, the second Chern class $c_2(\mathcal{E})$ of a rank r Ulrich bundle $\mathcal{E}$ on $(S, \mathcal{O}_S(H))$ with first Chern class $c_1(\mathcal{E}) = D$, if it exists, is fixed. We are going to denote it by $c_2^U(r, D)$. We have

$$c_2^U(r, D) = \frac{1}{2}(D^2 - DK_S) - r(H^2 - \chi(\mathcal{O}_S)).$$

In the particular case of *special* Ulrich bundles (see Definition 3.3.5), we have the following incarnation of the previous proposition.

Proposition 5.1.3. *Let $(S, \mathcal{O}_S(H))$ be a smooth projective surface and let $\mathcal{E}$ be a rank-two vector bundle on S. Then, $\mathcal{E}$ is a special Ulrich bundle if and only if $\mathcal{E}$ is initialized,*

$$c_1(\mathcal{E}) = 3H + K_S \quad \textit{and} \quad c_2(\mathcal{E}) = \frac{5H^2 + 3HK_S}{2} + 2\chi(\mathcal{O}_S).$$

In the toolkit of techniques to construct vector bundles on projective surfaces, there are two that have been thoroughly exploited to construct Ulrich bundles. On the one hand, elementary transformations (or, more specifically, the so-called Lazarsfeld–Mukai construction) and, on the other hand, Serre correspondence based on the Cayley–Bacharach property. In the next paragraphs we are going to recall them.

Definition 5.1.4. Let $(S, \mathcal{O}_S(H))$ be a smooth projective surface, $\mathcal{L}$ a line bundle on S, and $Z \subset S$ a finite set of distinct points. We say that the couple $(\mathcal{L}, Z)$ has the *Cayley–Bacharach property* (CB for short) or that Z satisfies CB with respect to $\mathcal{L}$ if and only if for any section $s \in \mathrm{H}^0(\mathcal{L})$ and any $Z' \subset Z$ such that $\mathrm{length}(Z') = \mathrm{length}(Z) - 1$, $s_{|Z'} = 0$ implies $s_{|Z} = 0$.

Example 5.1.5. (1) The most classical examples of sets of points on a surface having the Cayley–Bacharach property are those obtained from the following statement: if C and D are two curves on $\mathbb{P}^2$ of degree m and n, respectively, intersecting on a set Z of mn points, then any curve of degree $m + n - 3$ passing through any subset of Z of $mn - 1$ points should also pass through the whole of Z. In our terminology, it means that the couple $(\mathcal{O}_{\mathbb{P}^2}(m + n - 3), Z)$ has the Cayley–Bacharach property. For a proof of this statement and its relation to Residue Theory, see [103, p. 670 and ff.].

(2) As an immediate consequence, it is possible to state Chasles' theorem. Let $C, D \subset \mathbb{P}^2$ be two cubic curves intersecting on a set of nine distinct points $Z := \{p_1, \ldots, p_9\}$. Then any cubic curve $E \subset \mathbb{P}^2$ containing eight of the nine points of Z should contain the whole of Z.

(3) Even more classically, we can recover Pascal's theorem stating that the pairs of opposite sides of a hexagon Γ meet in three collinear points if Γ is inscribed in a conic in the projective plane. Indeed, this is a degenerate version of Chasles' theorem when C and D consist of the alternate edges of the hexagon Γ.

Finite subsets $Z \subset S$ satisfying CB with respect to a line bundle $\mathcal{L}$ can be used to construct rank-two vector bundles on S.

Theorem 5.1.6. *Let $(S, \mathcal{O}_S(H))$ be a smooth projective surface. Let $Z \subset S$ be a finite set of distinct points and let $\mathcal{L}$ and $\mathcal{M}$ be two line bundles on S. Then, there exists a (nontrivial) extension of the form*

$$0 \longrightarrow \mathcal{L} \longrightarrow \mathcal{E} \longrightarrow \mathcal{I}_Z \otimes \mathcal{M} \longrightarrow 0 \tag{5.4}$$

defining a rank 2 vector bundle $\mathcal{E}$ if and only if the couple $(\mathcal{L}^\vee \otimes \mathcal{M} \otimes \omega_S, Z)$ satisfies the Cayley–Bacharach property.

We are going to divide the proof of Theorem 5.1.6 into several steps. First of all, let us consider a subset $Z' \subsetneq Z$. Then $\mathcal{I}_Z \otimes \mathcal{M} \subset \mathcal{I}_{Z'} \otimes \mathcal{M}$. Hence starting with an extension of $\mathcal{I}_{Z'} \otimes \mathcal{M}$ by $\mathcal{L}$ as in (5.4), one can pull it back in an obvious way to get an extension of $\mathcal{I}_Z \otimes \mathcal{M}$ by $\mathcal{L}$. This defines a map

$$\phi : \operatorname{Ext}^1(\mathcal{I}_{Z'} \otimes \mathcal{M}, \mathcal{L}) \longrightarrow \operatorname{Ext}^1(\mathcal{I}_Z \otimes \mathcal{M}, \mathcal{L}). \tag{5.5}$$

Under this setting, we have the following:

Proposition 5.1.7. *Under the hypothesis of Theorem 5.1.6, given an element $e \in \operatorname{Ext}^1(\mathcal{I}_Z \otimes \mathcal{M}, \mathcal{L})$, the corresponding coherent sheaf $\mathcal{E}_e$ defined as an extension (5.4) fails to be a vector bundle if and only if there is a proper subset $Z' \subsetneq Z$ such that e belongs to the image of the map ϕ in (5.5).*

Proof. Assume that $\mathcal{E} := \mathcal{E}_e$ is not a vector bundle. In any case, $\mathcal{E}$ is a torsion-free sheaf, and therefore there exists a short exact sequence

$$0 \longrightarrow \mathcal{E} \longrightarrow \mathcal{F} := \mathcal{E}^{\vee\vee} \longrightarrow \tau \longrightarrow 0, \tag{5.6}$$

with $\mathcal{F}$ being a vector bundle, since it is a reflexive sheaf with support being a smooth surface. Moreover, τ is a coherent sheaf supported on a finite set of points (those where $\mathcal{E}$ fails to be free).

The extension (5.4) defining $\mathcal{E}$ plus the short exact sequence (5.6) give rise to a diagram

$$\begin{array}{ccccccccc} & & & & 0 & & 0 & & \\ & & & & \downarrow & & \downarrow & & \\ 0 & \longrightarrow & \mathcal{L} & \longrightarrow & \mathcal{E} & \longrightarrow & \mathcal{I}_Z \otimes \mathcal{M} & \longrightarrow & 0 \\ & & \| & & \downarrow & & \downarrow & & \\ 0 & \longrightarrow & \mathcal{L} & \longrightarrow & \mathcal{F} := \mathcal{E}^{\vee\vee} & \longrightarrow & \mathcal{I}_{Z'} \otimes \mathcal{M} & \longrightarrow & 0. \\ & & & & \downarrow & & \downarrow & & \\ & & & & \tau & = & \tau & & \\ & & & & \downarrow & & \downarrow & & \\ & & & & 0 & & 0 & & \end{array} \tag{5.7}$$

The map $\mathcal{L} \longrightarrow \mathcal{F}$ defines a global section $s \in \mathrm{H}^0(\mathcal{F} \otimes \mathcal{L}^\vee)$ which, in turn, defines the extension $e' \in \mathrm{Ext}^1(\mathcal{I}_{Z'} \otimes \mathcal{M}, \mathcal{L})$ displayed on the second row in (5.7). Moreover, under the map

$$\phi : \mathrm{Ext}^1(\mathcal{I}_{Z'} \otimes \mathcal{M}, \mathcal{L}) \longrightarrow \mathrm{Ext}^1(\mathcal{I}_Z \otimes \mathcal{M}, \mathcal{L}),$$

we have that $\phi(e') = e$. Finally, since $\tau \neq 0$, we have $Z' \subsetneq Z$ and obtain the "only if" part of the proposition.

In order to prove the converse implication, suppose that there exists $Z' \subsetneq Z$ such that $\mathcal{E} := \mathcal{E}_e$ is obtained by pulling back an extension $e' \in \mathrm{Ext}^1(\mathcal{I}_{Z'} \otimes \mathcal{M}, \mathcal{L})$, that is, suppose that $\phi(e') = e$. Then we have the following set of inclusions:

$$\mathcal{E} \subset \mathcal{F}_{e'} \subset \mathcal{F} := \mathcal{F}_{e'}^{\vee\vee},$$

with $\mathcal{F}$ a vector bundle and $\mathcal{F}/\mathcal{E}$ a coherent sheaf with finite support. Since a generically injective map between two vector bundles of the same rank is either an isomorphism or drops rank along a divisor (defined locally by the determinant map), we can conclude that $\mathcal{E}$ is not a vector bundle. □

Corollary 5.1.8. *Under the hypothesis of Theorem* 5.1.6, *there is an extension class* $e \in \mathrm{Ext}^1(\mathcal{I}_Z \otimes \mathcal{M}, \mathcal{L})$ *defining a rank-two vector bundle* $\mathcal{E}_e$ *on* S *if and only if for every proper subset* $Z' \subsetneq Z$ *the induced map*

$$\phi : \mathrm{Ext}^1(\mathcal{I}_{Z'} \otimes \mathcal{M}, \mathcal{L}) \longrightarrow \mathrm{Ext}^1(\mathcal{I}_Z \otimes \mathcal{M}, \mathcal{L})$$

is not surjective.

Proof. This follows immediately from Proposition 5.1.7 taking into account that, since we are considering Z to be reduced, there is only a finite number of subsets $Z' \subset Z$. □

Proof of Theorem 5.1.6: It will be a straightforward consequence of Corollary 5.1.8 and Serre's duality under the form

$$\mathrm{Ext}^1(\mathcal{G}, \omega_S) \cong \mathrm{H}^1(\mathcal{G})^\vee$$

for a coherent sheaf $\mathcal{G}$ supported on the surface S. First of all, notice that it is enough to check Corollary 5.1.8 for subsets $Z' \subset Z$ of length $|Z'| = |Z| - 1$. So, let $Z' := Z \backslash \{s\}$ for $s \in Z$. Then, tensoring the short exact sequence

$$0 \longrightarrow \mathcal{I}_Z \longrightarrow \mathcal{I}_{Z'} \longrightarrow \mathbf{k}(s) \longrightarrow 0$$

by $\mathcal{L}^\vee \otimes \mathcal{M} \otimes \omega_S$ and considering the associated long exact sequence of cohomology, we see that $(\mathcal{L}^\vee \otimes \mathcal{M} \otimes \omega_S, Z)$ satisfies the Caylay–Bacharach property if and only if the map

$$\mathrm{H}^1(\mathcal{I}_Z \otimes \mathcal{L}^\vee \otimes \mathcal{M} \otimes \omega_S) \longrightarrow \mathrm{H}^1(\mathcal{I}_{Z'} \otimes \mathcal{L}^\vee \otimes \mathcal{M} \otimes \omega_S)$$

is not injective. In turn, by Serre's duality, this is equivalent to saying that the map

$$\mathrm{Ext}^1(\mathcal{I}_{Z'} \otimes \mathcal{M}, \mathcal{L}) \longrightarrow \mathrm{Ext}^1(\mathcal{I}_Z \otimes \mathcal{M}, \mathcal{L})$$

is not surjective. □

Example 5.1.9. (1) Let $S \subset \mathbb{P}^n$ be a smooth surface and let $s \in S$ be a point. Let $\mathcal{L}$ and $\mathcal{M}$ be two line bundles such that s is a base point of the linear system associated to the line bundle $\mathcal{L}^\vee \otimes \mathcal{M} \otimes \omega_S$. Then trivially $(\mathcal{L}^\vee \otimes \mathcal{M} \otimes \omega_S, \{s\})$ satisfies CB and therefore we can define a rank 2 vector bundle as an extension of the form

$$0 \longrightarrow \mathcal{L} \longrightarrow \mathcal{E} \longrightarrow \mathcal{I}_{\{s\}|S} \otimes \mathcal{M} \longrightarrow 0.$$

For instance, a smooth surface $S_d \subset \mathbb{P}^3$ of degree d supports aCM rank-two vector bundles $\mathcal{E}$ with $c_1(\mathcal{E}) = 3 - d$ given as extensions of the form

$$0 \longrightarrow \mathcal{O}_{S_d} \longrightarrow \mathcal{E} \longrightarrow \mathcal{I}_{\{s\}|S_d}(3-d) \longrightarrow 0$$

where $s \in S_d$ is an arbitrary point.

(2) Analogously, for two distinct points $s, t \in S \subset \mathbb{P}^n$ not separated by the linear system associated to the line bundle $\mathcal{L}^\vee \otimes \mathcal{M} \otimes \omega_S$, we have rank 2 vector bundles of the form

$$0 \longrightarrow \mathcal{L} \longrightarrow \mathcal{E} \longrightarrow \mathcal{I}_{\{s,t\}|S_d} \otimes \mathcal{M} \longrightarrow 0.$$

Again, this allows us to construct rank 2 vector bundles $\mathcal{E}$ on a smooth surface $S_d \subset \mathbb{P}^3$ of degree d with $c_1(\mathcal{E}) = 4 - d$ as extensions of the form

$$0 \longrightarrow \mathcal{O}_{S_d} \longrightarrow \mathcal{E} \longrightarrow \mathcal{I}_{\{s,t\}|S}(4-d) \longrightarrow 0.$$

From Theorem 5.1.6 and Proposition 5.1.3, we obtain a method to construct special Ulrich bundles on a projective surface $(S, \mathcal{O}_S(H))$.

Corollary 5.1.10. *Let $(S, \mathcal{O}_S(H))$ be a smooth projective surface. Let $\mathcal{O}_S(D_1)$ and $\mathcal{O}_S(D_2)$ be two line bundles and $Z \subset S$ be a finite set of distinct points such that:*
1. *$(\mathcal{O}_S(D_2 - D_1 + K_S), Z)$ has the CB property;*
2. *$h^0(\mathcal{I}_Z(D_2 - H)) = h^0(\mathcal{O}_S(D_1 - H)) = 0$; and*
3. *$\mathcal{O}_S(D_1 + D_2) \cong \mathcal{O}_S(3H + K_S)$ and $D_1D_2 + |Z| = \frac{5H^2+3HK_S}{2} + 2\chi(\mathcal{O}_S)$.*

Then, any nontrivial extension of the form

$$0 \longrightarrow \mathcal{O}_S(D_1) \longrightarrow \mathcal{E} \longrightarrow \mathcal{I}_Z \otimes \mathcal{O}_S(D_2) \longrightarrow 0$$

defines a rank 2 special Ulrich bundle $\mathcal{E}$ on $(S, \mathcal{O}_S(H))$.

Another exploited method to construct special rank 2 Ulrich bundles $\mathcal{E}$ on a surface $(S, \mathcal{O}_S(H))$ is given by using the so-called Lazarsfeld–Mukai exact sequence. Let us recall it here.

Proposition 5.1.11. *Let $(S, \mathcal{O}_S(H))$ be a smooth projective surface.*

(1) Let $C \in |3H + K_S|$ be a smooth curve on S and let $\mathcal{L}$ be a line bundle on C with

$$\deg(\mathcal{L}) = \frac{1}{2}H(5H + 3K_S) + 2\chi(\mathcal{O}_S).$$

If $\sigma_1, \sigma_2 \in \mathrm{H}^0(\mathcal{L})$ define a base point free pencil and $\mathrm{H}^1(\mathcal{L} \otimes \mathcal{O}_S(H + K_S)) = 0$ then the vector bundle $\mathcal{E}$ defined by the Lazarsfeld–Mukai exact sequence

$$0 \longrightarrow \mathcal{E}^\vee \longrightarrow \mathcal{O}_S^2 \stackrel{\sigma_1,\sigma_2}{\longrightarrow} \mathcal{L} \longrightarrow 0$$

is a special rank 2 Ulrich bundle on S.

(2) Every special rank 2 Ulrich bundle $\mathcal{E}$ on S can be obtained from a Lazarsfeld–Mukai sequence as in (1).

Proof. (1) Let us first prove that the induced morphism

$$\mathrm{H}^0(\mathcal{O}_S(H + K_S))^2 \stackrel{\sigma_1,\sigma_2}{\longrightarrow} \mathrm{H}^0(\mathcal{L} \otimes \mathcal{O}_S(H + K_S))$$

is an isomorphism. Applying Riemann–Roch theorem on S and C, we obtain

$$\chi(\mathcal{L} \otimes \mathcal{O}_S(H + K_S)) = 2\chi(\mathcal{O}_S(H + K_S))$$

and

$$\chi(\mathcal{L} \otimes \mathcal{O}_S(2H + K_S)) = 2\chi(\mathcal{O}_S(2H + K_S)).$$

The hypothesis $\mathrm{H}^1(\mathcal{L} \otimes \mathcal{O}_S(H + K_S)) = 0$ implies $\mathrm{H}^1(\mathcal{L} \otimes \mathcal{O}_S(2H + K_S)) = 0$. These two last vanishings, together with the Kodaira vanishing theorem on S, give us $2\mathrm{h}^0(\mathcal{O}_S(H + K_S)) = \mathrm{h}^0(\mathcal{L} \otimes \mathcal{O}_S(H + K_S))$, and thus it suffices to show that the map

$$\mathrm{H}^0(\mathcal{O}_S(H + K_S))^2 \stackrel{\sigma_1,\sigma_2}{\longrightarrow} \mathrm{H}^0(\mathcal{L} \otimes \mathcal{O}_S(H + K_S))$$

is an injection. Since $C \in |3H + K_S|$, there is an exact sequence

$$0 \longrightarrow \mathcal{O}_S(-2H) \longrightarrow \mathcal{O}_S(H + K_S) \longrightarrow \mathcal{O}_C(H + K_S) \longrightarrow 0,$$

from which we deduce $\mathrm{H}^0(\mathcal{O}_S(H + K_S)) \cong \mathrm{H}^0(\mathcal{O}_C(H + K_S))$.

By the base point free pencil trick [11, p. 126], there is an exact sequence

$$0 \longrightarrow \mathrm{H}^0(\mathcal{L}^\vee \otimes \mathcal{O}_S(H + K_S)) \longrightarrow \mathrm{H}^0(\mathcal{O}_C(H + K_S))^2 \stackrel{\sigma_1,\sigma_2}{\longrightarrow} \mathrm{H}^0(\mathcal{L} \otimes \mathcal{O}_S(H + K_S)).$$

By the adjunction formula, $K_C = (3H + 2K_S)_{|C}$. So our hypothesis and Serre duality give $0 = \mathrm{H}^1(C, \mathcal{L} \otimes \mathcal{O}_S(H+K_S)) = \mathrm{H}^0(C, \mathcal{L}^\vee \otimes \mathcal{O}_S(2H+K_S))$, whence $\mathrm{H}^0(C, \mathcal{L}^\vee \otimes \mathcal{O}_S(H+K_S)) = 0$ and $\mathrm{H}^0(\mathcal{O}_S(H+K_S))^2 \xrightarrow{\sigma_1,\sigma_2} \mathrm{H}^0(\mathcal{L} \otimes \mathcal{O}_S(H+K_S))$ is an injection.

We will now check that $\mathcal{E}$ is a rank 2 special Ulrich bundle. Since by construction $c_1(\mathcal{E}) = 3H + K_S$, it suffices to show that $\mathcal{E}$ is 0-regular (see Proposition 5.1.3). Twisting the Lazarsfeld–Mukai sequence by $\mathcal{O}_S(H+K_S)$, we see that $\mathrm{H}^1(\mathcal{E}^\vee(H+K_S)) = \mathrm{H}^2(\mathcal{E}^\vee(H+K_S)) = 0$, and Serre duality gives $\mathrm{H}^1(\mathcal{E}(-H)) = \mathrm{H}^0(\mathcal{E}(-H)) = 0$. Since $c_1(\mathcal{E}) = 3H+K_S$, we have $\mathcal{E}(-H) = \mathcal{E}^\vee(2H+K_S)$. We apply Serre duality once more and obtain $\mathrm{H}^2(\mathcal{E}(-2H)) = \mathrm{H}^0(\mathcal{E}^\vee(2H+K_S)) = \mathrm{H}^0(\mathcal{E}(-H)) = 0$, which allows us to conclude that $\mathcal{E}$ is a rank 2 special Ulrich bundle.

(2) Let $\mathcal{E}$ be a rank 2 special Ulrich bundle on S. Since $\mathcal{E}$ is globally generated, two general sections $\sigma_1, \sigma_2 \in \mathrm{H}^0(\mathcal{E})$ become dependent on a smooth curve $C \in |3H + K_S|$. The cokernel of the induced map $0 \longrightarrow \mathcal{E}^\vee \longrightarrow \mathcal{O}_S^2$ is a line bundle $\mathcal{L}$ on C generated by global sections and we get the Lazarsfeld–Mukai sequence

$$0 \longrightarrow \mathcal{E}^\vee \longrightarrow \mathcal{O}_S^2 \xrightarrow{\sigma_1,\sigma_2} \mathcal{L} \longrightarrow 0.$$

Since $\mathcal{E}$ is an Ulrich bundle, by Serre duality, we have $\chi(\mathcal{E}(H+K_S)) = \chi(\mathcal{E}(-H)) = 0$. So $\chi(\mathcal{L} \otimes \mathcal{O}_S(H+K_S)) = 2\chi(\mathcal{O}_S(H+K_S))$. Applying the Riemann–Roch theorem on S and C, we get $\deg(\mathcal{L}) = \frac{1}{2}H(5H+3K_S) + 2\chi(\mathcal{O}_S)$. □

In this chapter we are going to study Ulrich bundles on different kinds of projective surfaces. This study is presented according to the complexity of the surface in terms of its Kodaira dimension. Let us recall it here.

Definition 5.1.12. Let S be a smooth projective surface with canonical line bundle $\mathcal{O}_S(K_S)$. The *Kodaira dimension* $\kappa(S)$ of S is defined as follows. If for any $n \geq 1$, the linear system $|nK_S|$ is empty, we define $\kappa(S) = -\infty$. Otherwise, let $\phi_{|nK_S|}$ be the rational map from S to the projective space defined by $|nK_S|$. Then

$$\kappa(S) := \max_n \{\dim \phi_{|nK_S|}(S)\}.$$

Therefore, we have that $\kappa(S)$ can be $-\infty, 0, 1, 2$. Let us recall briefly the classification of smooth projective surfaces accordingly to the value of $\kappa(S)$. To this end, let us define $B(S)$ as the set of isomorphism classes of surfaces birationally equivalent to S. If S_1 and S_2 belong to $B(S)$, we say that S_1 *dominates* S_2 if there is a birational morphism $S_1 \longrightarrow S_2$.

Recall that for any projective surface S' and a point $s \in S'$, there exists a surface S and a morphism $\epsilon : S \longrightarrow S'$ such that:
(1) $\epsilon^{-1}(s) = E \cong \mathbb{P}^1$.
(2) ϵ defines an isomorphism between $S \setminus E$ and $S' \setminus \{s\}$.
(3) S and ϵ are uniquely determined.

Then S is called the *blow-up* of S' at s and E is called the *exceptional curve* of the blow-up. The self-intersection of E turns out to be $E^2 = -1$. We have

Theorem 5.1.13. *Let S and S_0 be surfaces and let $f : S \longrightarrow S_0$ be a birational morphism. Then,*

(1) *There is a sequence of blow-ups $\epsilon_k : S_k \longrightarrow S_{k-1}$ for $k = 1, \dots, n$ and an isomorphism $u : S \longrightarrow S_n$ such that $f = \epsilon_1 \circ \cdots \circ \epsilon_n \circ u$.*
(2) *Let $f : S \longrightarrow S_0$ be a composition of n blow-ups and an isomorphism. Denote by $NS(S)$ the Néron–Severi group of divisors of S modulo algebraic equivalence. It holds that $NS(S) \cong NS(S_0) \oplus \mathbb{Z}^n$, where $\mathbb{Z}^n$ is freely generated by the exceptional curves of the respective blow-ups.*

Proof. See [21, Chapter II]. □

In view of the previous theorem, we see that any birational morphism from a surface to itself should be an isomorphism. In particular, the relation of dominance defines an order on $B(S)$, and we can introduce the following definition.

Definition 5.1.14. A surface S is *minimal* if its class in $B(S)$ is minimal, equivalently, if every birational morphism $S \longrightarrow S'$ is an isomorphism.

Definition 5.1.15. An integral curve $E \subset S$ on a surface S such that $E \cong \mathbb{P}^1$ and $E^2 = -1$ is called a (-1)-*curve*.

An integral curve $E \subset S$ on a surface is exceptional if it is the exceptional curve of a blow-up. Hence $E \cong \mathbb{P}^1$ and $E^2 = -1$, namely, it is a (-1)-curve. Conversely, a (-1)-curve E is the exceptional curve of a blow-up thanks to *Castelnuovo's contraction theorem.* By Theorem 5.1.13, S is minimal if and only if it does not contain exceptional curves. Enriques–Kodaira classification (see [21]) shows that any minimal smooth projective surface belongs to one of the following families:

(1) $\kappa(S) = -\infty$. Then S is either $\mathbb{P}^2$ or a *geometrically ruled surface* S, which means that there exists a surjective map $\pi : S \longrightarrow C$ on a smooth curve C with fibers isomorphic to $\mathbb{P}^1$.
(2) $\kappa(S) = 0$. They fall into four classes:
    (2.1) *Abelian surfaces*, $\mathrm{h}^1(\mathcal{O}_S) = 2$ and $K_S \equiv 0$; S is a 2-dimensional torus admitting an ample divisor.
    (2.2) *Enriques surfaces*, $\mathrm{h}^1(\mathcal{O}_S) = \mathrm{h}^2(\mathcal{O}_S) = 0$ and $2K_S \equiv 0$.
    (2.3) *K3 surfaces*, $K_S \equiv 0$ and $\mathrm{h}^1(\mathcal{O}_S) = 0$.
    (2.4) *Bielliptic surfaces*, $\mathrm{h}^2(\mathcal{O}_S) = 0$ and $\mathrm{h}^1(\mathcal{O}_S) = 1$; S is isomorphic to $C_1 \times C_2/G$ where C_1 and C_2 are elliptic curves and G is a finite group of translations of C_1 acting on C_2 such that $C_2/G \cong \mathbb{P}^1$.
(3) $\kappa(S) = 1$. They are proper elliptic surfaces. An *elliptic surface* S is a surface with a surjective map towards a smooth curve such that the general fiber is a smooth connected curve of genus one. An elliptic surface satisfies $\kappa(S) \leq 1$ and there are

examples with $\kappa(S) = -\infty$ (for example, $S \cong \mathbb{P}^1 \times C$, for C an elliptic curve) and $\kappa(S) = 0$ (e. g., a smooth quartic surface $S \subset \mathbb{P}^3$ containing a line). However, the majority of elliptic surfaces have Kodaira dimension one. They are the *proper elliptic surfaces.*

(4) $\kappa(S) = 2$. They are *surfaces of general type*, for which a finer satisfactory classification is out of reach for the time being.

5.2 Minimal surfaces with $\kappa = -\infty$

In this section, we will focus our attention on the existence of Ulrich bundles on smooth minimal surfaces with $\kappa = -\infty$. We will deal with low rank Ulrich bundles on any ruled surface and, in the case of Hirzebruch surfaces, we will describe Ulrich bundles of arbitrary rank. We will end the section determining whether the d-uple Veronese embedding of $\mathbb{P}^2$ supports a rank r Ulrich bundle.

Let us start discussing the existence of low rank Ulrich bundles on ruled surfaces.

Definition 5.2.1. A *geometrically ruled surface*, or simply a *ruled surface*, is a smooth surface S together with a surjective morphism $\pi : S \longrightarrow C$ to a smooth curve C such that for every point $p \in C$, the fiber $f := \pi^{-1}(p)$ is isomorphic to $\mathbb{P}^1$ and π admits a section $\sigma : C \longrightarrow S$, that is, $\pi \circ \sigma$ is the identity on C.

Recall that if C is a smooth curve, then a vector bundle $\mathcal{E}$ on C is called *normalized* if

$$h^0(C, \mathcal{E}) > 0 \quad \text{and} \quad h^0(C, \mathcal{E}(\mathfrak{v})) = 0 \tag{5.8}$$

for any divisor $\mathfrak{v}$ on C of negative degree.

It follows from [108, Proposition V.2.8] that any ruled surface $\pi : S \longrightarrow C$ can be written as $S \cong \mathbb{P}(\mathcal{E})$ where $\mathcal{E}$ is a normalized rank-two vector bundle on C. So, from now on, while dealing with ruled surfaces, we will assume that $S \cong \mathbb{P}(\mathcal{E})$ where $\mathcal{E}$ is a normalized rank-two vector bundle on a smooth curve C of genus g. We denote by $\mathfrak{e} := \bigwedge^2 \mathcal{E}$ and define *the invariant e of S* as $e := -\deg(\mathfrak{e})$. By Nagata's theorem, we have $e \geq -g$ (see [164, Theorem 1]).

The Picard group of a ruled surface S is generated by $\pi^* \operatorname{Pic}(C)$ and by the class of any effective divisor C_0 such that $\pi_* \mathcal{O}_S(C_0) \cong \mathcal{E}$. If $\mathfrak{b}$ is a divisor on C, we will write $\mathfrak{b}f$ instead of $\pi^* \mathfrak{b}$. Thus the class of any divisor D on S can be written uniquely as $aC_0 + \mathfrak{b}f$ with $a \in \mathbb{Z}$ and $\mathfrak{b} \in \operatorname{Pic}(C)$. Keeping this notation, according to [108, Corollary V.2.18, Proposition V.2.21], we have:

(1) In case $e \geq 0$, a divisor $aC_0 + \mathfrak{b}f$ on S is ample if and only if

$$a > 0 \quad \text{and} \quad \deg(\mathfrak{b}) > ae,$$

(2) In case $e < 0$, a divisor $aC_0 + \mathfrak{b}f$ on S is ample if and only if

$$a > 0 \quad \text{and} \quad \deg(\mathfrak{b}) > \frac{ae}{2}.$$

The canonical divisor K_S on S is in the class $-2C_0 + (\mathfrak{k} + \mathfrak{e})f$, where $\mathfrak{k}$ is the canonical divisor on C and the intersection pairing on S is given by $C_0^2 = -e$, $C_0 f = 1$, and $f^2 = 0$.

Now we introduce the numerical function $d(a, g, e)$ which will play an important role in the remanning part of this section. We fix integers a, g, and e and define

$$d(a, g, e) := g - 1 + \frac{(a-1)e}{2}. \tag{5.9}$$

To compute the cohomology of line bundles on S, we apply the projection formula

$$R^i\pi_*(\mathcal{O}_S(D) \otimes \pi^*\mathcal{E}) \cong R^i\pi_*(\mathcal{O}_S(D)) \otimes \mathcal{E}$$

(see [108, Lemma V.2.4, Exercises III.8.3 and III.8.4]) and get that, for any $a \geq 0$,

$$\mathrm{h}^i(S, \mathcal{O}_S(aC_0 + \mathfrak{b}f)) = \mathrm{h}^i(C, (S^a\mathcal{E})(\mathfrak{b})) \tag{5.10}$$

where $S^a\mathcal{E}$ stands for the ath symmetric power of $\mathcal{E}$. On the other hand, since $\mathcal{E}$ is normalized, by [108, Theorem V.2.12], there is an everywhere nonzero section in $\mathrm{H}^0(C, \mathcal{E})$ defining the exact sequence

$$0 \longrightarrow \mathcal{O}_C \longrightarrow \mathcal{E} \longrightarrow \mathcal{O}_C(\mathfrak{e}) \longrightarrow 0.$$

Thus, for any $t \geq 1$, there also exists an exact sequence of the form

$$0 \longrightarrow S^{t-1}\mathcal{E} \longrightarrow S^t\mathcal{E} \longrightarrow \mathcal{O}_C(t\mathfrak{e}) \longrightarrow 0. \tag{5.11}$$

Finally, for each divisor $\mathfrak{d}$ on C, the induction on t using the long exact cohomology sequence associated to (5.11) yields

$$\mathrm{h}^0(C, (S^t\mathcal{E})(\mathfrak{d})) \geq \mathrm{h}^0(C, \mathcal{O}_C(\mathfrak{d})), \tag{5.12}$$

$$\mathrm{h}^0(C, (S^t\mathcal{E})(\mathfrak{d})) \leq \sum_{i=0}^{t} \mathrm{h}^0(C, \mathcal{O}_C(\mathfrak{d} + i\mathfrak{e})). \tag{5.13}$$

In the following result we characterize Ulrich line bundles on ruled surfaces $S \cong \mathbb{P}(\mathcal{E})$. Later on this will be used to analyze their existence on $(S, \mathcal{O}_S(H))$ according to the invariant e of S and the very ample divisor H.

Proposition 5.2.2. *Let $(S, \mathcal{O}_S(H))$ be a ruled surface with $H = aC_0 + \mathfrak{b}f$. There is an Ulrich line bundle on S if and only if $d(a, g, e) \in \mathbb{Z}$ and there exist divisors $\mathfrak{u} \in \mathrm{Pic}^{d(a,g,e)}(C)$ satisfying*

$$\mathrm{h}^0(C, (S^{a-1}\mathcal{E})(\mathfrak{u})) = 0. \tag{5.14}$$

The Ulrich line bundles on S are exactly those of the form

$$\mathcal{O}_S((2a-1)C_0 + (\mathfrak{b}+\mathfrak{u})f),$$

and their Ulrich duals

$$\mathcal{O}_S((a-1)C_0 + (2\mathfrak{b}+\mathfrak{k}+\mathfrak{e}-\mathfrak{u})f),$$

for each $\mathfrak{u}$ *on C satisfying condition* (5.14) *above. In particular, if* $(a-1)e$ *is odd, then there are no Ulrich line bundles on* $(S, \mathcal{O}_S(H))$.

Proof. Assume that S supports an Ulrich line bundle $\mathcal{L}$ and write it as $\mathcal{L} \cong \mathcal{O}_S(a_1C_0 + \mathfrak{b}_1 f) \otimes \mathcal{O}_S(H)$. By Proposition 3.3.3, its Ulrich dual $\mathcal{M} := \mathcal{O}_S(3H - 2C_0 + (\mathfrak{k}+\mathfrak{e})f) \otimes \mathcal{L}^\vee$ is also an Ulrich line bundle.

In particular, if $\mathcal{M} \cong \mathcal{O}_S(a_2C_0 + \mathfrak{b}_2 f) \otimes \mathcal{O}_S(H)$, we have

$$(a_1+a_2)C_0 + (\mathfrak{b}_1+\mathfrak{b}_2)f = H - 2C_0 + (\mathfrak{k}+\mathfrak{e})f = (a-2)C_0 + (\mathfrak{b}+\mathfrak{k}+\mathfrak{e})f.$$

Since both $\mathcal{L}$ and $\mathcal{M}$ are assumed to be Ulrich bundles on $(S, \mathcal{O}_S(H))$, it follows that $\chi(\mathcal{L}(-H)) = \chi(\mathcal{M}(-H)) = 0$. Thus, by the Riemann–Roch theorem (Theorem 1.4.3), we get

$$(a_i+1)\big(\deg(\mathfrak{b}_i) - d(a_i+1, g, e)\big) = 0, \quad i = 1,2.$$

If $\deg(\mathfrak{b}_i) = d(a_i+1, g, e)$, for $i = 1,2$, then the equality $\mathfrak{b}_1 + \mathfrak{b}_2 = \mathfrak{b}+\mathfrak{k}+\mathfrak{e}$ yields $\deg(\mathfrak{b}) = ae/2$. Thus

$$H^2 = -a^2e + 2a\deg(\mathfrak{b}) = 0,$$

contradicting the ampleness of $\mathcal{O}_S(H)$. If $a_1 = a_2 = -1$, then $a = 0$, which again contradicts the ampleness of H. Therefore, we can assume $a_1 = -1$ and $a_2 = a-1$, whence $\deg(\mathfrak{b}_2) = d(a, g, e)$ and

$$\mathcal{L} \cong \mathcal{O}_S(-C_0 + (\mathfrak{b}+\mathfrak{k}+\mathfrak{e}-\mathfrak{b}_2)f) \otimes \mathcal{O}_S(H),$$
$$\mathcal{M} \cong \mathcal{O}_S((a-1)C_0 + \mathfrak{b}_2 f) \otimes \mathcal{O}_S(H).$$

Let $D := (a-1)C_0 + \mathfrak{b}_2 f + H$, so that $\mathcal{O}_S(D) \cong \mathcal{M}$ and $\mathcal{O}_S(3H + K_S - D) \cong \mathcal{L}$. By Proposition 5.1.1, $\mathcal{O}_S(D)$ is an Ulrich line bundle if and only if

$$\begin{aligned} DH &= \frac{1}{2}(3H^2 + HK_S), \\ D^2 &= 2(H^2 - \chi(\mathcal{O}_S)) + DK_S, \end{aligned} \tag{5.15}$$

and $\mathrm{H}^0(S, \mathcal{O}_S(D-H)) = \mathrm{H}^0(S, \mathcal{O}_S(2H+K_S-D)) = 0$. The divisor D satisfies the equalities required in (5.15). On the other hand, it always holds that

$$\mathrm{h}^0(S, \mathcal{O}_S(2H + K_S - D)) = \mathrm{h}^0(S, \mathcal{L}(-H)) = 0$$

by [108, Exercises III.8.2, III.8.3, and III.8.4] since the coefficient of C_0 is negative. Therefore, since, according to (5.10),

$$h^0(S, \mathcal{O}_S(D - H)) = h^0(C, (S^{a-1}\mathcal{E}) \otimes \mathcal{O}_C(\mathfrak{b}_2)),$$

we can take $\mathfrak{u} := \mathfrak{b}_2$, and the statement follows from Proposition 5.1.1. □

Remark 5.2.3. Notice that the result in Proposition 5.2.2 does not depend on the choice of the divisor $\mathfrak{b}$. So the only restriction on $\mathfrak{b}$ comes from H being very ample and, in particular, ample. Therefore we only should impose that $\deg(\mathfrak{b}) > ae$ if $e \geq 0$ and $\deg(\mathfrak{b}) > \frac{ae}{2}$ if $e < 0$.

Example 5.2.4. (1) When $g = 0$, $e = 0$, and $H = C_0 + f$, we are dealing with the case of the quadric surface $Q := \mathbb{P}^1 \times \mathbb{P}^1 \subset \mathbb{P}^3$. In this case, we saw in Theorem 2.3.2 that the only two Ulrich bundles are $\mathcal{O}_Q(C_0) = \mathcal{O}_Q(1, 0)$ and $\mathcal{O}_Q(f) = \mathcal{O}_Q(0, 1)$, in agreement with Proposition 5.2.2.

(2) When $g = 0$, $e = 1$, and $H = C_0 + 2f$, we are dealing with the case of the cubic scroll $S \subset \mathbb{P}^4$, namely, the rational normal scroll of degree 3. It was proved in [85] that this is one of the varieties of *finite representation type* (see Definition 2.1.14), or, in other words, it only supports a finite number of nonisomorphic, indecomposable aCM bundles. The only Ulrich line bundles are in correspondence with the two divisors corresponding to lines inside S. To see this, consider first the lines from the linear system $|f|$. Then $\mathcal{I}_{f|S}$ is an aCM line bundle and $h^0(\mathcal{I}_{f|S}(H)) = 3$ so $\mathcal{I}_{f|S}(H) = \mathcal{O}_S(C_0+f)$ is an Ulrich line bundle. The dual Ulrich line bundle $\mathcal{O}_S(2f)$, according to Proposition 5.2.2, corresponds to the ideal sheaf (twisted by H) of the exceptional line C_0 (since $C_0 \cong \mathbb{P}^1$ and $C_0^2 = -1$). Notice, however, that this is the only case of ruled surface that is not minimal.

Using this characterization, we will get the first result concerning the existence of Ulrich line bundles on ruled surfaces. To prove it, we need to recall the notion of theta divisors.

Denote by $M_C^{ss}(r; d)$ the moduli space of rank r semistable vector bundles $\mathcal{E}$ on C of degree $\deg(\mathcal{E}) = d$.

Definition 5.2.5. Let $\mathcal{F}$ be a rank r vector bundle of degree d on a smooth curve C. Set $j = \gcd(d, r)$ and write $d = jd_1$, $r = jr_1$. We define

$$\Theta_{\mathcal{F}} = \{\mathcal{F}_1 \in M_C^{ss}(r_1; r_1(g-1) - d_1) \mid h^0(C, \mathcal{F} \otimes \mathcal{F}_1) > 0\}.$$

If $\Theta_{\mathcal{F}}$ is a divisor on $M_C^{ss}(r_1; r_1(g-1) - d_1)$, we call it a *theta divisor*.

Lemma 5.2.6. *For a generic semistable vector bundle $\mathcal{F}$ of rank r and degree d, $\Theta_{\mathcal{F}}$ is a divisor on $M_C^{ss}(r_1; r_1(g-1) - d_1)$.*

Proof. See [182, Proposition 1.8.1]. □

We have seen that for a generic semistable vector bundle $\mathcal{F}$, $\Theta_{\mathcal{F}}$ is a divisor of the corresponding moduli space. Without the hypothesis that $\mathcal{F}$ is generic, the result is not necessarily true. In fact, it has been shown that for some values of r and d, there exist vector bundles, sometimes even infinite families of vector bundles $\mathcal{F}$, for which $\Theta_{\mathcal{F}} = M_C^{ss}(r_1; r_1(g-1) - d_1)$ (see [176]).

Theorem 5.2.7. *Let $(S, \mathcal{O}_S(H))$ be a ruled surface with $H = aC_0 + \mathfrak{b}f$.*

(1) If $e > 0$, then there are Ulrich line bundles on $(S, \mathcal{O}_S(H))$ if and only if $a = 1$. In this case, there exist exactly two families of dimension g of Ulrich line bundles on S. For any generic divisor $\mathfrak{u}$ of degree $g-1$,

$$\mathcal{O}_S(C_0 + (\mathfrak{b} + \mathfrak{u})f)$$

and its Ulrich dual are Ulrich line bundles.

(2) If $(a-1)e = 0$, then there also exist two families of dimension g of Ulrich line bundles.

Proof. (1) First of all, notice that since H is an ample divisor, $a \geq 1$. Assume that there exists an Ulrich line bundle on $(S, \mathcal{O}_S(H))$. By Proposition 5.2.2, there exists a divisor $\mathfrak{u} \in \operatorname{Pic}^{d(a,g,e)}(C)$ satisfying

$$h^0(C, (S^{a-1}\mathcal{E})(\mathfrak{u})) = 0. \tag{5.16}$$

On the other hand, if $e > 0$ and $a > 1$, by Riemann–Roch theorem (Theorem 1.4.3), for any divisor $\mathfrak{u} \in \operatorname{Pic}^{d(a,g,e)}(C)$,

$$\chi(\mathcal{O}_C(\mathfrak{u})) = \deg(\mathfrak{u}) - g + 1 > 0,$$

which contradicts the fact that by (5.12), $h^0(C, \mathcal{O}_C(\mathfrak{u})) \leq h^0(C, (S^{a-1}\mathcal{E})(\mathfrak{u})) = 0$. Hence, if there exists an Ulrich line bundle on $(S, \mathcal{O}_S(H))$, then necessarily $a = 1$. Thus, $d(a,g,e) = g - 1 \in \mathbb{Z}$ and since there are divisors $\mathfrak{u} \in \operatorname{Pic}(C)$ of degree $g-1$ such that $h^0(C, \mathcal{O}_C(\mathfrak{u})) = 0$, by Proposition 5.2.2, there are exactly two Ulrich line bundles as in the statement.

(2) Since $(a-1)e = 0$, we have $d(a,g,e) = g - 1$. Thus, the set $\mathfrak{U} \subseteq \operatorname{Pic}^{d(a,g,e)}(C)$ of line bundles $\mathcal{O}_S(\mathfrak{u})$ such that $h^0(C, \mathcal{O}_C(\mathfrak{u})) = 0$ is open and nonempty because it is the complement of the theta divisor of the trivial line bundle. Moreover, $\dim \mathfrak{U} = \dim \operatorname{Pic}^{g-1}(C) = g$.

In particular, if Ulrich line bundles $\mathcal{U}$ on S exist, their characterization in Proposition 5.2.2 means that they form two families according to whether $c_1(\mathcal{U})f$ is $2a-1$ or $a-1$. Both families have dimension $\dim \mathfrak{U} = g$.

In the rest of the proof, we show that a general divisor on $\mathfrak{U}$ satisfies condition (5.14).

The case $a = 1$ holds due to the nonemptiness of $\mathfrak{U}$. So, we can restrict ourselves to the case $a \geq 2$ and $e = 0$. In this case, if $g = 0$, then $S \cong \mathbb{P}^1 \times \mathbb{P}^1$ and any divisor on C of degree −1 verifies the condition stated in Proposition 5.2.2, so we are done. Therefore, it only remains to deal with the case $a \geq 2$, $e = 0$ and $g \geq 1$.

Notice that for each $i \geq 0$, we have $\deg(\mathfrak{u} + i\mathfrak{e}) = d(a, g, 0) = g - 1$. Thus, for all $0 \leq i \leq a-1$ and for each $\mathfrak{u}$ with $\mathcal{O}_S(\mathfrak{u} + i\mathfrak{e}) \in \mathfrak{U} = \mathrm{Pic}^{g-1}(C) \setminus \Theta_{\mathcal{O}_C}$,

$$\mathrm{h}^0(C, \mathcal{O}_C(\mathfrak{u} + i\mathfrak{e})) = 0.$$

Inequality (5.13) yields $\mathrm{h}^0(C, (S^{a-1}\mathcal{E})(\mathfrak{u})) = 0$ for such a $\mathfrak{u}$. Thus, once more, the statement follows from Proposition 5.2.2. □

Remark 5.2.8. The above theorem illustrates once more that the property of being Ulrich strongly depends on the very ample line bundle that is fixed (see also Example 3.2.3). We will come back to this point in Section 5.6.

When $a \geq 2$ and $e < 0$, the picture is much more intricate. In order to prove the existence of Ulrich line bundles in this setting, we will relate their existence to the existence of suitable theta divisors. In fact, in view of Proposition 5.2.2, we characterize the existence of Ulrich line bundles in terms of the existence of theta divisors.

Proposition 5.2.9. *Let $(S, \mathcal{O}_S(H))$ be a ruled surface with $H = aC_0 + \mathfrak{b}f$. There is an Ulrich line bundle on S if and only if $d(a, g, e) \in \mathbb{Z}$ and $\Theta_{S^{a-1}\mathcal{E}}$ is a proper divisor on $\mathrm{Pic}^{d(a,g,e)}(C)$.*

Proof. By definition, $d(a, g, e) \in \mathbb{Z}$ if and only if $(a-1)e$ is even. Since the rank r and the degree d of $S^{a-1}\mathcal{E}$ are $r = a$ and $d = -a(a-1)e/2$, respectively, we have

$$\Theta_{S^{a-1}\mathcal{E}} = \{\mathcal{L} \in \mathrm{Pic}^{d(a,g,e)}(C) \mid \mathrm{h}^0(C, (S^{a-1}\mathcal{E}) \otimes \mathcal{L}) > 0\}.$$

Then, the statement follows from Proposition 5.2.2. □

Therefore, our next goal is to study the existence of theta divisors of symmetric powers of normalized rank two vector bundles $\mathcal{E}$ on a curve C.

Lemma 5.2.10. *Let C be a smooth curve, $\mathcal{E}$ a rank-two vector bundle on C and $\mathcal{L}_1$ a line subbundle of maximal degree of $\mathcal{E}$. Then, $\mathcal{E} \otimes \mathcal{L}_1^{-1}$ is normalized.*

Proof. By assumption, there exists an injective map $0 \to \mathcal{L}_1 \to \mathcal{E}$ and therefore also a map $0 \to \mathcal{O}_C \to \mathcal{E} \otimes \mathcal{L}_1^{-1}$. Hence, $\mathrm{h}^0(\mathcal{E} \otimes \mathcal{L}_1^{-1}) > 0$. Assume now that there is a line bundle $\mathcal{L}$ of negative degree such that $\mathrm{h}^0(\mathcal{E} \otimes \mathcal{L}_1^{-1} \otimes \mathcal{L}) > 0$. Then $\mathcal{L}_1 \otimes \mathcal{L}^{-1}$ is a subsheaf of $\mathcal{E}$ of degree higher than the degree of $\mathcal{L}_1$ contradicting the assumption. □

Proposition 5.2.11. *Let C be a smooth curve. Fix integers d, r, and r', with $0 < r' < r$. For a rank r vector bundle $\mathcal{E}$ of degree d on C, we define*

$$s_{r'}(\mathcal{E}) := r'd - r\max\{\deg(\mathcal{E}') \mid \mathrm{rank}(\mathcal{E}') = r', \mathcal{E}' \subset \mathcal{E}\}.$$

For a fixed s with $0 < s \leq r'(r-r')(g-1)$, $s \equiv r'd\ (r)$, we define

$$M_{r',s}(r; d) = \{\mathcal{E} \in M_C^{ss}(r; d) \mid s_{r'}(\mathcal{E}) = s\}.$$

Then $M_{r',s}(r;d)$ is nonempty, irreducible of dimension $r^2(g-1)+1+s-r'(r-r')(g-1)$ and $M_{r',s}(r;d) \subset \overline{M_{r',s+r}(r;d)}$. Moreover, a generic $\mathcal{E} \in M_{r',s}(r;d)$ is given by an exact sequence

$$0 \longrightarrow \mathcal{E}' \longrightarrow \mathcal{E} \longrightarrow \mathcal{E}'' \longrightarrow 0$$

with both $\mathcal{E}'$ and $\mathcal{E}''$ stable and $\mathcal{E}'$ being the unique subbundle of $\mathcal{E}$ of rank r' and degree $d' := \frac{r'd-s}{r}$.

Proof. If $g = 0, 1$, there are no integers s satisfying the above restrictions. For $g \geq 2$, see [188, Theorem 0.1 and Corollary 1.12]. □

Corollary 5.2.12. *Let C be a smooth curve of genus g and $\mathcal{E}$ a rank-two vector bundle of degree d on C. If $\mathcal{E} \in M_{1,s}(2;d)$, then there exists a line bundle $\mathcal{L}$ on C of degree $\frac{d-s}{2}$ such that $\bar{\mathcal{E}} = \mathcal{E} \otimes \mathcal{L}^{-1}$ is normalized. In particular,*

$$0 < s = \deg(\bar{\mathcal{E}}) \leq g.$$

Proof. The statement follows from Lemma 5.2.10 and Proposition 5.2.11. □

Proposition 5.2.13. *Let C be a smooth curve of genus g, $\mathcal{F}$ a rank r vector bundle of degree d, and $\mathcal{L}$ a line bundle of degree d' on C. With the notations above, $\Theta_{\mathcal{F}}$ is a proper divisor of $M_C^{ss}(r_1; r_1(g-1)-d_1)$ if and only if $\Theta_{\mathcal{F}\otimes\mathcal{L}}$ is a proper divisor of $M_C^{ss}(r_1; r_1(g-1-d')-d_1)$.*

Proof. The map

$$\begin{array}{ccc} M_C^{ss}(r_1; r_1(g-1)-d_1) & \longrightarrow & M_C^{ss}(r_1; r_1(g-1-d')-d_1), \\ \mathcal{F}' & \mapsto & \mathcal{F}' \otimes \mathcal{L}^{-1} \end{array}$$

gives a bijection between the two moduli spaces (with inverse $\mathcal{F}'' \to \mathcal{F}'' \otimes \mathcal{L}$). From the definition of the theta locus, under this map, $\Theta_{\mathcal{F}}$ maps to $\Theta_{\mathcal{F}\otimes\mathcal{L}}$. So, one locus is a divisor if and only if the other is. □

Corollary 5.2.14. *Let C be a smooth curve of genus g, $\mathcal{E}$ a rank-two vector bundle of degree d on C, $\mathcal{L}$ a line bundle of degree d' on C, and β a positive integer.*

1. *If βd is even, $\Theta_{S^\beta\mathcal{E}}$ is a proper divisor of $\operatorname{Pic}^{g-1-\frac{\beta d}{2}}(C)$ if and only if $\Theta_{S^\beta(\mathcal{E}\otimes\mathcal{L})}$ is a proper divisor of $\operatorname{Pic}^{g-1-\frac{\beta d}{2}-\beta d'}(C)$.*
2. *If βd is odd, $\Theta_{S^\beta\mathcal{E}}$ is a proper divisor of $M_C^{ss}(2; 2(g-1)-\beta d)$ if and only if $\Theta_{S^\beta(\mathcal{E}\otimes\mathcal{L})}$ is a proper divisor of $M_C^{ss}(2; 2(g-1)-\beta d-2\beta d')$.*

Proof. This follows from Proposition 5.2.13 since $S^\beta(\mathcal{E}\otimes\mathcal{L}) = (S^\beta\mathcal{E})\otimes\mathcal{L}^\beta$ and $M_C^{ss}(1;d) = \operatorname{Pic}^{g-1-\frac{\beta d}{2}-\beta d'}(C)$. □

Now we are ready to state the first main result concerning the existence of proper theta divisors. To this end, we first consider a normalized rank-two vector bundle $\mathcal{E}$ of even degree $\deg(\mathcal{E}) = -e = 2f$ on a smooth curve C and an integer $a \geq 2$. In particular, the slope of $\mathcal{E}$ is f. The bundle $\mathcal{F} = S^{a-1}\mathcal{E}$ has rank a and slope $(a-1)f$. Therefore,

$$\Theta_{S^{a-1}\mathcal{E}} = \{\mathcal{L} \in \mathrm{Pic}^{d(a,g,e)}(C) \mid \mathrm{h}^0(C, (S^{a-1}\mathcal{E}) \otimes \mathcal{L}) > 0\}.$$

Proposition 5.2.15. *Let C be a smooth curve of genus g. Fix an even degree $-e = 2f$ and an integer s, $0 < s \leq g$. Let $\mathcal{E}$ be a normalized rank 2 vector bundle, generic in the stratum $M_{1,s}(2; 2f)$. Then, for all $a \geq 2$, $\Theta_{S^{a-1}\mathcal{E}}$ is a proper divisor of $\mathrm{Pic}^{d(a,g,e)}(C)$.*

Proof. The condition that $\Theta_{S^{a-1}\mathcal{E}}$ is a proper divisor is an open condition in the moduli space of vector bundles. Using Proposition 5.2.11, it suffices to prove the result for the smallest stratum $M_{1,2}(2; 2f)$ of the moduli space of vector bundles corresponding to those bundles with subbundles of the largest degree. Applying Proposition 5.2.11, these vector bundles are extensions of two line bundles of degree $f-1$ and $f+1$, respectively. They can be deformed to strictly semistable rank-two bundles and, in particular, to a direct sum $\mathcal{L}_1 \oplus \mathcal{L}_2$ of two generic line bundles of the same degree f. Then $S^{a-1}\mathcal{E}$ is the set of tensors in $(\mathcal{L}_1 \oplus \mathcal{L}_2)^{\otimes(a-1)}$ that are invariant under the action of the symmetric group. The set

$$\{\mathcal{L} \in \mathrm{Pic}^{d(a,g,e)}(C) \mid \mathrm{h}^0(C, \mathcal{L}_1^k \otimes \mathcal{L}_2^{a-1-k} \otimes \mathcal{L}) > 0\}$$

is a theta divisor on $\mathrm{Pic}^{d(a,g,e)}(C)$. Moreover, $\Theta_{S^{a-1}\mathcal{E}}$ is contained in the union of these theta divisors as k varies. Thus, it is still a divisor. □

Let us now consider a normalized rank two vector bundle $\mathcal{E}$ of odd degree $\deg(\mathcal{E}) = -e = 2f + 1$ on a curve C and an integer $a \geq 2$.

Remark 5.2.16. Notice that on any smooth curve C, normalized rank-two vector bundles $\mathcal{E}$ of $\deg(\mathcal{E}) > 0$ are stable. By a generic normalized rank-two vector bundle of positive degree on C we mean a vector bundle on an open subset of the corresponding moduli space. Similarly, a generic curve of genus g is a curve in an open set of the moduli space parameterizing curves of genus g.

Proposition 5.2.17. *Let C be a generic curve of genus g and let $\mathcal{E}$ be a generic normalized rank two vector bundle of degree $-e = 2f + 1$ on C. Then, for odd $a > 2$, $\Theta_{S^{a-1}\mathcal{E}}$ is a proper divisor of $\mathrm{Pic}^{d(a,g,e)}(C)$ and for even $a \geq 2$, $\Theta_{S^{a-1}\mathcal{E}}$ is a proper divisor of the moduli space $M_C^{ss}(2; 2g - 2 - (a-1)(2f+1))$.*

Proof. We start by deforming the curve C to a chain C_0 of elliptic curves as follows. Consider $C_1, \dots, C_g$ generic elliptic curves. Let P_i, Q_i be generic points on C_i. We construct a curve C_0 of arithmetic genus g by identifying P_i with Q_{i-1}, $i = 2, \dots, g$. Let us now determine a generic normalized rank-two vector bundle $\mathcal{E}_0$ of degree $2f+1$ on C_0. To this end, take a generic indecomposable rank-two vector bundle $\mathcal{E}_1$ of degree 1 on C_1, a direct sum of two generic line bundles of degree one $\mathcal{L}_1^i \oplus \mathcal{L}_1'^i$ on C_i, $i = 3, 5, \dots, 2f+1$; and a direct sum of two generic line bundles of degree zero $\mathcal{L}_0^i \oplus \mathcal{L}_0'^i$ on the remaining components C_i, $i = 2, 4, 6 \ldots 2f$ and $i = 2f+2, 2f+3, \ldots, g$. We take the gluing so that $\mathcal{L}_0^i$, $i = 2f+2, 2f+3, \ldots, g$ glue to each other but the gluing are generic otherwise. One can then find a degree zero line subbundle $\mathcal{L}_0^i$ of $\mathcal{L}_1^i \oplus \mathcal{L}_1'^i$ on C_i, $i = 3, 5, \ldots, 2f+1$ that glues with both $\mathcal{L}_0^{i-1}$ and $\mathcal{L}_0^{i+1}$ and a degree zero line subbundle of $\mathcal{E}_1$ that glues with $\mathcal{L}_0^2$. Gluing these degree zero line subbundles on each component, we get a degree zero line subbundle of the vector bundle $\mathcal{E}_0$ on the chain. One can, in fact, check that the largest degree of a subbundle of $\mathcal{E}_0$ on C_0 is zero (see [196] for details). Hence, $\mathcal{E}_0$ is a generic normalized rank-two vector bundle on C_0 of degree $2f+1$.

In order to show that the theta divisor of the symmetric power of $\mathcal{E}_0$ is a divisor, it suffices to find a line bundle $\mathcal{L}_0$ on C_0 (respectively vector bundle $\mathcal{F}_0$) of the appropriate degree such that $(S^{a-1}\mathcal{E}_0) \otimes \mathcal{L}_0$ (respectively $(S^{a-1}\mathcal{E}_0) \otimes \mathcal{F}_0$) does not have any limit linear section.

Notice that the vector bundle $\mathcal{E}_0$ that we built on C_0 has restriction to C_i, $i = 2, \ldots, g$ a direct sum of two generic line bundles $\mathcal{E}_i = \mathcal{L}_j^i \oplus \mathcal{L}_j'^i$ of the same degree j, $j = 0, 1$ and has restriction to C_1 a generic indecomposable rank-two vector bundle $\mathcal{E}_1$ of degree 1. The $(a-1)$-symmetric power of $\mathcal{E}_i$, $i > 1$ is a subsheaf of $\bigoplus_{m+l=a-1} (\mathcal{L}_j^i)^m \otimes (\mathcal{L}_j'^i)^l$. The $(a-1)$-symmetric power of $\mathcal{E}_1$ is a subsheaf of the $(a-1)$-tensor power of $\mathcal{E}_1$. Since $\mathcal{E}_1$ is a rank-two vector bundle, $\mathcal{E}_1^\vee \cong \mathcal{E}_1 \otimes \bar{\mathcal{L}}^{-1}$ where $\bar{\mathcal{L}} = \det(\mathcal{E}_1)$. On the other hand, it follows from [14, Lemma 22] that $\mathcal{E}_1 \otimes \mathcal{E}_1^\vee = \bigoplus_{i=0,\ldots,3} \mathcal{M}_i$ where the $\mathcal{M}_i$ are the elements in C_1 of order 2.

Therefore,

$$\otimes^{a-1} \mathcal{E}_1 = \begin{cases} (\bigoplus_{i=0,\ldots,3} \mathcal{M}_i)^{\otimes k} \otimes \bar{\mathcal{L}}^k & \text{if } a = 2k+1, \\ (\bigoplus_{i=0,\ldots,3} \mathcal{M}_i)^{\otimes k} \otimes \mathcal{E}_1 \otimes \bar{\mathcal{L}}^k & \text{if } a = 2(k+1). \end{cases} \tag{5.17}$$

A limit linear series of slope $g-1$ of the tensor product of $S^{a-1}\mathcal{E}_0$ with an arbitrary line bundle $\mathcal{L}_0$ of degree $g - 1 - \frac{(2f+1)(a-1)}{2}$ (respectively an arbitrary rank-two vector bundle $\mathcal{F}_0$ of degree $2(g-1) - (2f+1)(a-1)$) would give rise to a section on each component with proper vanishing at the nodes.

Assume first that a is odd. According to (5.17) and the construction of $\mathcal{E}_0$,

$$(\otimes^{a-1}\mathcal{E}_0 \otimes \mathcal{L}_0)_{|C_i} = \begin{cases} (\bigoplus_{i=0,\ldots,3} \mathcal{M}_i)^{\otimes k} \otimes \bar{\mathcal{L}}^k \otimes \mathcal{L}_{01} & \text{if } i = 1, \\ \bigoplus_{m+l=a-1} (\mathcal{L}_j^i)^m \otimes (\mathcal{L}_j'^i)^l \otimes \mathcal{L}_{0i} & \text{if } i > 1, \end{cases}$$

where $\mathcal{L}_{0i}$ denotes the restriction of $\mathcal{L}_0$ to C_i.

On the other hand, the restriction $\mathcal{L}_{01}$ has degree $g-1-\frac{a-1}{2}$ on C_1, the restriction $\mathcal{L}_{0i}$ has degree $g-a$ on the components $C_3, C_5, \ldots, C_{2f+1}$ and degree $g-1$ on the remaining components $C_2, C_4, C_6, \ldots, C_{2f}$ and C_{2f+2}, C_{2f+3}, C_g. Moreover, it is generic of the stated degree on each component. Notice also that $\mathcal{L}_j^i, \mathcal{L}_j'^i, \bar{\mathcal{L}}$ are fixed and determined by the generic choice of $\mathcal{E}_0$. In addition, the line bundles $\mathcal{M}_i$ form a subgroup, so their product is another element $\mathcal{M}_j$ in this subgroup and their degree is zero. Hence, the degree of $\bar{\mathcal{L}}^k \otimes \mathcal{L}_{01}$ and of $\mathcal{L}_j^i \otimes \mathcal{L}_j'^i \otimes \mathcal{L}_{0i}$ is $g-1$. By the generality of $\mathcal{L}_{0i}$, the sum of the orders of vanishing of a section of the line bundles $\mathcal{M}_i \otimes \bar{\mathcal{L}}^k \otimes \mathcal{L}_{01}, \mathcal{L}_j^i \otimes \mathcal{L}_j'^i \otimes \mathcal{L}_{0i}$ at the two nodes of the elliptic curve is at most $g-2$. In order to have a limit linear series, the order of vanishing at Q_i of the sections on the component C_i and the order of vanishing at P_{i+1} of the sections on the component C_{i+1} needs to be at least $g-1$. Therefore, the sum O of vanishing orders at the nodes satisfies

$$(g-1)^2 \le O \le (g-2)g,$$

which is impossible.

Consider now the case in which a is even and then $\mathcal{F}_0$ is an arbitrary rank-two vector bundle of degree $2g-2-(2f+1)(a-1)$. We take $\mathcal{F}_{0,1}$ on C_1 to be a generic vector bundle of degree $2g-1-a$, $\mathcal{F}_{0i}$ the direct sum of two generic line bundles of degree $g-a$ on the components $C_3, C_5, \dots, C_{2f+1}$ and the direct sum of two generic line bundles of degree $g-1$ on the remaining components $C_2, C_4, C_6, \dots, C_{2f}$ and on $C_{2f+2}, C_{2f+3}, \dots, C_g$. Any two indecomposable vector bundles of rank two and odd degree differ by a product with a line bundle (see [14, Corollary to Theorem 7]). Therefore, there exists a line bundle $\mathcal{L}_{0,1}$ of degree $g-1$ such that $\mathcal{F}_{0,1} = \mathcal{E}_{0,1} \otimes \mathcal{L}_{0,1}$ Thus, we have:

$$(\otimes^{a-1}\mathcal{E}_0 \otimes \mathcal{F}_0)_{|C_i} = \begin{cases} (\bigoplus_{i=0,\dots,3} \mathcal{M}_i)^{\otimes k+1} \otimes \bar{\mathcal{L}}^k \otimes \mathcal{L}_{01} & \text{if } i = 1, \\ \bigoplus_{m+l=a-1} (\mathcal{L}_j^i)^m \otimes (\mathcal{L}_j'^i)^l \otimes (\mathcal{L}_{0i} \oplus \mathcal{L}_{0i}') & \text{if } i > 1. \end{cases}$$

The same argument as before shows that this cannot have a limit section. □

Theorem 5.2.18. *Let $(S, \mathcal{O}_S(H))$ be a ruled surface over C, $S \cong \mathbb{P}(\mathcal{E})$, with $\mathcal{E}$ generic among those of degree $e < 0$ and write $H = aC_0 + \mathfrak{b}f$. It holds:*
1. *If e is even, then there is an Ulrich line bundle on $(S, \mathcal{O}_S(H))$.*
2. *If e is odd, a is odd (which implies that $d(a, g, e) \in \mathbb{Z}$) and C is a generic smooth curve, then there is an Ulrich line bundle on $(S, \mathcal{O}_S(H))$.*

Proof. (1) From Proposition 5.2.9, we need to check that $\Theta_{S^{a-1}\mathcal{E}}$ is a proper divisor. From Corollary 5.2.12, it suffices to check it for a generic point of $M_{1,2}(2; 2f)$ where $-e = 2f$. From Proposition 5.2.15, this holds.

(2) The result follows from Proposition 5.2.9 and Proposition 5.2.17. □

Remark 5.2.19. Because the proof we gave for Proposition 5.2.17 uses a deformation argument to a special kind of curve, we cannot conclude that Theorem 5.2.18 (2) is true for *every* curve. It is likely though that, as in the case of even degree, the result holds for every curve.

As a consequence, we have the following result:

Corollary 5.2.20. *Let a, g, and e be integers such that $-g \le e < 0$, $a \ge 2$, and $\frac{(a-1)e}{2} \in \mathbb{Z}$. Then, there exists a geometrically ruled surface $S \longrightarrow C$ over a curve C of genus g with invariant e such that $(S, \mathcal{O}_S(aC_0 + \mathfrak{b}f))$ supports an Ulrich line bundle.*

Proof. The claim follows from Theorem 5.2.18. □

Now we will turn our attention to the existence of special rank-two Ulrich bundles. Recall that, according to Definition 3.3.5, a rank-two Ulrich bundle $\mathcal{E}$ on a surface $(S, \mathcal{O}_S(H))$ is *special* if and only if $c_1(\mathcal{E}) = 3H + K_S$ where, as usual, K_S denotes the canonical divisor on S. Since later on we will analyze the existence of rank $r \geq 2$ Ulrich bundles on Hirzebruch surfaces (that is, surfaces ruled over $\mathbb{P}^1$), now we will assume $g \geq 1$.

Proposition 5.2.21. *Let $(S, \mathcal{O}_S(H))$ be a ruled surface over a smooth curve of genus $g \geq 1$, invariant e and $H = C_0 + \mathfrak{b}f$. Then, there exists a family of dimension $2\deg(\mathfrak{b}) - e + 3g - 3$ of rank-two simple strictly semistable special Ulrich bundles on S.*

Proof. It follows from Proposition 5.2.2 and Theorem 5.2.7 that for a general divisor $\mathfrak{u} \in \mathrm{Pic}^{g-1}(C)$,

$$\mathcal{L}_1 = \mathcal{O}_S((2\mathfrak{b} + \mathfrak{k} + \mathfrak{e} - \mathfrak{u})f) \quad \text{and} \quad \mathcal{L}_2 = \mathcal{O}_S(C_0 + (\mathfrak{b} + \mathfrak{u})f)$$

are Ulrich line bundles on $(S, \mathcal{O}_S(H))$. Moreover, since $S \cong \mathbb{P}(\mathcal{E})$ with $\mathcal{E}$ normalized, by (5.10) and Serre's duality, we have

$$\mathrm{H}^0(\mathcal{L}_2 \otimes \mathcal{L}_1^\vee) = \mathrm{H}^2(\mathcal{L}_2 \otimes \mathcal{L}_1^\vee) = 0,$$

and hence, by Riemann–Roch theorem (Theorem 1.4.3),

$$\mathrm{ext}^1(\mathcal{L}_1, \mathcal{L}_2) = \mathrm{h}^1(\mathcal{L}_2 \otimes \mathcal{L}_1^\vee) = -\chi(\mathcal{L}_2 \otimes \mathcal{L}_1^\vee) = 2\deg(\mathfrak{b}) - e + 2g - 2 > 0.$$

Therefore, we can consider rank-two vector bundles $\mathcal{F}$ on S given by a nontrivial extension of the following type:

$$0 \longrightarrow \mathcal{O}_S(C_0 + (\mathfrak{b} + \mathfrak{u})f) \longrightarrow \mathcal{F} \longrightarrow \mathcal{O}_S(C_0 + (2\mathfrak{b} + \mathfrak{k} + \mathfrak{e} - \mathfrak{u})f) \longrightarrow 0.$$

Since $\mathcal{L}_1$ and $\mathcal{L}_2$ are Ulrich duals, $c_1(\mathcal{F}) = c_1(\mathcal{L}_1) + c_1(\mathcal{L}_2) = 3H + K_S$ and, since $\mathcal{F}$ is an extension of Ulrich line bundles, by Proposition 3.3.1, $\mathcal{F}$ is also an Ulrich bundle. Hence, $\mathcal{F}$ is a rank-two special Ulrich bundle. Moreover, by Proposition 3.3.2 (1) it is simple and, in particular, indecomposable. On the other hand, since Ulrich bundles are semistable by Proposition 3.3.14 and the slopes of $\mathcal{L}_2$ and $\mathcal{F}$ coincide, it is strictly semistable. Finally, since $\mathrm{h}^0(\mathcal{F} \otimes \mathcal{L}_2^\vee) = 1$, we have a family of the desired dimension. □

Now we will consider ruled surfaces $(S, \mathcal{O}_S(H))$ over a curve C of genus $g \geq 1$ with $H = aC_0 + \mathfrak{b}f$, $a \geq 2$. Since we do not know if all of them support Ulrich line bundles, the strategy to get special rank-two Ulrich bundles will be different. To this end, we set

$$u_e := \begin{cases} ae & \text{if } e > 0, \\ e & \text{if } e \leq 0. \end{cases}$$

With this notation we have the following result:

Theorem 5.2.22. *Let $(S, \mathcal{O}_S(H))$ be a ruled surface over a curve of genus $g \geq 1$, invariant e and $H = aC_0 + \mathfrak{b}f$. If $a = 2$ or $a \geq 3$ and*

$$\deg(\mathfrak{b}) > \max\left\{ \frac{(a-3)(g-1)+ea}{2}, (g-1) + u_e, \frac{e(3a+1)}{6} + \frac{2g}{3} \right\},$$

then, for any general finite set $Z \subset S$ of $(a-1)(\deg(\mathfrak{b}) - ea/2)$ points and any general $\mathfrak{v} \in \operatorname{Pic}^{g-1}(C)$, there exist rank 2 μ-stable special Ulrich bundles $\mathcal{F}$ on $(S, \mathcal{O}_S(H))$ fitting into the exact sequence

$$0 \longrightarrow \mathcal{O}_S(aC_0 + (\mathfrak{b} + \mathfrak{v})f) \longrightarrow \mathcal{F} \longrightarrow \mathcal{I}_Z((2a-2)C_0 + (2\mathfrak{b} + \mathfrak{k} + \mathfrak{e} - \mathfrak{v})f) \longrightarrow 0. \quad (5.18)$$

Proof. We proceed in several steps.

Step 1. We first prove that there are vector bundles in the extension (5.18). To this end, we check that the couple

$$(\mathcal{O}_S(K_S + (a-2)C_0 + (\mathfrak{k} + \mathfrak{e})f), Z)$$

verifies the Cayley–Bacharach property (see Definition 5.1.4). If $a = 2, 3$, then $h^0(\mathcal{O}_S(K_S + (a-2)C_0 + (\mathfrak{k} + \mathfrak{e})f)) = 0$, and the Cayley–Bacharach property is automatically satisfied. If $a \geq 4$, the divisor $(a-2)C_0 + (\mathfrak{k} + \mathfrak{e})f$ is ample. Hence, by the Kodaira vanishing theorem,

$$\begin{aligned} h^0(\mathcal{O}_S(K_S + (a-2)C_0 + (\mathfrak{k} + \mathfrak{e})f)) &= \chi(\mathcal{O}_S(K_S + (a-2)C_0 + (\mathfrak{k} + \mathfrak{e})f)) \\ &= (a-3)\left(g - 1 + \deg(\mathfrak{b}) - \frac{ea}{2}\right). \end{aligned}$$

Thus, the hypothesis $2\deg(\mathfrak{b}) > (\alpha - 3)(g-1) + e\alpha$ implies

$$h^0(\mathcal{O}_S(K_S + (a-2)C_0 + (\mathfrak{k} + \mathfrak{e})f)) \leq \ell(Z) - 1,$$

and it follows that for a general Z and any $x \in \operatorname{supp}(Z)$,

$$h^0(\mathcal{I}_{Z\setminus\{x\}}(K_S + (a-2)C_0 + (\mathfrak{k} + \mathfrak{e})f)) = 0.$$

Hence $(\mathcal{O}_X(K_S + (a-2)C_0 + (\mathfrak{k} + \mathfrak{e})f), Z)$ satisfies the Cayley–Bacharach property and, therefore, there are rank-two vector bundles $\mathcal{F}$ given by an exact sequence of the following type:

$$0 \longrightarrow \mathcal{O}_S(aC_0 + (\mathfrak{b} + \mathfrak{v})f) \longrightarrow \mathcal{F} \longrightarrow \mathcal{I}_Z((2a-2)C_0 + (2\mathfrak{b} + \mathfrak{k} + \mathfrak{e} - \mathfrak{v})f) \longrightarrow 0.$$

It is clear that the determinant of such a bundle equals to $\mathcal{O}_S(K_S + 3H)$. Moreover, a direct computation shows that the second Chern class $c_2(\mathcal{F})$ is

$$-\frac{a(5a-3)}{2}e + \deg(\mathfrak{b})(5a-3) + (3a-2)(g-1) = \frac{1}{2}H(5H + 3K_S) + 2\chi(\mathcal{O}_S).$$

So, according to Proposition 5.1.3, $\mathcal{F}$ is an Ulrich bundle if and only if it is initialized.

Step 2. We are going to prove that the rank-two vector bundles we have constructed are initialized. Consider the exact sequence

$$0 \longrightarrow \mathrm{H}^0(\mathcal{O}_S(\mathfrak{v}f)) \longrightarrow \mathrm{H}^0(\mathcal{F}(-H)) \longrightarrow \mathrm{H}^0(\mathcal{I}_Z((a-2)C_0 + (\mathfrak{b} + \mathfrak{k} + \mathfrak{e} - \mathfrak{v})f)). \tag{5.19}$$

The genericity of $\mathfrak{v}$ ensures that $\mathrm{H}^0(\mathcal{O}_S(\mathfrak{v}f)) = 0$. On the other hand, $aC_0 + (\mathfrak{b} - \mathfrak{v})f$ is ample. In fact, since $\deg(\mathfrak{b}) > (g-1) + u_e$, $\deg(\mathfrak{b}) - g + 1 > ae/2$ if $e \le 0$ and $\deg(\mathfrak{b}) - g + 1 > ae$ if $e > 0$. Thus, Kodaira vanishing theorem implies

$$\begin{aligned} \mathrm{h}^0((a-2)C_0 + (\mathfrak{b} + \mathfrak{k} + \mathfrak{e} - \mathfrak{v})f) &= \chi((a-2)C_0 + (\mathfrak{b} + \mathfrak{k} + \mathfrak{e} - \mathfrak{v})f) \\ &= -\frac{a(a-1)e}{2} + (a-1)\deg(\mathfrak{b}) = \ell(Z). \end{aligned}$$

Therefore, for a general Z, we have

$$\mathrm{h}^0(\mathcal{I}_Z((a-2)C_0 + (\mathfrak{b} + \mathfrak{k} + \mathfrak{e} - \mathfrak{v})f)) = 0$$

and, by the exact sequence (5.19), get that $\mathcal{F}$ is initialized and hence it is a special Ulrich bundle.

Step 3. The vector bundle $\mathcal{F}$ is μ-stable.

They are certainly μ-stable in the range $e > 0$ and $a \ge 2$. In fact, rank-two Ulrich bundles are always μ-semistable and they can only be destabilized by Ulrich line bundles (see Corollary 3.3.17), which do not exist in this range by Theorem 5.2.7.

In order to prove the statement for $a \ge 2$ and $e \le 0$, we compute the dimension of the family $\mathbb{P}$ of vector bundles defined in the statement and estimate the dimension of the subfamily of strictly μ-semistable ones.

First, we check that a general vector bundle $\mathcal{F}$ belongs to one single extension. To this end, it suffices to prove that

$$\mathrm{h}^0(S, \mathcal{I}_Z((a-2)C_0 + (\mathfrak{b} + \mathfrak{k} - 2\mathfrak{v} + \mathfrak{e}))) = 0,$$

for general Z and $\mathfrak{v}$. Observe that

$$\begin{aligned} \mathrm{h}^0(S, \mathcal{O}_S((a-2)C_0 + (\mathfrak{b} + \mathfrak{k} - 2\mathfrak{v} + \mathfrak{e}))) &= \mathrm{h}^0(C, (S^{a-2}\mathcal{E})(\mathfrak{b} + \mathfrak{k} - 2\mathfrak{v} + \mathfrak{e})) \\ &\le \sum_{i=0}^{a-2} \mathrm{h}^0(C, \mathcal{O}_C(\mathfrak{b} + \mathfrak{k} - 2\mathfrak{v} + (i+1)\mathfrak{e})). \end{aligned}$$

On the other hand, $\deg(\mathfrak{b} + \mathfrak{k} - 2\mathfrak{v} + (i+1)\mathfrak{e}) = \deg(\mathfrak{b}) - (i+1)e > g - 1$ by the assumption $\deg(\mathfrak{b}) > (g-1) + e$. Hence $\mathrm{h}^1(C, \mathcal{O}_C(\mathfrak{b} + \mathfrak{k} - 2\mathfrak{v} + (i+1)\mathfrak{e}))$ will be zero for all i with $0 \le i \le a-2$, and for a generic choice of $\mathfrak{v}$. Thus, applying the Riemann–Roch theorem, we obtain

$$\begin{aligned} \mathrm{h}^0(S, \mathcal{O}_S((a-2)C_0 + (\mathfrak{b} + \mathfrak{k} - 2\mathfrak{v} + \mathfrak{e}))) &\le (a-1)\deg(\mathfrak{b}) - \frac{a(a-1)e}{2} + (a-1)(1-g) \\ &\le \deg(Z). \end{aligned}$$

Hence, for a generic choice of Z, we have $\mathrm{h}^0(S, \mathcal{I}_Z((a-2)C_0 + (\mathfrak{b} + \mathfrak{k} - 2\mathfrak{v} + \mathfrak{e}))) = 0$.

The vector bundles $\mathcal{F}$ are parameterized by a projective bundle $\mathbb{P}$ with typical fiber the projectivization of

$$\begin{aligned}\mathrm{Ext}^1(\mathcal{I}_Z((2a-2)C_0 + (2\mathfrak{b} + \mathfrak{k} + \mathfrak{e} - \mathfrak{v})f), \mathcal{O}_S(aC_0 + (\mathfrak{b} + \mathfrak{v})f))\\ \cong \mathrm{H}^1(S, \mathcal{I}_Z((a-4)C_0 + (\mathfrak{b} + 2\mathfrak{k} + 2\mathfrak{e} - 2\mathfrak{v})f))\end{aligned}$$

over an open subset of $\mathrm{Hilb}^l(S) \times \mathrm{Pic}^{g-1}(C)$ with $l = (a-1)(\deg(\mathfrak{b}) - ea/2)$.

Assume $a \geq 3$. We have already seen that

$$\begin{aligned}&\mathrm{h}^0(S, \mathcal{O}_S((a-4)C_0 + (\mathfrak{b} + 2\mathfrak{k} + 2\mathfrak{e} - 2\mathfrak{v})f))\\ &\qquad = (a-3)\left(g - 1 + \deg(\mathfrak{b}) - \frac{ea}{2}\right) \quad \text{and}\\ &\mathrm{h}^1(S, \mathcal{O}_S((a-4)C_0 + (\mathfrak{b} + 2\mathfrak{k} + 2\mathfrak{e} - 2\mathfrak{v})f))\\ &\qquad = \mathrm{h}^2(S, \mathcal{O}_S((a-4)C_0 + (\mathfrak{b} + 2\mathfrak{k} + 2\mathfrak{e} - 2\mathfrak{v})f)) = 0.\end{aligned} \tag{5.20}$$

It follows from the first equality (5.20) and the hypothesis $2\deg(\mathfrak{b}) > (a-3)(g-1) + ea$ that

$$\mathrm{h}^0(S, \mathcal{I}_Z((a-4)C_0 + (\mathfrak{b} + 2\mathfrak{k} + 2\mathfrak{e} - 2\mathfrak{v})f)) = 0, \tag{5.21}$$

for a general choice of Z.

Now assume that $a = 2$. Trivially, $\mathrm{h}^0(S, \mathcal{O}_S(-2C_0 + (\mathfrak{b} + 2\mathfrak{k} + 2\mathfrak{e} - 2\mathfrak{v})f)) = 0$. Moreover, for $i \geq 1$,

$$\begin{aligned}\mathrm{h}^i(S, \mathcal{O}_S(-2C_0 + (\mathfrak{b} + 2\mathfrak{k} + 2\mathfrak{e} - 2\mathfrak{v})f)) &= \mathrm{h}^{2-i}(S, \mathcal{O}_S((-\mathfrak{b} - \mathfrak{k} - \mathfrak{e} + 2\mathfrak{v})f))\\ &= \mathrm{h}^{2-i}(C, \mathcal{O}_C(-\mathfrak{b} - \mathfrak{k} - \mathfrak{e} + 2\mathfrak{v}))\\ &= \mathrm{h}^{i-1}(C, \mathcal{O}_C(\mathfrak{b} + 2\mathfrak{k} + \mathfrak{e} - 2\mathfrak{v})).\end{aligned}$$

We have $\deg(\mathfrak{b} + 2\mathfrak{k} + \mathfrak{e} - 2\mathfrak{v}) = \deg(\mathfrak{b}) - e + 2g - 2$, and thus

$$\begin{aligned}\mathrm{h}^1(S, \mathcal{O}_S(-2C_0 + (\mathfrak{b} + 2\mathfrak{k} + 2\mathfrak{e} - 2\mathfrak{v})f)) = \deg(\mathfrak{b}) - e + g - 1,\\ \mathrm{h}^2(S, \mathcal{O}_S(-2C_0 + (\mathfrak{b} + 2\mathfrak{k} + 2\mathfrak{e} - 2\mathfrak{v})f)) = 0.\end{aligned} \tag{5.22}$$

In both cases ($a = 2$ and $a \geq 3$), the exact cohomology sequence associated to

$$0 \longrightarrow \mathcal{I}_Z \longrightarrow \mathcal{O}_S \longrightarrow \mathcal{O}_Z \longrightarrow 0$$

tensored by $\mathcal{O}_S((a-4)C_0 + (\mathfrak{b} + 2\mathfrak{k} + 2\mathfrak{e} - 2\mathfrak{v})f)$ and the equalities (5.20)–(5.22) yield

$$\mathrm{h}^1(S, \mathcal{I}_Z((a-4)C_0 + (\mathfrak{b} + 2\mathfrak{k} + 2\mathfrak{e} - 2\mathfrak{v})f)) = 2\deg(\mathfrak{b}) - (a-3)(g-1) - ae.$$

Therefore, since $\mathbb{P}$ depends on the sets $Z \in \mathrm{Hilb}^l(S)$, which has dimension $2l$, the line bundles $\mathfrak{u}$, and the extension classes, it follows that

$$\begin{aligned} \dim \mathbb{P} &= 2(a-1)\left(\deg(\mathfrak{b}) - \frac{ea}{2}\right) + g + 2\deg(\mathfrak{b}) - (a-3)(g-1) - ae - 1 \\ &= 2a\deg(\mathfrak{b}) - ea^2 - (a-4)(g-1). \end{aligned} \tag{5.23}$$

From now on we will assume that $\mathcal{F}$ is strictly μ-semistable. So it contains an Ulrich line bundle and, in particular, by Proposition 5.2.2 this implies that $(a-1)e$ is even. Recall also that we are under the assumption $e \le 0$. Ulrich line bundles are given by Proposition 5.2.2. If $\mathcal{L} := \mathcal{O}_S((2a-1)C_0 + (\mathfrak{b}+\mathfrak{u})f) \subset \mathcal{F}$, then we have the diagram

$$\begin{array}{ccccccccc} & & & & \mathcal{L} & & & & \\ & & & & \downarrow & & & & \\ 0 & \longrightarrow & \mathcal{O}_S(H + \mathfrak{v}f) & \longrightarrow & \mathcal{F} & \longrightarrow & \mathcal{I}_Z(2H + K_S - \mathfrak{v}f) & \longrightarrow & 0 \end{array}$$

which implies that there is a nonzero morphism from $\mathcal{O}_S((2a-1)C_0 + (\mathfrak{b}+\mathfrak{u})f)$ to either $\mathcal{O}_S(aC_0 + (\mathfrak{b}+\mathfrak{v})f)$ or $\mathcal{O}_S((2a-2)C_0 + (2\mathfrak{b}+\mathfrak{k}+\mathfrak{e}-\mathfrak{v})f)$, which is a contradiction since $\mathrm{H}^0(\mathcal{O}_S((1-a)C_0 + (\mathfrak{v}-\mathfrak{u})f)) = \mathrm{H}^0(\mathcal{O}_S(-C_0 + (\mathfrak{b}+\mathfrak{k}+\mathfrak{e}-\mathfrak{v}-\mathfrak{u})f)) = 0$.

Thus, we necessarily have $\mathcal{O}_S((a-1)C_0 + (2\mathfrak{b}+\mathfrak{k}+\mathfrak{e}-\mathfrak{u})f) \subset \mathcal{F}$ and, since its quotient is also an Ulrich line bundle, get an exact sequence of the form

$$0 \longrightarrow \mathcal{O}_S((a-1)C_0 + (2\mathfrak{b}+\mathfrak{k}+\mathfrak{e}-\mathfrak{u})f) \longrightarrow \mathcal{F} \longrightarrow \mathcal{O}_S((2a-1)C_0 + (\mathfrak{b}+\mathfrak{u})f) \longrightarrow 0. \tag{5.24}$$

By the projection formula and Serre's duality,

$$\begin{aligned} &\mathrm{Ext}^1(\mathcal{O}_S((2a-1)C_0 + (\mathfrak{b}+\mathfrak{u})f), \mathcal{O}_S((a-1)C_0 + (2\mathfrak{b}+\mathfrak{k}+\mathfrak{e}-\mathfrak{u})f)) \\ &\qquad \cong \mathrm{H}^1(S, \mathcal{O}_S(-aC_0 + (\mathfrak{b}+\mathfrak{k}+\mathfrak{e}-2\mathfrak{u})f)) \\ &\qquad \cong \mathrm{H}^1(S, \mathcal{O}_S((a-2)C_0 + (2\mathfrak{u}-\mathfrak{b})f))^\vee \\ &\qquad \cong \mathrm{H}^1(C, (S^{a-2}\mathcal{E})(2\mathfrak{u}-\mathfrak{b}))^\vee. \end{aligned}$$

Hence, extensions as in the exact sequence (5.24) are parameterized by a space of dimension

$$\begin{aligned} &\mathrm{h}^1(C, (S^{a-2}\mathcal{E})(2\mathfrak{u}-\mathfrak{b})) - 1 \\ &\qquad = (a-1)\left(\deg(\mathfrak{b}) - \frac{ae}{2} + 1 - g\right) - 1 + \mathrm{h}^0(C, (S^{a-2}\mathcal{E})(2\mathfrak{u}-\mathfrak{b})) \\ &\qquad \le (a-1)\left(\deg(\mathfrak{b}) - \frac{ae}{2} + 1 - g\right) - 1 + \sum_{i=0}^{a-2} \mathrm{h}^0(C, \mathcal{O}_C(2\mathfrak{u}-\mathfrak{b}+i\mathfrak{e})), \end{aligned}$$

where the inequality follows from the inequality (5.13).

Notice that $(a-1-i)e \le 0$. Hence, the Riemann–Roch theorem and the Clifford theorem (see [108, Proposition IV.5.4]) imply

$$\begin{aligned} &h^0(C, \mathcal{O}_C(2\mathfrak{u} - \mathfrak{b} + i\mathfrak{e})) \\ &\le g - 1 + \max\left\{ (a-1-i)e - \deg(\mathfrak{b}), \frac{(a-1-i)e - \deg(\mathfrak{b})}{2} + 1 \right\}, \end{aligned}$$

for each general choice of $\mathfrak{u}$.

Since $e \le 0$, $\deg(\mathfrak{b}) \ge 0$, and thus

$$h^0(C, \mathcal{O}_C(2\mathfrak{u} - \mathfrak{b} + i\mathfrak{e})) \le \frac{(a-1-i)e - \deg(\mathfrak{b})}{2} + g.$$

Hence the strictly μ-semistable vector bundles we are interested in are parameterized by a family of dimension at most

$$\frac{a-1}{2}\left(\deg(\mathfrak{b}) + 2 - \frac{ae}{2}\right) - 1.$$

The dimension of the family of strictly μ-semistable bundles is smaller than the value $\dim \mathbb{P}$ given by the equality (5.23) if

$$\frac{a}{2}\left(3\deg(\mathfrak{b}) - \frac{e}{2}(3a+1) - 2g\right) + 4(g-1) + \frac{b}{2} + 2 > 0,$$

a condition which is automatically satisfied, since by hypothesis $3\deg(\mathfrak{b}) \ge \frac{e(3a+1)}{2} + 2g$.

Since a general $\mathcal{F}$ is uniquely determined by an extension in $\mathbb{P}$, we conclude that there are μ-stable Ulrich bundles. □

Now we will turn our attention to *Hirzebruch surfaces*. Our aim is to give a complete classification of rank r Ulrich bundles on them. This will be achieved by means of techniques quite common in the context of derived categories. The reader not familiar with these notions has a short introduction to the needed material in Appendix A.

Denote by $\pi : \mathbb{F}_e \longrightarrow \mathbb{P}^1$ a Hirzebruch surface with invariant $e \ge 0$, that is, $\mathbb{F}_e \cong \mathbb{P}(\mathcal{O}_{\mathbb{P}^1} \oplus \mathcal{O}_{\mathbb{P}^1}(-e))$. Since $\mathrm{Pic}(\mathbb{P}^1) \cong \mathbb{Z}$, we simplify the notation and denote any line bundle $\mathcal{O}_{\mathbb{F}_e}(D) \in \mathrm{Pic}(\mathbb{F}_e)$ by $\mathcal{O}_{\mathbb{F}_e}(D) = \mathcal{O}_{\mathbb{F}_e}(aC_0 + bf)$ with $a, b \in \mathbb{Z}$. Furthermore, the canonical divisor is given by $K_{\mathbb{F}_e} = -2C_0 - (e+2)f$. In this particular case, in addition to (5.10), we have

$$H^i(\mathbb{F}_e, \mathcal{O}_{\mathbb{F}_e}(-C_0 + bf)) = 0 \quad \text{for all } i \ge 0,\ b \in \mathbb{Z}. \tag{5.25}$$

On $\mathbb{F}_e$ there are two well-known short exact sequences that we will use later on. The first is

$$0 \longrightarrow \mathcal{O}_{\mathbb{F}_e}(-f) \longrightarrow \mathcal{O}_{\mathbb{F}_e}^2 \longrightarrow \mathcal{O}_{\mathbb{F}_e}(f) \longrightarrow 0, \tag{5.26}$$

and the second is

$$0 \longrightarrow \mathcal{O}_{\mathbb{F}_e}(-C_0 - ef) \longrightarrow \mathcal{O}_{\mathbb{F}_e} \oplus \mathcal{O}_{\mathbb{F}_e}(-ef) \longrightarrow \mathcal{O}_{\mathbb{F}_e}(C_0) \longrightarrow 0. \tag{5.27}$$

Our next goal is, given a rank r Ulrich bundle $\mathcal{F}$ on $\mathbb{F}_e$, to see that it can be written as the cokernel of a certain map between direct sum of line bundles. To this end, we will first determine the cohomology of $\mathcal{F}$.

Lemma 5.2.23. *Let $\mathcal{F}$ be a rank r Ulrich bundle on $(\mathbb{F}_e, \mathcal{O}_{\mathbb{F}_e}(aC_0 + bf))$. Then, it holds:*
(1) $h^i(\mathcal{F}(tC_0 + sf)) = 0$ *for* $i = 0, 2$ *and all* $-2a \le t \le -a$ *and* $-2b \le s \le -b$.
(2) $h^1(\mathcal{F}(-aC_0 + sf)) = h^2(\mathcal{F}(-aC_0 + sf)) = 0$ *for all* $s \ge -b$.
(3) $h^1(\mathcal{F}(tC_0 + sf)) = h^2(\mathcal{F}(tC_0 + sf)) = 0$ *for all* $t \ge -a$ *and* $s \ge -b + e$.
(4) $h^0(\mathcal{F}(tC_0 + sf)) = h^1(\mathcal{F}(tC_0 + sf)) = 0$ *for all* $t \le -2a$ *and* $s \le -2b - e$.

Proof. It can be proven by induction on s and t and using the cohomology exact sequences associated to the exact sequence (5.26) and (5.27) twisted by $\mathcal{F}$. □

Using a *Beilinson-type spectral sequence* (see Theorem A.0.27), we get the following resolution:

Theorem 5.2.24. *Let $\mathcal{F}$ be a rank r Ulrich bundle on $(\mathbb{F}_e, \mathcal{O}_{\mathbb{F}_e}(aC_0 + bf))$ with the first Chern class $c_1(\mathcal{F}) = \alpha C_0 + \beta f$. Then, $\mathcal{F}$ fits into the following short exact sequence:*

$$0 \longrightarrow \mathcal{O}^{\gamma}_{\mathbb{F}_e}(H - C_0 - (e+1)f) \longrightarrow \mathcal{O}^{\delta}_{\mathbb{F}_e}(H - C_0 - ef) \oplus \mathcal{O}^{\tau}_{\mathbb{F}_e}(H - f) \longrightarrow \mathcal{F} \longrightarrow 0 \tag{5.28}$$

where $\gamma = \alpha + \beta - r(a + b - 1) - e(\alpha - ar)$, $\delta = \beta - r(b - 1) - e(\alpha - ar)$, *and* $\tau = \alpha - r(a - 1)$.

Proof. Let $\mathcal{F}$ be a rank r Ulrich bundle on $\mathbb{F}_e$ and denote by $\overline{\mathcal{F}} = \mathcal{F} \otimes \mathcal{O}_{\mathbb{F}_e}(-aC_0 - bF)$. According to Example A.0.26 (2) and by Theorem A.0.27 applied to $\overline{\mathcal{F}}$, we have a complex $L^\bullet$ with terms

$$\begin{aligned}
L^{-3} &= H^{-1}(\overline{\mathcal{F}} \otimes \mathcal{O}_{\mathbb{F}_e}(-C_0 - f)) \otimes \mathcal{O}_{\mathbb{F}_e}(-C_0 - (e+1)f),\\
L^{-2} &= H^{0}(\overline{\mathcal{F}} \otimes \mathcal{O}_{\mathbb{F}_e}(-C_0 - f)) \otimes \mathcal{O}_{\mathbb{F}_e}(-C_0 - (e+1)f)\\
&\quad \oplus H^{-1}(\overline{\mathcal{F}} \otimes \mathcal{O}_{\mathbb{F}_e}(-C_0)) \otimes \mathcal{O}_{\mathbb{F}_e}(-C_0 - ef),\\
L^{-1} &= H^{1}(\overline{\mathcal{F}} \otimes \mathcal{O}_{\mathbb{F}_e}(-C_0 - f)) \otimes \mathcal{O}_{\mathbb{F}_e}(-C_0 - (e+1)f)\\
&\quad \oplus H^{0}(\overline{\mathcal{F}} \otimes \mathcal{O}_{\mathbb{F}_e}(-C_0)) \otimes \mathcal{O}_{\mathbb{F}_e}(-C_0 - ef) \oplus H^{0}(\overline{\mathcal{F}} \otimes \mathcal{O}_{\mathbb{F}_e}(-f)) \otimes \mathcal{O}_{\mathbb{F}_e}(-f),\\
L^{0} &= H^{2}(\overline{\mathcal{F}} \otimes \mathcal{O}_{\mathbb{F}_e}(-C_0 - f)) \otimes \mathcal{O}_{\mathbb{F}_e}(-C_0 - (e+1)f)\\
&\quad \oplus H^{1}(\overline{\mathcal{F}} \otimes \mathcal{O}_{\mathbb{F}_e}(-C_0)) \otimes \mathcal{O}_{\mathbb{F}_e}(-C_0 - ef)\\
&\quad \oplus H^{1}(\overline{\mathcal{F}} \otimes \mathcal{O}_{\mathbb{F}_e}(-f)) \otimes \mathcal{O}_{\mathbb{F}_e}(-f) \oplus H^{0}(\overline{\mathcal{F}}) \otimes \mathcal{O}_{\mathbb{F}_e},\\
L^{1} &= H^{2}(\overline{\mathcal{F}} \otimes \mathcal{O}_{\mathbb{F}_e}(-C_0)) \otimes \mathcal{O}_{\mathbb{F}_e}(-C_0 - ef)\\
&\quad \oplus H^{2}(\overline{\mathcal{F}} \otimes \mathcal{O}_{\mathbb{F}_e}(-f)) \otimes \mathcal{O}_{\mathbb{F}_e}(-f) \oplus H^{1}(\overline{\mathcal{F}}) \otimes \mathcal{O}_{\mathbb{F}_e},\\
L^{2} &= H^{3}(\overline{\mathcal{F}} \otimes \mathcal{O}_{\mathbb{F}_e}(-f)) \otimes \mathcal{O}_{\mathbb{F}_e}(-f) \oplus H^{2}(\overline{\mathcal{F}}) \otimes \mathcal{O}_{\mathbb{F}_e},
\end{aligned}$$

$$L^3 = \mathrm{H}^3(\overline{\mathcal{F}}) \otimes \mathcal{O}_{\mathbb{F}_e},$$

exact everywhere except at zero and $\mathrm{H}^0(L^\bullet) = \overline{\mathcal{F}}$.

It follows from the general cohomological properties and Lemma 5.2.23 that

$$L^{-3} = L^{-2} = L^1 = L^2 = L^3 = 0$$

and

$$\begin{aligned}
&\mathrm{H}^0(\overline{\mathcal{F}} \otimes \mathcal{O}_{\mathbb{F}_e}(-C_0)) = \mathrm{H}^0(\overline{\mathcal{F}} \otimes \mathcal{O}_{\mathbb{F}_e}(-f)) = 0,\\
&\mathrm{H}^2(\overline{\mathcal{F}} \otimes \mathcal{O}_{\mathbb{F}_e}(-C_0 - f)) = \mathrm{H}^0(\overline{\mathcal{F}}) = 0,\\
&\gamma := \mathrm{h}^1(\overline{\mathcal{F}} \otimes \mathcal{O}_{\mathbb{F}_e}(-C_0 - f)) = -\chi(\overline{\mathcal{F}} \otimes \mathcal{O}_{\mathbb{F}_e}(-C_0 - f)),\\
&\delta := \mathrm{h}^1(\overline{\mathcal{F}} \otimes \mathcal{O}_{\mathbb{F}_e}(-C_0)) = -\chi(\overline{\mathcal{F}} \otimes \mathcal{O}_{\mathbb{F}_e}(-C_0)),\\
&\tau := \mathrm{h}^1(\overline{\mathcal{F}} \otimes \mathcal{O}_{\mathbb{F}_e}(-f)) = -\chi(\overline{\mathcal{F}} \otimes \mathcal{O}_{\mathbb{F}_e}(-f)).
\end{aligned}$$

All these values can be computed using the Riemann–Roch theorem, and we get $\overline{\mathcal{F}}$ as the following cokernel:

$$0 \longrightarrow \mathcal{O}_{\mathbb{F}_e}^{\gamma}(-C_0 + (-e-1)f) \longrightarrow \mathcal{O}_{\mathbb{F}_e}^{\delta}(-C_0 - ef) \oplus \mathcal{O}_{\mathbb{F}_e}^{\tau}(-f) \longrightarrow \overline{\mathcal{F}} \longrightarrow 0$$

where $\gamma = \alpha + \beta - r(a+b-1) - e(\alpha - ar)$, $\delta = \beta - r(b-1) - e(\alpha - ar)$, and $\tau = \alpha - r(a-1)$, which proves the claim. □

Let us now focus on the converse: given $\mathcal{F}$ a rank r vector bundle as a cokernel of a map as in (5.28), we want to determine when it is an Ulrich bundle.

Theorem 5.2.25. *Let $\mathcal{F}$ be a rank r vector bundle on $(\mathbb{F}_e, \mathcal{O}_{\mathbb{F}_e}(H))$, $H = aC_0 + bf$, with $c_1(\mathcal{F}) = \alpha C_0 + \beta f$. Assume that $c_1(\mathcal{F})H = \frac{r}{2}(3H^2 + HK_{\mathbb{F}_e})$ and that $\mathcal{F}$ is defined as the cokernel of an injective map ϕ:*

$$\mathcal{O}_{\mathbb{F}_e}^{\gamma}(H - C_0 - (e+1)f) \xrightarrow{\ \phi\ } \mathcal{O}_{\mathbb{F}_e}^{\delta}(H - C_0 - ef) \oplus \mathcal{O}_{\mathbb{F}_e}^{\tau}(H - f) \tag{5.29}$$

where $\gamma = \alpha + \beta - r(a+b-1) - e(\alpha - ar)$, $\delta = \beta - r(b-1) - e(\alpha - ar)$ and $\tau = \alpha - r(a-1)$ are nonnegative integers. Then, $\mathcal{F}$ is an Ulrich bundle if and only if $\mathrm{H}^2(\mathcal{F}(-2H)) = 0$.

Proof. By construction, $\mathcal{F}$ is given by an exact sequence of the following type:

$$0 \longrightarrow \mathcal{O}_{\mathbb{F}_e}^{\gamma}(H - C_0 - (e+1)f) \xrightarrow{\ \phi\ } \mathcal{O}_{\mathbb{F}_e}^{\delta}(H - C_0 - ef) \oplus \mathcal{O}_{\mathbb{F}_e}^{\tau}(H - f) \longrightarrow \mathcal{F} \longrightarrow 0. \tag{5.30}$$

Twisting this exact sequence by $\mathcal{O}_{\mathbb{F}_e}(-aC_0 - bf) = \mathcal{O}_{\mathbb{F}_e}(-H)$, taking cohomology, using (5.25) and (5.10), we get that

$$\mathrm{H}^0(\mathcal{F}(-H)) = \mathrm{H}^1(\mathcal{F}(-H)) = \mathrm{H}^2(\mathcal{F}(-H)) = 0.$$

In particular, $\chi(\mathcal{F}(-H)) = 0$, which is equivalent to the fact that $c_2(\mathcal{F}) = \frac{1}{2}(c_1(\mathcal{F})^2 - c_1(\mathcal{F})K_{\mathbb{F}_e}) - r(H^2 - 1)$. Hence, according to Proposition 5.1.1, $\mathcal{F}$ is an Ulrich bundle if and only if $\mathrm{H}^2(\mathcal{F}(-2H)) = 0$. □

Definition 5.2.26. Let $(\mathbb{F}_e, \mathcal{O}_{\mathbb{F}_e}(H))$, $H = aC_0 + bf$, be a Hirzebruch surface, r a positive integer, and $c_1 = \alpha C_0 + \beta f$. We say that (r, c_1) is an *Ulrich admissible pair* on $(\mathbb{F}_e, \mathcal{O}_{\mathbb{F}_e}(H))$ if

$$\begin{aligned} c_1 H &= \frac{r}{2}(3H^2 + HK_{\mathbb{F}_e}), \\ &\alpha + \beta - r(a + b - 1) - e(\alpha - ar) \geq 0, \\ &\beta - r(b - 1) - e(\alpha - ar) \geq 0, \\ &\alpha - r(a - 1) \geq 0. \end{aligned}$$

As a consequence of Theorem 5.2.25, there exists a rank r Ulrich bundle on a Hirzebruch surface $(\mathbb{F}_e, \mathcal{O}_{\mathbb{F}_e}(H))$, $H = aC_0 + bf$, for any Ulrich admissible pair (r, c_1). In fact, we have

Theorem 5.2.27. *Let $(\mathbb{F}_e, \mathcal{O}_{\mathbb{F}_e}(H))$, $H = aC_0 + bf$, be a Hirzebruch surface and $(r, \alpha C_0 + \beta f)$ an Ulrich admissible pair. Then, there exists a rank r Ulrich bundle $\mathcal{F}$ with $c_1(\mathcal{F}) = \alpha C_0 + \beta f$.*

Proof. Consider the nonnegative integers $\gamma = \alpha + \beta - r(a + b - 1) - e(\alpha - ar)$, $\delta = \beta - r(b - 1) - e(\alpha - ar)$, and $\tau = \alpha - r(a - 1)$, and take

$$\begin{aligned} \phi \in \operatorname{Hom}(\mathcal{O}^{\gamma}_{\mathbb{F}_e}((a-1)C_0 + (b-e-1)f), \mathcal{O}^{\delta}_{\mathbb{F}_e}((a-1)C_0 + (b-e)f) \oplus \mathcal{O}^{\tau}_{\mathbb{F}_e}(aC_0 + (b-1)f)) \\ \cong \mathrm{H}^0(\mathcal{O}^{\gamma\delta}_{\mathbb{F}_e}(0, 1) \oplus \mathcal{O}^{\gamma\tau}_{\mathbb{F}_e}(1, e)) \end{aligned}$$

a general injective morphism. Define $\mathcal{F} := \operatorname{coker}(\phi)$. We have the short exact sequence

$$0 \longrightarrow \mathcal{O}^{\gamma}_{\mathbb{F}_e}(H - C_0 - (e+1)f) \xrightarrow{\phi} \mathcal{O}^{\delta}_{\mathbb{F}_e}(H - C_0 - ef) \oplus \mathcal{O}^{\tau}_{\mathbb{F}_e}(H - f) \longrightarrow \mathcal{F} \longrightarrow 0$$

and, in particular, $c_1(\mathcal{F}) = \alpha C_0 + \beta f$. Since $(r, \alpha C_0 + \beta f)$ is an Ulrich admissible pair, $c_1(\mathcal{F})H = \frac{r}{2}(3H^2 + HK_{\mathbb{F}_e})$. By Riemann–Roch theorem (Theorem 1.4.3), this equality is equivalent to the fact that

$$\begin{aligned} &\chi(\mathcal{O}^{\gamma}_{\mathbb{F}_e}((-a-1)C_0 + (-b-e-1)f)) \\ &= \chi(\mathcal{O}^{\delta}_{\mathbb{F}_e}((-a-1)C_0 + (-b-e)f) \oplus \mathcal{O}^{\tau}_{\mathbb{F}_e}(-aC_0 - (b+1)f)), \end{aligned}$$

which implies that $\chi(\mathcal{F}(-2H)) = 0$. On the other hand, by [62, Theorem 1.1], $\mathcal{F}(-2H)$ has at most one nonzero cohomology group. These facts, together with $\mathrm{H}^0(\mathcal{F}(-2H)) = 0$, imply that $\mathrm{H}^1(\mathcal{F}(-2H)) = \mathrm{H}^2(\mathcal{F}(-2H)) = 0$, and the result follows from Theorem 5.2.25. □

In the last part of this section, we determine whether the d-uple Veronese embedding of $\mathbb{P}^2$ supports a rank r Ulrich bundle. More precisely, we explicitly describe the integers d and r such that rank r Ulrich bundles on $\mathbb{P}^2$ for the Veronese embedding given by $\mathcal{O}_{\mathbb{P}^2}(d)$ exist. Ulrich line bundles on $(\mathbb{P}^2, \mathcal{O}_{\mathbb{P}^2}(d))$ exist only for $d = 1$. For $r \geq 2$, we have the following characterization.

Proposition 5.2.28. *Let $\mathcal{E}$ be a rank r vector bundle on $(\mathbb{P}^2, \mathcal{O}_{\mathbb{P}^2}(d))$. It holds:*
(1) *If $\mathcal{E}$ is an Ulrich bundle, then it has a locally free resolution of the following type:*

$$0 \longrightarrow \mathcal{O}_{\mathbb{P}^2}^{\frac{r}{2}(d-1)}(-2) \longrightarrow \mathcal{O}_{\mathbb{P}^2}^{\frac{r}{2}(d+1)}(-1) \longrightarrow \mathcal{E}(-d) \longrightarrow 0.$$

(2) *If $\mathcal{E}$ has natural cohomology and a locally free resolution of the following type:*

$$0 \longrightarrow \mathcal{O}_{\mathbb{P}^2}^{\frac{r}{2}(d-1)}(-2) \longrightarrow \mathcal{O}_{\mathbb{P}^2}^{\frac{r}{2}(d+1)}(-1) \longrightarrow \mathcal{E}(-d) \longrightarrow 0, \tag{5.31}$$

then $\mathcal{E}$ is an Ulrich bundle.

Proof. (1) See [60, Corollary 4.3] or [143, Corollary 4.6].

(2) It follows from [143, Corollary 4.6], taking into account that for vector bundles with resolution (5.31) and natural cohomology, the induced morphism

$$\mathrm{H}^2(\mathcal{O}_{\mathbb{P}^2}^{\frac{r}{2}(d-1)}(-d-2)) \longrightarrow \mathrm{H}^2(\mathcal{O}_{\mathbb{P}^2}^{\frac{r}{2}(d+1)}(-d-1))$$

has maximal rank, which means that it is either injective or surjective. □

Theorem 5.2.29. *Assume $d, r \geq 2$. A rank r Ulrich bundle $\mathcal{E}$ on $(\mathbb{P}^2, \mathcal{O}_{\mathbb{P}^2}(d))$ exists if and only if $r(d-1) \equiv 0 \pmod 2$. In addition, for $d \geq 3$ and $r(d-1) \equiv 0 \pmod 2$, there is a stable (hence indecomposable) rank r Ulrich bundle $\mathcal{E}$ on $(\mathbb{P}^2, \mathcal{O}_{\mathbb{P}^2}(d))$, with a minimal free resolution of the following type:*

$$0 \longrightarrow \mathcal{O}_{\mathbb{P}^2}^{\frac{r}{2}(d-1)}(-2) \longrightarrow \mathcal{O}_{\mathbb{P}^2}^{\frac{r}{2}(d+1)}(-1) \longrightarrow \mathcal{E}(-d) \longrightarrow 0.$$

Proof. We analyze separately the different values of d.

For $d = 2$, r is necessarily even and, by [60, Theorems 5.2 and 5.4], the tangent bundle $T_{\mathbb{P}^2}$ is the unique rank 2 Ulrich bundle on $(\mathbb{P}^2, \mathcal{O}_{\mathbb{P}^2}(2))$ and $T_{\mathbb{P}^2}^s$ is the unique rank $2s$ Ulrich bundle on $(\mathbb{P}^2, \mathcal{O}_{\mathbb{P}^2}(2))$.

For even $d > 2$, again r is necessarily even and, by [60, Theorem 6.1], there exist rank $2s$ stable (hence indecomposable) Ulrich bundles on $(\mathbb{P}^2, \mathcal{O}_{\mathbb{P}^2}(d))$ for all $s \geq 1$.

Let us assume that d is odd. By [60, Theorem 6.2], for $d \geq 3$ odd, there exist stable Ulrich bundles of any rank $r \geq 2$ on $(\mathbb{P}^2, \mathcal{O}_{\mathbb{P}^2}(d))$ if and only if there exists a rank 3 stable Ulrich bundle on $(\mathbb{P}^2, \mathcal{O}_{\mathbb{P}^2}(d))$. So our goal will be to prove the existence of rank 3 Ulrich bundles on $(\mathbb{P}^2, \mathcal{O}_{\mathbb{P}^2}(d))$ for all odd integers $d \geq 3$. To this end, we write $d = 2k+1$ with $k \geq 1$ and denote by $M_{\mathbb{P}^2}^s(3; 0, \frac{3}{2}k(k+1))$ the moduli space of μ-stable rank 3 vector bundles $\mathcal{F}$ on $\mathbb{P}^2$ with Chern classes $c_1(\mathcal{F}) = 0$ and $c_2(\mathcal{F}) = \frac{3}{2}k(k+1)$.

By [147, Corollary 2.4.2], a generic vector bundle $\mathcal{F}$ of the moduli space $M_{\mathbb{P}^2}^s(3; 0, \frac{3}{2}k(k+1))$ has natural cohomology and a locally free resolution of the following type:

$$0 \longrightarrow \mathcal{O}_{\mathbb{P}^2}^{t-\alpha_-}(-s-1) \oplus \mathcal{O}_{\mathbb{P}^2}^{\alpha_-}(-s) \longrightarrow \mathcal{O}_{\mathbb{P}^2}^{t+3-\alpha_+}(-s+1) \oplus \mathcal{O}_{\mathbb{P}^2}^{\alpha_+}(-s) \longrightarrow \mathcal{F} \longrightarrow 0$$

where

$$\begin{aligned} s &= \max\{\rho \in \mathbb{Z} \mid 3\rho^2 + 2c_1(\mathcal{F})\rho - 3\rho \leq 2c_2(\mathcal{F}) - c_1(\mathcal{F})^2 + c_1(\mathcal{F}) - 1\}, \\ \alpha_+ &= \max\{0, 2c_2(\mathcal{F}) - c_1(\mathcal{F})^2 + 3 - 3s^2 - 2c_1(\mathcal{F})s\}, \\ \alpha_- &= \max\{0, -2c_2(\mathcal{F}) + c_1(\mathcal{F})^2 - 3 + 3s^2 + 2c_1(\mathcal{F})s\}, \\ t &= \frac{1}{2}(3s + c_1(\mathcal{F}) - 3 + 2c_2(\mathcal{F}) - c_1(\mathcal{F})^2 + 3 - 3s^2 - 2c_1(\mathcal{F})s). \end{aligned}$$

We easily check that

$$s = k, \quad \alpha_+ = 3(k+1), \quad \alpha_- = 0, \quad \text{and} \quad t = 3k.$$

Therefore, the resolution of $\mathcal{F}$ has the following shape:

$$0 \longrightarrow \mathcal{O}_{\mathbb{P}^2}^{3k}(-k-1) \longrightarrow \mathcal{O}_{\mathbb{P}^2}^{3k+3}(-k) \longrightarrow \mathcal{F} \longrightarrow 0,$$

and, applying Proposition 5.2.28, we conclude that $\mathcal{E} := \mathcal{F}(k-1+d)$ is a rank 3 Ulrich bundle on $(\mathbb{P}^2, \mathcal{O}_{\mathbb{P}^2}(d))$ for any odd $d = 2k+1$. □

It is natural to ask whether the above result generalizes to $\mathbb{P}^n$. It is not difficult to see that, for any $n \geq 2$ and $d \geq 1$, $(\mathbb{P}^n, \mathcal{O}_{\mathbb{P}^n}(d))$ carries an Ulrich bundle of rank r for $r \gg 0$. More precisely, we have seen in Corollary 3.3.13 that $(\mathbb{P}^n, \mathcal{O}_{\mathbb{P}^n}(d))$ carries an Ulrich bundle of rank $n!$. Nevertheless, the following problem is quite open:

Problem 5.2.30. Determine the triples $(n, d, r) \in \mathbb{Z}^3$ such that $(\mathbb{P}^n, \mathcal{O}_{\mathbb{P}^n}(d))$ carries a rank r indecomposable Ulrich bundle.

Remark 5.2.31. The reader could look at [71], [106] and [143] for recent contributions to the aforementioned problem in the particular case of $n = 3$, i.e., for an update about triples $(3, d, r) \in \mathbb{Z}^3$ such that $(\mathbb{P}^3, \mathcal{O}_{\mathbb{P}^3}(d))$ carries a rank r indecomposable Ulrich bundle.

5.3 Nonminimal surfaces with $\kappa = -\infty$

In this section, we are going to continue the study of Ulrich bundles on surfaces with $\kappa = -\infty$. However, whereas in the previous section we focused our attention on minimal surfaces, now we are going to deal with the case of nonminimal ones. We focus on the most interesting and studied case, namely, when S is a blow-up of $\mathbb{P}^2$ such that $-K_S$ is still an ample divisor. They are the so-called *del Pezzo surfaces*.

Definition 5.3.1. A *del Pezzo surface* is a smooth surface S whose anticanonical divisor $-K_S$ is ample. Its degree is defined as K_S^2.

Remark 5.3.2. Any del Pezzo surface is rational. Indeed, according to Castelnuovo's criterion [133, Chapter III, Theorem 2.4], a smooth surface S is rational if and only if $h^0(\mathcal{O}_S(2K_S)) = h^1(\mathcal{O}_S) = 0$. In the case of del Pezzo surfaces, the former cohomology group is zero because $-2K_S$ is ample and therefore $2K_S$ is not effective. The later cohomology group is zero thanks to the Kodaira vanishing theorem.

Definition 5.3.3. A set of points $\{p_1, \dots, p_r\}$ on $\mathbb{P}^2$ with $r \leq 9$ is in *general position* if no three of them lie on a line and no six of them lie on a conic.

The following theorem characterizes all del Pezzo surfaces.

Theorem 5.3.4. *Let S be a del Pezzo surface of degree d. Then, $1 \leq d \leq 9$ and*
(1) *If $d = 9$, then S is isomorphic to $\mathbb{P}^2$ and $-K_{\mathbb{P}^2} = 3H_{\mathbb{P}^2}$ gives the usual Veronese embedding of $\mathbb{P}^2$ in $\mathbb{P}^9$.*
(2) *If $d = 8$, then S is isomorphic to either $\mathbb{P}^1 \times \mathbb{P}^1$ or to a blow-up of $\mathbb{P}^2$ at one point.*
(3) *If $1 \leq d \leq 7$, then S is isomorphic to a blow-up of $9 - d$ points in general position.*

Conversely, any surface described under (1), (2), (3) for $d \geq 1$ is a del Pezzo surface of the corresponding degree.

Proof. See [148, Chapter IV, Theorem 24.4]. □

Since we defined Ulrich bundles with respect to a very ample line bundle, from now on we are going to focus our attention on those del Pezzo surfaces with very ample anticanonical divisor. They are classically called *strong del Pezzo surfaces*. Nevertheless, to simplify the terminology, we will keep calling them del Pezzo surfaces. The following theorem characterizes them:

Theorem 5.3.5. *If the surface S is obtained from $\mathbb{P}^2$ by blowing up $r \leq 6$ points in general position, then $-K_S$ is very ample and its global sections yield a closed embedding of S in a projective space of dimension*

$$\dim H^0(S, \mathcal{O}_S(-K_S)) - 1 = K_S^2 = 9 - r.$$

The set of exceptional curves is identified under this embedding with the set of lines in the projective space which lie on S. The image of S has degree $9 - r$.

Proof. See [148, Chapter IV, Theorem 24.5]. □

Corollary 5.3.6. *Let S be a strong del Pezzo surface. Then, S is isomorphic either to $\mathbb{P}^1 \times \mathbb{P}^1$ or to the blow-up of $\mathbb{P}^2$ at r points in general position, $r = 0, \dots, 6$.*

From now on, since Veronese embeddings of $\mathbb{P}^2$ in $\mathbb{P}^{\binom{d+2}{2}-1}$ and quadric surfaces have been already discussed in Section 5.2, we are going to focus on strong del Pezzo surfaces defined as the blow-up of $\mathbb{P}^2$ at r, $1 \leq r \leq 6$, points in general position.

Remark 5.3.7. If S is a strong del Pezzo surface and $\pi : S \longrightarrow S'$ is the map blowing down a line, then S' is a del Pezzo surface with $K_{S'}^2 = K_S^2 + 1$ (see [148, Chapter IV, Corollary 24.5.2]).

In the following theorem we summarize the well-known results about the Picard group and the intersection product on blow-ups.

Theorem 5.3.8. *Let $\{p_1, \ldots, p_r\}$ be a set of points in $\mathbb{P}^2$ and let $\pi : S \longrightarrow \mathbb{P}^2$ be the blow-up of $\mathbb{P}^2$ at these points. Let $l \in \operatorname{Pic}(S)$ be the pullback of a line in $\mathbb{P}^2$, let E_i be the exceptional curves and let $e_i \in \operatorname{Pic}(S)$ be their linear equivalence classes. Then:*
(1) *$\operatorname{Pic}(S) \cong \mathbb{Z}^{r+1}$ generated by $l, e_1, \ldots, e_r$.*
(2) *The intersection pairing on S is given by $l^2 = 1$, $e_i^2 = -1$, $le_i = 0$, and $e_ie_j = 0$ for $i \neq j$.*
(3) *The canonical class is $K_S = -3l + \sum_{i=1}^r e_i$.*

Moreover, if $0 \leq r \leq 6$ and the points are in general position, the following holds:
(4) *If D is any effective divisor on S, $D \sim al - \sum_{i=1}^r b_ie_i$ then the degree of D as a curve embedded in $\mathbb{P}^{9-r}$ by $-K_S$ is $\deg(D) = 3a - \sum_{i=1}^r b_i$ and its self-intersection is $D^2 = a^2 - \sum_{i=1}^r b_i^2$.*
(5) *The arithmetic genus of D is*

$$p_a(D) = \frac{1}{2}(D^2 - \deg(D)) + 1 = \frac{1}{2}(a-1)(a-2) - \frac{1}{2}\sum_{i=1}^r b_i(b_i - 1).$$

Proof. See [108, Chapter V, Proposition 4.8]. □

Remark 5.3.9. Using the same notation as in the previous theorem, if C is any irreducible curve on S, other than the exceptional E_i, then $C_0 := \pi(C)$ is an irreducible plane curve and C is the strict transform of C_0. Let C_0 have degree a and multiplicity b_i at each p_i. Then $\pi^* C_0 = C + \sum_{i=1}^r b_iE_i$. Since C_0 is linearly equivalent to a times the class of a line on $\mathbb{P}^2$, we get $C \sim al - \sum_{i=1}^r b_ie_i$ with $a > 0$ and $b_i \geq 0$ (see [108, Chapter V, Remark 4.8.1]).

In order to have a good understanding of the properties of the del Pezzo surfaces, it is important to keep track of the (-1)-lines present on them. Let us collect some results on their behavior.

Proposition 5.3.10. (1) *$\mathbb{P}^1 \times \mathbb{P}^1$ and $\mathbb{P}^2$ have no (-1)-lines.*

(2) *Let S be a strong del Pezzo surface which is the blow-up of $\mathbb{P}^2$ of r points in general position with $1 \leq r \leq 6$. Then, the (-1)-lines of S are*
(2.1) *the r exceptional divisors $e_1, \ldots, e_r$;*
(2.2) *for $r \geq 2$, $F_{i,j} = l - e_i - e_j$ with $1 \leq i < j \leq r$;*
(2.3) *for $r = 5$, $G = 2l - e_1 - e_2 - e_3 - e_4 - e_5$;*
(2.4) *for $r = 6$, $G_j = 2l - \sum_{i \neq j} e_i$.*

In particular, S has exactly $r + \binom{r}{2} + \binom{r}{5}$ (-1)-lines.

Proof. See [108, Chapter V, Theorem 4.9]. □

Proposition 5.3.11. *Let $(S, \mathcal{O}_S(-K_S))$ be a del Pezzo surface of degree d and set $r = 9-d$. If $L_1, \dots, L_r$ are mutually disjoint (-1)-lines of S, then there exists a blow-up map $\pi : S \longrightarrow \mathbb{P}^2$ at r points in general position in $\mathbb{P}^2$ such that $L_1, \dots, L_r$ are the exceptional divisors.*

Proof. See [108, Chapter V, Proposition 4.10]. □

Now we are ready to study Ulrich bundles on del Pezzo surfaces.

Theorem 5.3.12. *Let $\mathcal{E}$ be a rank $r \geq 1$ Ulrich bundle on a del Pezzo surface $(S, \mathcal{O}_S(-K_S))$ of degree d and take $D = c_1(\mathcal{E})$. Then,*
(1) $\deg(D) = dr$.
(2) $D^2 = 2c_2(\mathcal{E}) + (d-2)r > 0$.
(3) $0 \leq DL \leq 2r$ *for all the lines L on S.*
(4) *D can be represented by an irreducible nonsingular curve.*

Proof. (1) and (2) are immediate consequences of Proposition 5.1.1.

In order to prove (3) and (4), let us first assume that $(S, \mathcal{O}_S(-K_S))$ is a cubic surface and let $\mathcal{E}$ be a rank r Ulrich bundle on $(S, \mathcal{O}_S(-K_S))$. By Theorem 3.2.9 (5), $\mathcal{E}$ has a minimal resolution of the form

$$0 \longrightarrow \mathcal{O}_{\mathbb{P}^3}(-1)^{3r} \longrightarrow \mathcal{O}_{\mathbb{P}^3}^{3r} \longrightarrow \mathcal{E} \longrightarrow 0.$$

Therefore, tensoring it by $\mathcal{O}_L$ for an arbitrary line $L \subset S$, we get

$$0 \longrightarrow \mathcal{A} \longrightarrow \mathcal{O}_L^{3r} \longrightarrow \mathcal{E}_L \longrightarrow 0.$$

By Grothendieck's theorem, $\mathcal{A}$ completely splits as a sum of line bundles on $L \cong \mathbb{P}^1$. On the other hand, $\mathcal{A}$ is a subbundle of $\mathcal{O}_L^{3r}$ and at the same time a quotient of $\mathcal{O}_L(-1)$. Therefore, $\mathcal{A} \cong \mathcal{O}_L(-1)^a \oplus \mathcal{O}_L^b$ with $a + b = \operatorname{rank}(\mathcal{A}) = 2r$ and $a = \deg(\mathcal{E}_L) = DL$. So $0 \leq DL \leq 2r$. Finally, (4) follows from (2), (3) and [108, V, Exercise 4.8].

Now, assume that $(S, \mathcal{O}_S(-K_S))$ is a del Pezzo surface of degree $4 \leq d \leq 9$ and consider a blow-up $\pi : (S', \mathcal{O}_{S'}(-K_{S'})) \longrightarrow (S, \mathcal{O}_S(-K_S))$ where $(S', \mathcal{O}_{S'}(-K_{S'}))$ is a cubic surface. Then, for any line $L \subset S$, $\pi^{-1}(L)$ is also a line on the cubic surface. Therefore $DL = \pi^{-1}(D)\pi^{-1}(L)$, and we have the same inequalities. The analogous argument shows that D can be represented by an irreducible smooth curve. □

Proposition 5.3.13. *On any del Pezzo surface $(S, \mathcal{O}_S(-K_S))$ of degree d, any rank r stable vector bundle $\mathcal{E}$ with Hilbert polynomial $P_{\mathcal{E}}(t) = dr\binom{t+2}{2}$ is an Ulrich bundle.*

Proof. This is an immediate consequence of Proposition 5.1.1 applied to del Pezzo surfaces. Indeed, if $\mathcal{E}$ is a stable vector bundle with $P_{\mathcal{E}}(t) = dr\binom{t+2}{2}$, it holds that $h^0(\mathcal{E}(-1)) = 0$. On the other hand, by Serre's duality (see Theorem 1.3.2),

$$\chi(\mathcal{E}^\vee(t+2)) = \chi(\mathcal{E}(-t-3)) = dr\binom{-t-1}{2} = dr\binom{t+2}{2} = \chi(\mathcal{E}(t)).$$

So $\mathcal{E}^\vee(2)$ is also a stable vector bundle with the same Hilbert polynomial and therefore $h^2(\mathcal{E}(-2)) = h^0(\mathcal{E}^\vee(1)) = 0$. Moreover, the Chern classes of $\mathcal{E}$ satisfy the two conditions (5.2), and hence we can conclude by invoking Proposition 5.1.1. □

In the case of del Pezzo surfaces, the local dimension of moduli spaces of stable rank r Ulrich bundles are computed thanks to the following results:

Lemma 5.3.14. *Let $(S, \mathcal{O}_S(-K_S))$ be a del Pezzo surface of degree d. Let $\mathcal{E}$ and $\mathcal{F}$ be Ulrich bundles of ranks r and s, and with first Chern classes C and D, respectively. Then*

$$\chi(\mathcal{E} \otimes \mathcal{F}^\vee) = (d-1)rs - CD. \tag{5.32}$$

In particular, $\chi(\mathcal{E} \otimes \mathcal{E}^\vee) = (d-1)r^2 - C^2$.

Proof. It is a straightforward application of Riemann–Roch theorem to the particular case of del Pezzo surfaces. □

Corollary 5.3.15. *If $\mathcal{E}$ is a stable (respectively simple) rank r Ulrich bundle with $c_1(\mathcal{E}) = D$ on a del Pezzo surface $(S, \mathcal{O}_S(-K_S))$ of degree d, then the moduli space (respectively the modular family) of stable (respectively simple) vector bundles is smooth at $[\mathcal{E}]$ of dimension $D^2 - (d-1)r^2 + 1$.*

Proof. Let $\mathcal{E}$ be a stable (respectively simple) Ulrich bundle on S of rank r. By Serre's duality (Theorem 1.3.2) and the simplicity of $\mathcal{E}$, $H^2(\mathcal{E} \otimes \mathcal{E}^\vee) \cong H^0(\mathcal{E} \otimes \mathcal{E}^\vee(-1)) \cong \{0\} \subset H^0(\mathcal{E} \otimes \mathcal{E}^\vee) \cong \mathbf{k}$. Therefore, by Proposition 1.5.15, the moduli space (respectively the modular family) is smooth at the point $[\mathcal{E}]$ of dimension

$$h^1(\mathcal{E} \otimes \mathcal{E}^\vee) = 1 - \chi(\mathcal{E} \otimes \mathcal{E}^\vee) = D^2 - (d-1)r^2 + 1 \tag{5.33}$$

where the last equality follows from Corollary 5.3.14. □

Indeed, we can obtain more detailed information about the moduli space of Ulrich bundles.

Proposition 5.3.16. *(1) Let $\mathcal{E}$ be a rank r Ulrich bundle with the first Chern class $c_1(\mathcal{E}) = D$ on a del Pezzo surface $(S, \mathcal{O}_S(H))$ with $H = -K_S$. Then, there exists an exact sequence*

$$0 \longrightarrow \mathcal{O}_S^{r-1} \longrightarrow \mathcal{E} \longrightarrow \mathcal{I}_Z(D) \longrightarrow 0 \tag{5.34}$$

where Z is a set of points of length $c_2(\mathcal{E})$ and $h^0(\mathcal{I}_Z(D-H)) = 0$.

(2) In addition, fixing a divisor D of positive degree, $r \geq 2$ and $n \geq 1$, the family of coherent sheaves $\mathcal{E}$ on S obtained as extensions of the form (5.34) *with Z a set of points of length n satisfying $h^0(\mathcal{I}_Z(D-H)) = 0$ is irreducible (or empty).*

Proof. (1) Since $\mathcal{E}$ is globally generated, the quotient of a general map $\mathcal{O}_S^{r-1} \longrightarrow \mathcal{E}$ will be a rank-one torsion-free sheaf. This sheaf should be of the form $\mathcal{I}_Z(D)$ for a finite set of points Z of length $c_2(\mathcal{E})$. Since $h^0(\mathcal{E}(-H)) = h^1(\mathcal{O}_S(-H)) = 0$, we have $h^0(\mathcal{I}_Z(D-H)) = 0$.

(2) First of all, recall that the Hilbert scheme $\mathrm{Hilb}^n(S)$ which parameterizes 0-dimensional subschemes of S of length n is irreducible of dimension $2n$ [96]. The condition $\mathrm{h}^0(\mathcal{I}_Z(D-H)) = 0$ defines an open irreducible subset $U \subset \mathrm{Hilb}^n(S)$. Over the elements of U, the extensions of the form (5.34) are parameterized by a choice of $r-1$ $\mathbf{k}$-independent elements of $\mathrm{Ext}^1(\mathcal{I}_Z(D), \mathcal{O}_S) \cong \mathrm{H}^1(\mathcal{I}_Z(D-H))$. On the other hand, there is an exact sequence of the form

$$0 \longrightarrow \mathrm{H}^0(\mathcal{O}_S(D-H)) \longrightarrow \mathrm{H}^0(\mathcal{O}_Z) \longrightarrow \mathrm{H}^1(\mathcal{I}_Z(D-H)) \longrightarrow \mathrm{H}^1(\mathcal{O}_S(D-H)) \longrightarrow 0.$$

Finally, since $\mathrm{H}^2(\mathcal{O}_S(D-H)) \cong \mathrm{H}^0(\mathcal{O}_S(-D)) = 0$, we see that $\dim \mathrm{Ext}^1(\mathcal{I}_Z(D), \mathcal{O}_S) = n - \chi(\mathcal{O}_S(D-H))$ is constant once we fix n and D. So these coherent sheaves are parameterized by a Grassmannian of a vector bundle over the irreducible open subset U, and hence they form an irreducible family. □

Corollary 5.3.17. *The Ulrich bundles of fixed rank and first Chern class on a del Pezzo surface (if they exist) form an irreducible family. More precisely, there exists a smooth irreducible variety T and a vector bundle $\mathcal{E}$ on $S \times T$ such that any Ulrich bundle on S with those invariants is of the form $\mathcal{E}_t$ for some $t \in T$.*

Proof. By Remark 5.1.2, rank r Ulrich bundles $\mathcal{E}$ with the first Chern class $D := c_1(\mathcal{E})$, have a fixed second Chern class $c_2^U(r, D)$. Then we can consider the irreducible family T' obtained in Proposition 5.3.16 (2) with respect to the data $r, D, c_2^U(r, D)$. Being Ulrich is an open condition on T'. So, there exists an open subset $T \subset T'$ giving the parameter space of the statement. □

Corollary 5.3.18. *Let $\mathcal{E}$ be a rank r Ulrich bundle on a del Pezzo surface $(S, \mathcal{O}_S(-K_S))$, with the first Chern class $c_1(\mathcal{E}) = D$ and second Chern class $c_2(\mathcal{E}) = n$. Then, the subvariety $M^U_{S,-K_S}(r; D, n) \subset M_{S,-K_S}(r; D, n)$ of Ulrich bundles with these invariants is irreducible. Therefore, the open part $M^s_{S,-K_S}(r; D, n) \subset M_{S,-K_S}(r; D, n)$ of stable vector bundles, whenever nonempty, is smooth irreducible of dimension $D^2-(d-1)r^2+1$ and consists exclusively of Ulrich bundles.*

Proof. Since Ulrich bundles are in any case semistable, by the general properties of moduli functors, there exists a map from the irreducible variety T constructed in Corollary 5.3.17 towards the moduli space $M_{S,-K_S}(r; D, n)$ whose image is exactly $M^U_{S,-K_S}(r; D, n)$, and therefore is irreducible as well. Since, as we saw in Proposition 5.3.13, any stable vector bundle with these invariants (and therefore with Hilbert polynomial $P_{\mathcal{E}}(t) = dr\binom{t+2}{2}$) is an Ulrich bundle, we have that the open subvariety $M^s_{S,-K_S}(r; D, n) \subset M^U_{S,-K_S}(r; D, n)$, whenever nonempty, is smooth irreducible of dimension $D^2 - (d-1)r^2 + 1$. □

The rest of this section will be devoted to giving a precise characterization of the pairs (D, r) for which there exist (stable) rank r Ulrich bundles on a del Pezzo surface $(S, \mathcal{O}_S(-K_S))$ with the first Chern class D. Let us start first with Ulrich line bundles. It turns out that they correspond to a basic class of projective curves.

Definition 5.3.19. A *rational normal curve* of degree d is a nondegenerate rational smooth curve of degree d in some $\mathbb{P}^d$.

Remark 5.3.20. If D is a rational normal curve of degree d then

$$\mathrm{H}^0(\mathcal{O}_{\mathbb{P}^d}(1)) \longrightarrow \mathrm{H}^0(\mathcal{O}_D(1)) \cong \mathrm{H}^0(\mathcal{O}_{\mathbb{P}^1}(d))$$

is injective and therefore an isomorphism. This means that D is embedded through the complete linear system $|\mathcal{O}_{\mathbb{P}^1}(d)|$ and so, up to an automorphism of $\mathbb{P}^d$, is unique. Moreover, D is an aCM subvariety in $\mathbb{P}^d$ (Example 2.1.5).

Proposition 5.3.21. *Let $(S, \mathcal{O}_S(-K_S))$ be a del Pezzo surface of degree $d = K_S^2$. Then $\mathcal{O}_S(D)$ is an Ulrich line bundle if and only if D can be represented by a rational normal curve $C \subset S \subset \mathbb{P}^d$ of degree d.*

Proof. Let $\mathcal{O}_S(D)$ be an Ulrich line bundle on a del Pezzo surface $S \subset \mathbb{P}^d$. By Theorem 5.3.12 (1), D can be represented by a smooth irreducible curve C of degree d. Moreover, thanks to Theorem 5.3.8 (5), we can compute the arithmetic genus of C as

$$p_a(C) = \frac{1}{2}(D^2 - \deg(D)) + 1 = 0$$

again by Theorem 5.3.12, (2). Namely, C is a rational normal curve of degree d.

Conversely, if C is such a curve, we are going to rely on the characterization of Ulrich bundles on surfaces introduced in Proposition 5.1.1. To prove that $\mathcal{O}_S(D)$ is an Ulrich line bundle on $(S, \mathcal{O}_S(H)) = (S, \mathcal{O}_S(-K_S))$, we consider the short exact sequence

$$0 \longrightarrow \mathcal{O}_S \longrightarrow \mathcal{O}_S(C) \longrightarrow \mathcal{O}_C(C) \cong \mathcal{O}_{\mathbb{P}^1}(C^2) \longrightarrow 0.$$

Since $p_a(C) = 0$, applying again Theorem 5.3.8 (5), we obtain that $C^2 - d-2$. Tensoring the previous short exact sequence by $\mathcal{O}_S(-H)$ and considering the associated long exact sequence of global sections, we get $\mathrm{h}^0(\mathcal{O}_S(C - H)) = 0$ since $\mathrm{h}^0(\mathcal{O}_S(-H)) = 0$ and $\mathcal{O}_C(C - H) = \mathcal{O}_{\mathbb{P}^1}(-2)$ has no nonzero global sections. On the other hand, $\mathrm{h}^2(\mathcal{O}_S(C - 2H)) = \mathrm{h}^0(\mathcal{O}_S(H - C))$ by Serre's duality. The latter is zero since $H - C$ is a nontrivial divisor of degree zero (let us observe that the hyperplane section has arithmetic genus one). Since $\deg(\mathcal{O}_S(C)) = d$, we can assert that $\mathcal{O}_S(C)$ is an Ulrich line bundle due to Proposition 5.1.1. □

Corollary 5.3.22. *Let $(S, \mathcal{O}_S(-K_S))$ be a del Pezzo surface of degree $3 \leq d \leq 9$ given as the blow-up of $\mathbb{P}^2$ at $r = 9 - d$ points. Then, the list of Ulrich line bundles on it, up to a permutation of the exceptional divisors, in terms of the standard basis of $\mathrm{Pic}(S)$ is contained in the following table:*

$d = K_S^2$	u(D,d)	D
3	1	l
	20	$2l - e_1 - e_2 - e_3$
	30	$3l - 2e_1 - e_2 - e_3 - e_4 - e_5$
	20	$4l - 2e_1 - 2e_2 - 2e_3 - e_4 - e_5 - e_6$
	1	$5l - 2e_1 - 2e_2 - 2e_3 - 2e_4 - 2e_5 - 2e_6$
4	10	$2l - e_1 - e_2$
	20	$3l - 2e_1 - e_2 - e_3 - e_4$
	10	$4l - 2e_1 - 2e_2 - 2e_3 - e_4 - e_5$
5	4	$2l - e_1$
	12	$3l - 2e_1 - e_2 - e_3$
	4	$4l - 2e_1 - 2e_2 - 2e_3 - e_4$
6	1	$2l$
	6	$3l - 2e_1 - e_2$
	1	$4l - 2e_1 - 2e_2 - 2e_3$
7	2	$3l - 2e_1$
8, 9	0	

where $u(D,d)$ is the number of divisor classes of that shape arising as a permutation of the exceptional divisors in the writing of D. In particular, a del Pezzo surface $(S, \mathcal{O}_S(-K_S))$ of degree $3 \leq d \leq 9$ given as the blow-up of $\mathbb{P}^2$ at $r = 9 - d$ points respectively supports the following number of Ulrich line bundles: 72, 40, 20, 8, 2, 0, 0.

We move forward to Ulrich bundles of higher rank on del Pezzo surfaces. First, we are going to study the case of the cubic surface $S \subset \mathbb{P}^3$. Then we are going to rely on the behavior of Ulrich bundles under pullbacks and pushforwards by the blow-down maps to obtain analogous results on the other del Pezzo surfaces.

For any divisor D on the cubic surface $S \subset \mathbb{P}^3$, we define

$$a(D) := \min\{DT \mid T \text{ rational normal cubic curve}\}.$$

Lemma 5.3.23. *Let $S \subset \mathbb{P}^3$ be a cubic surface. Let D be a divisor of degree $3r$, for $r \geq 2$, such that $0 \leq DL \leq 2r$ for all lines L. Then there exists a rational normal cubic curve T with $a(D) = DT$ such that:*

1. *The divisor $D' := D - T$ satisfies $0 \leq D'L \leq 2r - 2$ for all lines L.*
2. *If $r \geq 3$, $a(D) \geq 2r$ and D is not a multiple of $D_0 := 4l - 2e_1 - e_2 - e_3 - e_4 - e_5$ then D' is not a multiple of D_0 and $a(D') \geq 2(r-1)$.*

Proof. See [41, Proposition 3.8]. □

Theorem 5.3.24. *Let $S \subset \mathbb{P}^3$ be a cubic surface, D a divisor on S, and $r \geq 1$ an integer. Then the following are equivalent:*

1. *D is linearly equivalent to the sum of r rational normal cubic curves $\sum_{i=1}^{r} T_i$.*
2. *There exists a rank r Ulrich bundle $\mathcal{E}$ with the first Chern class D.*

Moreover, if $r \geq 2$, the two previous conditions are equivalent to
(3) $\deg(D) = 3r$ *and* $0 \leq DL \leq 2r$ *for all lines* L *in* S.

Proof. The case of $r = 1$ has been dealt with in Proposition 5.3.21. So let us assume that $r \geq 2$. Then, that (1) implies (2) follows from the fact that, since $\mathcal{O}_S(T_i)$ is an Ulrich line bundle, by Proposition 3.3.1, the direct sum $\bigoplus_{i=1}^r \mathcal{O}_S(T_i)$ is a rank r Ulrich bundle with the first Chern class $\sum_{i=1}^r T_i$. That (2) implies (3) was stated in Theorem 5.3.12.

Finally, we give a sketch of the proof that (3) implies (1) and address the interested reader to [41, Theorem 3.9]. When $r = 2$, a case-by-case analysis of the Picard group $\mathrm{Pic}(S)$ shows that any effective divisor D of degree 6 such $DL \leq 4$ for any line L can be written as the sum of two rational normal cubic curves. On the other hand, if $r > 2$, we use Lemma 5.3.23 to write $D = D' + T$ with T a rational normal cubic curve and D' a divisor of degree $3(r-1)$ satisfying the condition (3). Then, by induction hypothesis, D' can be written as a sum of rational normal cubic curves, so the same holds for D. □

Theorem 5.3.25. *Let $S \subset \mathbb{P}^3$ be a cubic surface, D a divisor on S, and $r \geq 2$ an integer satisfying the equivalent conditions of Theorem 5.3.24. Then the following are equivalent:*
(1) $DT \geq 2r$ *for all rational normal cubic curves* T *and* $D \neq mD_0$ *for any* $m \geq 2$ *and* $D_0 = 4l - 2e_1 - e_2 - e_3 - e_4 - e_5$.
(2) *There exist stable rank r Ulrich bundles $\mathcal{E}$ with $c_1(\mathcal{E}) = D$.*

In this case, the moduli space $M^s_{S,-K_S}(r; D, n)$ of rank r stable bundles $\mathcal{E}$ with $c_1(\mathcal{E}) = D$ and $c_2(\mathcal{E}) = \frac{D^2-r}{2}$ is smooth, irreducible of dimension $D^2 - 2r^2 + 1$, and consists exclusively of stable Ulrich bundles.

Proof. The proof that (1) implies (2) is by induction on r. The case $r = 2$ is done by a case-by-case analysis of the divisors D of degree six satisfying the hypothesis. So let us suppose that $r \geq 3$. By Lemma 5.3.23, we can choose a rational normal cubic curve T such that $D' := D - T$ satisfies the same hypothesis. By induction, there exists a stable Ulrich bundle $\mathcal{F}$ with $c_1(\mathcal{F}) = D'$. By Lemma 5.3.14, $\chi(\mathcal{F}^\vee(T)) = 2(r-1) - D'.T$. Since $DT = a(D)$, it holds that $D'T = a(D) - 1$. Moreover, $\mathrm{h}^0(\mathcal{F}^\vee(T)) = 0$ and $\mathrm{h}^2(\mathcal{F}^\vee(T)) = 0$ by stability of $\mathcal{F}$. So

$$\mathrm{ext}^1(\mathcal{F}, \mathcal{O}_S(T)) = \mathrm{h}^1(\mathcal{F}^\vee(T)) = -\chi(\mathcal{F}^\vee(T)) = a(D) + 1 - 2r > 0$$

because $a(D) \geq 2r$. Thus, there are nonsplitting extensions

$$0 \longrightarrow \mathcal{O}_S(T) \longrightarrow \mathcal{E} \longrightarrow \mathcal{F} \longrightarrow 0, \tag{5.35}$$

with $\mathcal{E}$ simple, by Proposition 1.5.10. Therefore, the modular family of simple bundles with these invariants has dimension $D^2 - 2r^2 + 1$ by Corollary 5.3.15. The general element of this modular family is either stable or given by an extension of the form (5.35) (see [119, Proposition 2.3.1]). But the vector bundles that are extensions of the form (5.35) form a family of dimension

$$\dim\{\mathcal{F}\} + \mathrm{ext}^1(\mathcal{F}, \mathcal{O}_S(T)) - 1 = D'^2 - 2(r-1) + 1 + a - 2r = D^2 - 2r^2 - a(D) + 2r$$

because $D = D' + T$. Since $a(D) \geq 2r$, this number is smaller than $D^2 - 2r^2 + 1$, and we can conclude that there exist stable Ulrich bundles with those invariants.

For the proof that (2) implies (1), one shows that if $a(D) < 2r$ or $D = mD_0$ for some $m \geq 2$, then there exists a family of properly semistable Ulrich bundles of dimension greater than or equal to $D^2 - 2r^2 + 1$. Now, since the moduli space $M^s_{S,-K_S}(r; D, n)$ of stable Ulrich bundles by Corollary 5.3.18 would be irreducible of dimension exactly that number, we conclude that they cannot exist. See [41, Theorem 4.3] for the details of this part.

Once the equivalence between (1) and (2) has been established, the last part of the statement follows from Corollary 5.3.18. □

Let $(S, \mathcal{O}_S(-K_S))$ be a del Pezzo surface of degree $4 \leq d \leq 9$, given as the blow-up of $\mathbb{P}^2$ at $r = 9 - d$ points. The exceptional divisors are $e_1, \ldots, e_r$. We can take general $p_{r+1}, \ldots, p_6$ points on S and consider the cubic surface $(S', -K_{S'})$, jointly with the blow-up map $\pi : S' \longrightarrow S$ associated to these $d - 3$ points. Let us call the new exceptional divisors $e_{r+1}, \ldots, e_6$. Recall that $-K_{S'} = -K_S - e_{r+1} - \cdots - e_6$. In the next theorem we are going to see that there is a close relation between Ulrich bundles on $(S, \mathcal{O}_S(-K_S))$ and on $(S', \mathcal{O}_{S'}(-K'_S))$. Recall that in Remark 5.1.2 it was defined $c_2^U(r, D)$ as the second Chern of a rank r Ulrich bundle with the first Chern class D on a projective surface $(S, \mathcal{O}_S(-K_S))$.

Theorem 5.3.26. *With the above notation, let $r \geq 2$ be an integer and D a divisor. Set $D' := \pi^* D - r\sum_{i=r+1}^{6} e_i$. Then there exists a rational map*

$$\begin{array}{llll} \Phi : & M^U_{S',-K_{S'}}(r; D', c_2^U(r, D')) & \dashrightarrow & M^U_{S,-K_S}(r; D, c_2^U(r, D)), \\ & [\mathcal{F}] & \mapsto & \left[\pi_*\mathcal{F}\left(\sum_{i=r+1}^{6} e_i\right)\right] \end{array}$$

and a map

$$\begin{array}{llll} \Psi : & M^U_{S,-K_S}(r; D, c_2^U(r, D)) & \longrightarrow & M^U_{S',-K_{S'}}(r; D', c_2^U(r, D')), \\ & [\mathcal{E}] & \mapsto & \left[\pi^*\mathcal{E}\left(-\sum_{i=r+1}^{6} e_i\right)\right] \end{array}$$

which are inverses of each other. Moreover, Φ sends stable Ulrich bundles on $(S', \mathcal{O}_{S'}(-K'_S))$ to stable Ulrich bundles on $(S, \mathcal{O}_S(-K_S))$.

Proof. In order to define Φ, suppose first that $M^U_{S',-K_{S'}}(r;D',c_2^U(r,D'))$ is nonempty. Then D' can be written as the sum $\sum_{i=1}^r T_i$ for T_i rational normal cubic curves. Notice that any such T_i has the coefficients of the e_i between 0 and 2. Now since the coefficient on the sum should be r, it means that for any T_i with coefficient 2 there should be a T_j with coefficient 0. Checking the list (5.3.22) of rational normal cubic curves, we can confirm, by a case-by-case analysis, that we can find two other rational normal cubic curves T_i' and T_j' such that $T_i + T_j \equiv T_i' + T_j'$. In other words, we can express D' as a sum $\sum_{i=1}^r T_i$ where all the T_i have one coefficient for all the new exceptional divisors e_i, $i = 1, \ldots, d-3$. Now we can consider the completely splitting Ulrich bundle $\mathcal{E} := \bigoplus_{i=1}^r \mathcal{O}_S(T_i)$. We have $\mathcal{E}(e_i)_{|e_i} \cong \mathcal{O}_{\mathbb{P}^1}(1)$ for all e_i, $i = 1, \ldots, d-3$. Since $[\mathcal{E}] \in M^U_{S',-K_{S'}}(r;D',c_2^U(r,D'))$ is irreducible, by semicontinuity there exists an open subset $U \subset \Gamma$ such that for any $[\mathcal{E}] \in U$, $\mathcal{E}(e_i)_{e_i} \cong \mathcal{O}_{\mathbb{P}^1}(1)^r$. Therefore, by Theorem 3.3.9, the map Φ is well-defined on U. On the other hand, by Theorem 3.3.7, the map Ψ is defined everywhere, and clearly they are inverses of each other.

To conclude, let us assume that there exists an Ulrich bundle $\mathcal{E}$ on S' such that $\tilde{\mathcal{E}} := \pi_*\mathcal{E}(\sum_{i=r+1}^6 e_i)$ is not stable. Then, by Corollary 3.3.17, there exists a short exact sequence

$$0 \longrightarrow \mathcal{E}_1 \longrightarrow \tilde{\mathcal{E}} \longrightarrow \mathcal{E}_2 \longrightarrow 0$$

of Ulrich bundles on $(S, \mathcal{O}_S(-K_S))$. Applying the map Ψ to this short exact sequence (which remains exact under pullbacks), we see that $\mathcal{E}$ was already properly semistable. □

Corollary 5.3.27. *Let $(S, \mathcal{O}_S(-K_S))$ be a del Pezzo surface of degree $3 \le d \le 9$, given as the blow-up of $\mathbb{P}^2$ at $r = 9 - d$ points. Let T_i, $i = 1, \ldots, r$, be divisor classes of rational normal curves of degree d. Then there exists an Ulrich bundle $\mathcal{E}$ with $c_1(\mathcal{E}) = \sum_{i=1}^r T_i$.*

Proof. It is immediate from Theorem 5.3.26 once we observe from the list of Ulrich line bundles on del Pezzo surfaces from Corollary 5.3.22 that if T is the divisor class of a rational normal curve on $(S, \mathcal{O}_S(-K_S))$ of degree d, then $\pi^*T - \sum_{i=1}^{d-3} e_i$ is the divisor class of a rational normal cubic curve on the cubic surface $(S, \mathcal{O}_S(-K_S))$. □

Remark 5.3.28. Observe that for $d = 8, 9$ the del Pezzo surfaces $(S, \mathcal{O}_S(-K_S))$ of degree d given as the blow-up at $9 - d$ points do not contain any rational normal curve of degree d. So Corollary 5.3.27 does not apply. On the other hand, it is easy to check that if we call $H := -K_S$ the anticanonical divisor, the divisor $r(\pi^*H - \sum_{i=r+1}^6 e_i)$ can be written as the sum of rational normal cubic curves. Therefore we prove that the moduli space of stable (Ulrich vector bundles) $M^s_{S,-K_S}(r;rH,c_2^U(r,rH))$ is a smooth irreducible quasiprojective variety of dimension r^2+1.

5.4 Surfaces with $\kappa = 0$

The goal of this section is to prove the existence of Ulrich bundles on smooth surfaces of Kodaira dimension 0. We first consider minimal smooth surfaces S with Kodaira dimension 0. So S will be either a K3 surface, or an Enriques surface, or a bielliptic surface, or an abelian surface (see the end of Section 5.1).

Let us observe that we cannot expect, in general, to find Ulrich line bundles on an arbitrary surface S with $\kappa = 0$ (for instance, they do not exist when $\mathrm{Pic}(S) \cong \mathbb{Z}$ which is the case, for example, for a general K3 surface). Whence we will focus on rank 2 Ulrich bundles. We will first deal with smooth K3 surfaces. Our approach follows an idea of Faenzi [92]. We start building a rank 2 aCM bundle $\mathcal{E}$ on a smooth K3 surface S using Serre's construction. Next we perform a sort of elementary modification of $\mathcal{E}$ along a single generic point $p \in S$ and get a simple nonreflexive sheaf $\mathcal{E}_0$ on S. Finally, flatly deforming $\mathcal{E}_0$, we get the desired Ulrich bundle. This method is based on a result of Artamkin. To introduce it, let us recall the following definition.

Definition 5.4.1. Let $(X, \mathcal{O}_X(H))$ be a smooth projective surface and $\mathcal{F}$ a torsion-free sheaf on X. For any point $p \in X$ and any surjection $\varphi : \mathcal{F} \longrightarrow \mathcal{O}_p$, we define the *elementary modification* $\mathcal{F}_\varphi$ of $\mathcal{F}$ as the kernel of φ. So, we have an exact sequence

$$0 \longrightarrow \mathcal{F}_\varphi \longrightarrow \mathcal{F} \longrightarrow \mathcal{O}_p \longrightarrow 0. \tag{5.36}$$

It is easy to check that elementary modifications preserve μ-stability. Moreover, we have:

$$\mathrm{rank}(\mathcal{F}) = \mathrm{rank}(\mathcal{F}_\varphi), \quad c_1(\mathcal{F}) = c_1(\mathcal{F}_\varphi), \quad \text{and} \quad \Delta(\mathcal{F}_\varphi) = \Delta(\mathcal{F}) + \frac{1}{r}.$$

Since $\mathrm{H}^i(X, \mathcal{O}_p) = 0$ for $i \geq 1$, the long exact sequence of cohomology associated to (5.36) implies that $\mathrm{H}^2(X, \mathcal{F}) = \mathrm{H}^2(X, \mathcal{F}_\varphi)$. Furthermore, as long as $p \in X$ and $\varphi : \mathcal{F} \longrightarrow \mathcal{O}_p$ are general and $\mathrm{h}^0(X, \mathcal{F}) > 0$, the map $\mathrm{H}^0(X, \mathcal{F}) \longrightarrow \mathrm{H}^0(X, \mathcal{O}_p)$ is surjective. Hence, if $\mathrm{h}^0(X, \mathcal{F}) > 0$, then the long exact sequence of cohomology associated to (5.36) implies that $\mathrm{h}^0(X, \mathcal{F}_\varphi) = \mathrm{h}^0(X, \mathcal{F}) - 1$ and $\mathrm{h}^1(X, \mathcal{F}_\varphi) = \mathrm{h}^1(X, \mathcal{F})$. Similarly, if $\mathrm{h}^0(X, \mathcal{F}) = 0$, then $\mathrm{h}^0(X, \mathcal{F}_\varphi) = \mathrm{h}^0(X, \mathcal{F})$ and $\mathrm{h}^1(X, \mathcal{F}_\varphi) = \mathrm{h}^1(X, \mathcal{F}) + 1$.

Proposition 5.4.2. *Let $(S, \mathcal{O}_S(H))$ be a smooth surface, $p \in S$ a point, and $\mathcal{E}$ a vector bundle on S of rank $r \geq 2$ such that $\mathrm{Ext}^2(\mathcal{E}, \mathcal{E}) = \mathrm{H}^2(\mathcal{O}_S)$. For any $0 \neq \varphi \in \mathrm{Hom}(\mathcal{E}, \mathcal{O}_p)$, set $\mathcal{E}_\varphi = \ker(\varphi)$. Then, $\mathcal{E}_\varphi$ has a universal deformation whose general sheaf is locally free at p.*

Proof. See [13, Theorem 1.4 and Corollary 1.5]. □

Theorem 5.4.3. *Any smooth K3 surface $(S, \mathcal{O}_S(H))$ admits a rank 2 special simple Ulrich bundle.*

Proof. A general hyperplane section C of S is a smooth curve of genus g and degree $H^2 = 2g - 2$ with $\omega_C \cong \mathcal{O}_C(H)$ and there is a short exact sequence

$$0 \longrightarrow \mathcal{O}_S(-H) \longrightarrow \mathcal{O}_S \longrightarrow \mathcal{O}_C \longrightarrow 0.$$

First, we apply Kodaira vanishing theorem and get that $\mathrm{H}^1_*(\mathcal{O}_S(tH)) = 0$. Therefore, $\mathcal{O}_S$ is an aCM line bundle on S. Let $Z \subset S$ be a set of $g + 2$ points in general linear position (meaning that no $g + 1$ of them are on a hyperplane). The set Z satisfies the CB property with respect to $\mathcal{O}_S(H)$ and by Theorem 5.1.6, Z gives rise to a rank 2 vector bundle $\mathcal{E}$ on S with $c_1(\mathcal{E}) = H$ (and, hence, $\mathcal{E}^\vee \cong \mathcal{E}(-H)$) which fits into an exact sequence

$$0 \longrightarrow \mathcal{O}_S \longrightarrow \mathcal{E} \longrightarrow \mathcal{I}_Z(H) \longrightarrow 0. \tag{5.37}$$

Let us check that $\mathcal{E}$ is aCM and simple. We first observe that $\mathrm{H}^1(\mathcal{E}) \cong \mathrm{Ext}^1(\mathcal{E}, \mathcal{O}_S)^\vee = 0$ and by Serre duality $\mathrm{h}^1(\mathcal{E}(-H)) = \mathrm{h}^1(\mathcal{E}^\vee(H)) = \mathrm{h}^1(\mathcal{E}) = 0$. We restrict the exact sequence (5.37) to a hyperplane section C that avoids Z and obtain the exact sequence

$$0 \longrightarrow \mathcal{O}_C \longrightarrow \mathcal{E}_{|C} \longrightarrow \mathcal{O}_C(H) \longrightarrow 0.$$

From $\mathrm{H}^1(\mathcal{E}(-H)) = \mathrm{H}^0(\mathcal{E}(-H)) = 0$ we deduce that $\mathrm{h}^0(\mathcal{E}_{|C}) = \mathrm{h}^0(\mathcal{E})$. From the fact that Z is a set of $g+2$ points in general linear position, we deduce that $\mathrm{h}^0(\mathcal{E}) = \mathrm{h}^0(\mathcal{O}_S) = 1$, and hence $\mathrm{H}^0(\mathcal{E}_{|C}(-H)) = 0$. This implies that $\mathrm{H}^1(\mathcal{E}(-2H)) = 0$ and actually $\mathrm{H}^1(\mathcal{E}(-tH)) = 0$ for all $t \geq 1$. Serre duality allows us to show that $\mathrm{H}^1(\mathcal{E}(tH)) = 0$ for all $t \geq 1$, and we conclude that $\mathcal{E}$ is an aCM bundle.

To see that $\mathcal{E}$ is simple, we apply the functor $\mathrm{Hom}(\mathcal{E}, -)$ to the exact sequence

$$0 \longrightarrow \mathcal{I}_Z(H) \longrightarrow \mathcal{O}_S(H) \longrightarrow \mathcal{O}_Z \longrightarrow 0$$

and get $0 \neq \mathrm{Hom}(\mathcal{E}, \mathcal{I}_Z(H)) \subset \mathrm{Hom}(\mathcal{E}, \mathcal{O}_S(H)) \cong \mathrm{H}^0(\mathcal{E}) \cong \mathbf{k}$, so $\mathrm{Hom}(\mathcal{E}, \mathcal{I}_Z(H)) \cong \mathbf{k}$. Applying $\mathrm{Hom}(\mathcal{E}, -)$ to the exact sequence (5.37) and taking into account that $\mathrm{Hom}(\mathcal{E}, \mathcal{O}_S) \cong \mathrm{H}^2(\mathcal{E})^\vee = 0$, we obtain $0 \neq \mathrm{Hom}(\mathcal{E}, \mathcal{E}) \subset \mathrm{Hom}(\mathcal{E}, \mathcal{I}_Z(H)) \cong \mathbf{k}$ and conclude that $\mathcal{E}$ is simple. By Serre duality, we also have $\mathrm{Ext}^2(\mathcal{E}, \mathcal{E}) \cong \mathbf{k}$.

We now choose a point $p \in S \setminus Z$ and a surjective map $\varphi : \mathcal{E} \longrightarrow \mathcal{O}_p$ such that the induced map on global sections $\mathrm{H}^0(\varphi) : \mathrm{H}^0(\mathcal{E}) \longrightarrow \mathrm{H}^0(\mathcal{O}_p)$ is an isomorphism. According to Definition 5.4.1, we perform an elementary modification along the point p and define $\mathcal{E}_\varphi := \ker(\varphi)$. It holds that $c_1(\mathcal{E}_\varphi) = H$, $c_2(\mathcal{E}_\varphi) = g + 3$ and $\mathrm{Hom}(\mathcal{E}_\varphi, \mathcal{E}_\varphi) \cong \mathrm{Ext}^2(\mathcal{E}_\varphi, \mathcal{E}_\varphi) \cong \mathbf{k}$.

Our final aim is to flatly deform $\mathcal{E}_\varphi$ to an Ulrich bundle. By Proposition 5.4.2, there exist a smooth connected variety X of dimension

$$\mathrm{ext}^1(\mathcal{E}_\varphi, \mathcal{E}_\varphi) = \mathrm{hom}(\mathcal{E}_\varphi, \mathcal{E}_\varphi) + \mathrm{ext}^2(\mathcal{E}_\varphi, \mathcal{E}_\varphi) - \chi(\mathcal{E}_\varphi \otimes \mathcal{E}_\varphi^\vee) = 2g + 8$$

and a flat family of simple sheaves $\mathcal{F}$ on $S \times X$ such that $\mathcal{F}_x$ is locally free for any point x in an open dense subset $X_1 \subset X$ and $\mathcal{F}_{x_0} = \mathcal{E}_\varphi$ for some distinguished point $x_0 \in X$.

It only remains to prove that $\mathcal{F}_x$ is an initialized rank 2 special Ulrich bundle for all generic $x \in X$. By construction $H^0(\mathcal{F}_{x_0}) = 0$ and by semicontinuity $H^0(\mathcal{F}_x) = 0$ for all x in an open dense subset $X_2 \subset X$. Hence, the isomorphism $\mathcal{F}_x^\vee \cong \mathcal{F}_x(-H)$ and Serre duality imply $H^i(\mathcal{F}_x(-H)) \cong H^{2-i}(\mathcal{F}_x^\vee(H))^\vee \cong H^{2-i}(\mathcal{F}_x)^\vee = 0$. We conclude that $\mathcal{F}_x(H)$ is a rank 2 initialized special Ulrich bundle for all x in the open dense subset $X_1 \cap X_2 \subset X$ which proves the claim. □

Theorem 5.4.4. *Any minimal smooth surface $(S, \mathcal{O}_S(H))$ of Kodaira dimension 0 supports a rank 2 special Ulrich bundle.*

Proof. The case of K3 surfaces was treated in Theorem 5.4.3, and we will give a uniform proof for the remaining three cases using Serre's construction. If $S \subset \mathbb{P}^n$ is an Enriques surface, we choose a set Z of $n+2$ general points of S. They obviously satisfy the CB property with respect to $\mathcal{O}_S(H)$ since no $n+1$ of them lie on a hyperplane. By Serre's construction, there exists a rank 2 vector bundle $\mathcal{E}$ on S given by an extension as follows:

$$0 \longrightarrow \mathcal{O}_S(K_S) \longrightarrow \mathcal{E} \longrightarrow \mathcal{I}_Z(H) \longrightarrow 0.$$

We have

$$\begin{aligned} \chi(\mathcal{E}) &= \chi(\mathcal{O}_S(K_S)) + \chi(\mathcal{I}_Z(H)) \\ &= \chi(\mathcal{O}_S(K_S)) + \chi(\mathcal{O}_S(H)) - n - 2 = 0, \\ h^0(\mathcal{E}) &= h^0(\mathcal{I}_Z(H)) + h^0(\mathcal{O}_S(K_S)) = 0, \quad \text{and} \\ \det(\mathcal{E}) &= \mathcal{O}_S(K_S + H). \end{aligned}$$

Therefore, by the characterization of Ulrich bundles on surfaces given in Proposition 5.1.3, $\mathcal{E}(H)$ is an initialized rank 2 special Ulrich bundle on S.

Finally, let $S \subset \mathbb{P}^n$ be an abelian surface or a bielliptic surface. We take a general hyperplane section $C = S \cap H$ of S. Then C is a smooth curve of genus $n+2$ (Riemann–Roch's theorem), and we take a set Z of $n+1$ general points on C. Any subset of n points of Z span H. Hence Z satisfies the CB property with respect to $\mathcal{O}_S(H)$, and, by Theorem 5.1.6, we can build a rank 2 vector bundle on S:

$$0 \longrightarrow \mathcal{O}_S(K_S) \longrightarrow \mathcal{E} \longrightarrow \mathcal{I}_Z(H) \longrightarrow 0. \tag{5.38}$$

Let $\eta \in \mathrm{Pic}^0(S) \setminus \{\mathcal{O}_S, \mathcal{O}_S(K_S)\}$ be an element of order 2. Since $\chi(\mathcal{O}_S(K_S)) = 0$ and $\chi(\mathcal{I}_Z(H)) = \chi(\mathcal{O}_S(H)) - n - 1 = 0$, from the exact sequence (5.38), we get $\chi(\mathcal{E} \otimes \eta) = \chi(\mathcal{E}) = 0$. We twist the exact sequence (5.38) by η and obtain $H^i(\mathcal{O}_S(K_S) \otimes \eta) = H^i(\eta) = 0$ for $0 \le i \le 2$. Let us check that $H^0(\mathcal{I}_Z(H) \otimes \eta) = 0$. Since the restriction map

$$H^0(S, \eta \otimes \mathcal{O}_S(H)) \longrightarrow H^0(C, \eta \otimes \mathcal{O}_S(H)_{|C})$$

is an isomorphism, it suffices to prove that $H^0(C, \eta \otimes \mathcal{O}_S(H)_{|C}(-Z)) = 0$. We have $\eta \otimes \mathcal{O}_S(H)_{|C} = \mathcal{O}_C(K_C) \otimes (\eta \otimes \mathcal{O}_S(-K_S))_{|C}$; since $\eta \otimes \mathcal{O}_S(-K_S)$ is nontrivial, so is its restriction to the curve C. Therefore, $h^0(C, \eta \otimes \mathcal{O}_S(H)_{|C}) = g(C) - 1 = n + 1$ and $H^0(C, \eta \otimes \mathcal{O}_S(H)_{|C}(-Z)) = 0$ because Z is a set of $n + 1$ general points on C. Finally, we apply Proposition 5.1.3 to $\mathcal{E} \otimes \eta \otimes \mathcal{O}_S(H)$ and get that $\mathcal{E} \otimes \eta \otimes \mathcal{O}_S(H)$ is a rank 2 special Ulrich bundle on S. □

Theorem 5.4.5. *For any smooth surface S of Kodaira dimension 0 there exists a rank 2 Ulrich bundle on $(S, \mathcal{O}_S(H))$ for a certain very ample line bundle $\mathcal{O}_S(H)$.*

Proof. By Theorem 5.4.4, the minimal model of any smooth projective surface with Kodaira dimension 0 carries a rank 2 special Ulrich bundle. Applying Proposition 3.3.7, we get the claim. □

We want to finish this section giving an application of the Lazarsfeld–Mukai exact sequence explained in the first section of this chapter (see Proposition 5.1.11). Indeed, we have:

Example 5.4.6. Let $S \subset \mathbb{P}^4$ be a smooth complete intersection surface of type $(2, 3)$. We have $\deg(S) = 6$, $q(S) = 0$, $K_S = 0$, and hence S is a K3 surface. In addition, $\frac{1}{2}H(5H + 3K_S) + 2\chi(\mathcal{O}_S) = 19$. Take a smooth curve $C \in |3H|$. By [9, Lemma 2.4], there is a line bundle $\mathcal{L}$ on C of degree $\deg(\mathcal{L}) = 19$ and two sections $\sigma_1, \sigma_2 \in H^0(\mathcal{L})$ which define a base point free pencil. In addition, $H^1(\mathcal{L} \otimes \mathcal{O}_S(H + K_S)) = H^1(\mathcal{L} \otimes \mathcal{O}_S(H)) = 0$ and, applying Proposition 5.1.11, we conclude that the vector bundle $\mathcal{E}$ defined by the Lazarsfeld–Mukai exact sequence

$$0 \longrightarrow \mathcal{E}^\vee \longrightarrow \mathcal{O}_S^2 \xrightarrow{\sigma_1, \sigma_2} \mathcal{L} \longrightarrow 0$$

is a special rank 2 Ulrich bundle on the complete intersection $S \subset \mathbb{P}^4$ of type $(2, 3)$.

Notice that the rank of the Ulrich bundle that we construct with this approach is lower than the rank 6 Ulrich bundle that we obtained in Theorem 4.3.2 applying matrix factorization tools.

5.5 Examples of surfaces with positive κ carrying an Ulrich bundle

In the previous sections of this chapter, we have worked on smooth projective surfaces S with $\kappa(S) \leq 0$, and in this section we deal with smooth projective surfaces of positive Kodaira dimension where the picture is still unclear. We will start with a broad class of elliptic regular surfaces S ($q(S) = 0$) with Kodaira dimension $\kappa(S) = 1$. Let us start fixing the results and notation from the theory of elliptic surfaces and Weierstrass fibrations needed in the sequel.

Definition 5.5.1. An *elliptic surface* is a smooth surface S together with a holomorphic map $\pi : S \longrightarrow C_0$ from S to a smooth curve C_0 such that the general fiber of π is a smooth connected curve of genus one.

An elliptic surface $\pi : S \longrightarrow C_0$ is *a minimal elliptic surface* if there are no (-1)-curves in the fibers of π. We say that $\pi : S \longrightarrow C_0$ is an elliptic surface with section C if a section $s : C_0 \longrightarrow S$ of π is given (this means that $\pi \circ s = Id_{C_0}$) and the image of s is the curve C on S.

Definition 5.5.2. Let S be a smooth surface and let C_0 be a smooth curve. A *Weierstrass fibration* $\pi : S \longrightarrow C_0$ is a flat and proper map such that every geometric fiber has arithmetic genus one, with general fiber smooth and such that there is a section of π not passing through the singular point of any fiber.

Definition 5.5.3. Let $\pi : S \longrightarrow C_0$ be a Weierstrass fibration with section $C \subset S$ and let $\mathcal{N}_{C|S}$ be the normal bundle of C in S. We define the *fundamental line bundle* $\mathcal{L}$ of π as the line bundle $\pi_* \mathcal{N}_{C|S}^\vee$ on C_0. We set $e := \deg(\mathcal{L})$.

It holds:

$$C^2 = -e, \quad F^2 = 0, \quad \text{and} \quad CF = 1 \quad \text{for a general fiber } F.$$

Moreover, it turns out that $S \cong E \times C$ for an elliptic curve E if and only if $\mathcal{L} \cong \mathcal{O}_{C_0}$.

In the next two propositions, we summarize the main results that we will need concerning Weierstrass fibrations.

Proposition 5.5.4. *Let $\pi : S \longrightarrow C_0$ be a Weierstrass fibration with section C, general fiber F and fundamental line bundle $\mathcal{L}$, with $e = \deg(\mathcal{L}) \geq 1$ and $g := g(C_0)$. We have:*
(1) $K_S \equiv (2g - 2 + e)F$, $q(S) = h^1(\mathcal{O}_S) = g$, $p_g(S) = h^0(\mathcal{O}_S(K_S)) = e - 1 + g$, $\chi(\mathcal{O}_S) = e$, *and* $h^{1,1}(S) = 10e + 2g$.
(2) $\kappa(S) = -\infty$ *if and only if S is a rational surface if and only if* $g = 0$ *and* $e = 1$.
(3) $\kappa(S) = 0$ *if and only if S is a K3 surface if and only if* $g = 0$ *and* $e = 2$.
(4) $\kappa(S) = 1$ *if and only if* $(g, e) \notin \{(0, 1), (0, 2)\}$.
(5) *Let* $\mathcal{O}_S(A) \in \operatorname{Pic}(S)$. *Then* $A \equiv aC + bF$ *if and only if* $\mathcal{O}_S(A) \cong \mathcal{O}_S(aC) \otimes \pi^* \mathcal{M}$ *for some* $\mathcal{M} \in \operatorname{Pic}^b(C_0)$.

Proof. See [145, Proposition 3.6]. □

So, the Picard rank $\rho(S)$ of a Weierstrass fibration $\pi : S \longrightarrow C_0$ satisfies $2 \leq \rho(S) \leq 10e + 2g$. Moreover, if $\pi : S \longrightarrow C_0$ is a Weierstrass fibration with $\rho(S) = 2$ then the group $N^1(S)$ of numerical equivalent classes of divisors on S satisfies $N^1(S) \cong \mathbb{Z}[C] \oplus \mathbb{Z}[F]$ (see [145, Remark 3.8]).

Proposition 5.5.5. *Let $\pi : S \longrightarrow C_0$ be a Weierstrass fibration with section C, general fiber F and fundamental line bundle $\mathcal{L}$, with $e = \deg(\mathcal{L}) \geq 1$ and $g = g(C_0)$. Let $D \equiv aC + bF$. We have:*
(1) *D is very ample if* $a \geq 3$ *and* $b \geq ae + 2g + 1$.
(2) *If D is very ample then* $a \geq 3$ *and* $b \geq ae + 1$.
(3) $|D|$ *is base point free if* $a \geq 2$ *and* $b \geq ae + 2g$.
(4) $H^1(\mathcal{O}_S(D)) = 0$ *if either* $a = 1$ *and* $b \geq 2g - 1$ *or* $a \geq 2$ *and* $b \geq ae + 2g - 1$.

Proof. See [145, Proposition] or [154, Sections III, IV, and VII]. □

Lemma 5.5.6. *Let $\pi : S \longrightarrow C_0$ be a Weierstrass fibration with section C and general fiber F. Let $D \equiv aC + bF$, with $a < 0$. Then $\mathrm{H}^0(\mathcal{O}_S(D)) = 0$.*

Proof. Let $H \equiv 3C + \beta F$ be a divisor, with $\beta \gg 0$. By Proposition 5.5.5 (1), H is very ample and

$$(aC + bF)H = -3ae + a\beta + 3b < 0.$$

Thus, we can conclude using Nakai–Moishezon criterion (see [108, Theorem V.1.10]). □

Proposition 5.5.7. *Let $\pi : S \longrightarrow C_0$ be a Weierstrass fibration with section C, $e \geq 1$, and let $H \equiv \alpha C + \beta F$ be a very ample divisor. Then $(S, \mathcal{O}_S(H))$ does not support Ulrich line bundles $\mathcal{O}_S(D)$ with D numerically equivalent to a linear combination of C and F.*

Proof. Let us suppose, on the contrary, that there exists a divisor $D \equiv aC + bF$ such that $\mathcal{O}_S(D + H)$ is Ulrich. Then its Ulrich dual $\mathcal{O}_S(2H + K_S - D)$ is also an Ulrich bundle. Set $H + K_S - D \equiv cC + dF$. Then $a + c = \alpha$ and $b + d = \beta + 2g - 2 + e$. Since $\mathcal{O}_S(D + H)$ is an Ulrich line bundle, we have $\chi(\mathcal{O}_S(D)) = 0$ and $\chi(\mathcal{O}_S(H + K_S - D)) = 0$. So, by Riemann–Roch theorem, we have

$$\begin{aligned} 2e &= a(ae - 2b + 2g - 2 + e), \\ 2e &= c(ce - 2d + 2g - 2 + e). \end{aligned} \tag{5.39}$$

Therefore

$$\begin{aligned} -2e &\leq ae - 2b + 2g - 2 + e \leq 2e, \\ -2e &\leq ce - 2d + 2g - 2 + e \leq 2e. \end{aligned}$$

Adding these two inequalities, we get

$$-4e \leq \alpha e - 2\beta \leq 4e.$$

From this we deduce $\alpha = 3$ and $\beta \leq 7e/2$. On the other hand, since $\mathrm{H}^0(\mathcal{O}_S(D + H)) \neq 0$ and $\mathrm{H}^0(\mathcal{O}_S(2H + K_S - D)) \neq 0$, we have $a + 3 \geq 0$ and $c + 3 \geq 0$. Therefore, $-3 \leq a \leq 6$. Now, the cases $a = 0, 3$ are immediately excluded from (5.39). To rule out the rest of cases, since $a + c = 3$, it is enough to show that a cannot be positive. In order to do that, we observe that for $\mathcal{O}_S(D + H)$ being Ulrich, we have

$$(D + H)H = \frac{1}{2}(3H^2 + HK_S)$$

or, equivalently,

$$b = (e - \beta/3)a + \beta + g - 1 - e.$$

Now suppose that $a > 0$. Since $0 = \mathrm{H}^0(\mathcal{O}_S(aC + bF))$ and since every linear system in $\mathrm{Pic}^b(C_0)$, for $b \geq g$, contains effective divisors, we can conclude, using Proposition 5.5.4 (5), that $b \leq g - 1$. Therefore

$$(e - \beta/3)a + \beta - e \leq 0,$$

which implies

$$(a - 1)e \leq (a - 3)\beta/3 \leq 7(a - 3)e/6.$$

Since $e \geq 1$, the latter yields $a \geq 15$, a contradiction. □

Let $H \equiv \alpha C + \beta F$ be a very ample divisor. We are going to construct rank 2 vector bundles $\mathcal{E}$ on $(S, \mathcal{O}_S(H))$ with Chern classes

$$c_1(\mathcal{E}) = 3H + K_S, \quad c_2(\mathcal{E}) = \frac{1}{2}(5H^2 + 3HK_S) + 2\chi(\mathcal{O}_S)$$

and such that $\mathrm{H}^0(\mathcal{E}(-H)) = 0$. By Proposition 5.1.3, any such bundle is a special Ulrich bundle. Let us consider the following divisors: A will be a divisor numerically equivalent to $(\alpha + 1)C + (\beta - 1)F$ and B a divisor such that $\mathcal{O}_S(B) \cong \mathcal{O}_S(3H + K_S - A)$. Then we have

$$A \equiv (\alpha + 1)C + (\beta - 1)F \quad \text{and} \quad B \equiv (2\alpha - 1)C + (2\beta + 2g - 1 + e)F. \tag{5.40}$$

We will construct such vector bundles via Serre's correspondence as extensions of the form

$$0 \longrightarrow \mathcal{O}_S(A) \longrightarrow \mathcal{E} \longrightarrow \mathcal{I}_Z(B) \longrightarrow 0, \tag{5.41}$$

using the Cayley–Bacharach property. Let us compute explicitly the cardinality of $|Z|$. We know that

$$\begin{aligned} c_2(\mathcal{E}) &= \frac{1}{2}(5H^2 + 3HK_S) + 2\chi(\mathcal{O}_S) \\ &= \frac{1}{2}(-5e\alpha^2 + 10\alpha\beta + 6\alpha g - 6\alpha + 3e\alpha) + 2e, \\ AB &= -2\alpha^2 e + 2e + 4\alpha\beta + 2g\alpha - 3\alpha + \beta + 2g. \end{aligned}$$

Therefore

$$z := |Z| = c_2(\mathcal{E}) - AB = \frac{\alpha e}{2}(3 - \alpha) + (\alpha - 1)\beta + g\alpha - 2g.$$

Since $\alpha \geq 3$ and $\beta \geq \alpha e + 1$, z is positive:

$$z \geq \frac{\alpha e}{2}(\alpha + 1) + (\alpha - 1) + g(\alpha - 2) > 0.$$

Proposition 5.5.8. *For arbitrary $\alpha \geq 3$ and $\beta \geq (2e+2g-1)(\alpha-2)+1$, and a general set Z of z points, the pair $(\mathcal{O}_S(B-A+K_S), Z)$ has the CB property.*

Proof. Using the Riemann–Roch formula, we compute the Euler characteristic of $\mathcal{O}_S(B-A+K_S) \cong \mathcal{O}_S((\alpha-2)C+(\beta+4g-2+2e)F)$ and obtain that

$$\begin{aligned}\chi(\mathcal{O}_S(B-A+K_S)) &= \frac{1}{2}((B-A+K_S)^2-(B-A+K_S)K_S)+\chi(\mathcal{O}_S)\\ &= -e\alpha^2/2+7e\alpha/2-4e+\alpha\beta+3g\alpha-\alpha-2\beta-6g+2.\end{aligned}$$

On the other hand, applying Proposition 5.5.5 (4), we see that $h^1(\mathcal{O}_S(B-A+K_S))=0$. Moreover, by Serre's duality and Lemma 5.5.6, we have

$$H^2(\mathcal{O}_S(B-A+K_S)) \cong H^0(\mathcal{O}_S(A-B))=0.$$

Gathering the previous computations, we obtain

$$h^0(\mathcal{O}_S(B-A+K_S))-z=\chi(\mathcal{O}_S(B-A+K_S))-z=(2e+2g-1)\alpha-\beta-4e-4g+2<0$$

for $\beta \geq (2e+2g-1)(\alpha-2)+1$, and hence CB property is trivially satisfied. □

Proposition 5.5.9. *Let $\pi : S \longrightarrow C_0$ be a Weierstrass fibration with section C, general fiber F, $e \geq 1$, and let $H \equiv \alpha C+\beta F$ be a very ample divisor. Let us assume that $\beta \geq \alpha e+2g$. Then any vector bundle $\mathcal{E}$ constructed as an extension* (5.41) *has the following cohomology table:*

	−3	−2	−1	0	1	2
h^2	$2H^2$	a	0	0	0	0
h^1	0	a	a	0	0	0
h^0	0	0	a	$2H^2$	$6H^2$	$12H^2$

where $a \leq h^0(\mathcal{I}_Z(B-H))=g$. In particular, $\mathcal{E}$ is a weakly Ulrich bundle on $(S, \mathcal{O}_S(H))$.

Proof. Let us start the table by obtaining the values $h^i(\mathcal{E}(-H))$ using the long exact cohomology sequence associated to

$$0 \longrightarrow \mathcal{O}_S(A-H) \longrightarrow \mathcal{E}(-H) \longrightarrow \mathcal{I}_Z(B-H) \longrightarrow 0. \tag{5.42}$$

We first show that there are no effective divisors linearly equivalent to $A-H \equiv C-F$. Let $yC+(ey+2g+1)F$, with $y \geq \max\{3, 2g+1\}$ a very ample divisor. Then

$$(C-F)(yC+(ey+2g+1)F) \leq 0$$

and therefore $H^0(\mathcal{O}_S(C-F))=0$. Analogously, using Serre's duality and Lemma 5.5.6, we see that $H^2(\mathcal{O}_S(C-F))=0$. Therefore, $-h^1(\mathcal{O}_S(C-F))=\chi(\mathcal{O}_S(C-F))=-g$.

Let us now focus on the cohomology groups $H^i(\mathcal{I}_Z(B-H))$. Notice that $B - H \equiv (\alpha-1)C + (\beta + 2g - 1 + e)F$. We have the structural sequence

$$0 \longrightarrow \mathcal{I}_Z(B-H) \longrightarrow \mathcal{O}_S(B-H) \longrightarrow \mathcal{O}_Z \longrightarrow 0.$$

Again, $H^1(\mathcal{O}_S(B-H)) = 0$ by Proposition 5.5.5 (4) and $H^2(\mathcal{O}_S(B-H)) = 0$ by Lemma 5.5.6. Therefore,

$$h^0(\mathcal{I}_Z(B-H)) = h^0(\mathcal{O}_S(B-H)) - |Z| = \chi(\mathcal{O}_S(B-H)) - |Z| = g,$$

for a general set of points Z. Moreover, $H^i(\mathcal{I}_Z(B-H)) = 0$ for $i = 1, 2$. Plugging in this information in the long exact sequence associated to (5.42), we obtain

$$a := h^0(\mathcal{E}(-H)) = h^1(\mathcal{E}(-H)) \leq g \quad \text{and} \quad H^2(\mathcal{E}(-H)) = 0.$$

One more application of Lemma 5.5.6 shows that $H^2(\mathcal{O}_S(A)) = H^2(\mathcal{I}_Z(B)) = 0$. This implies that $H^2(\mathcal{E}) = 0$. The hypothesis on β ensures that $H^1(\mathcal{O}_S(A)) = H^1(\mathcal{I}_Z(B)) = 0$, from where we get $H^1(\mathcal{E}) = 0$. Finally, applying Riemann–Roch theorem, we obtain

$$h^0(\mathcal{E}) = \chi(\mathcal{E}) = 2H^2.$$

The remaining entries of the table are obtained once we observe

$$H^i(\mathcal{E}(-tH)) \cong H^{2-i}(\mathcal{E}^\vee(K_S + tH))^\vee \cong H^{2-i}(\mathcal{E}((t-3)H))^\vee,$$

by Serre's duality and the fact that $\mathcal{E}^\vee \cong \mathcal{E}(-3H - K_S)$. Thus, according to Definition 3.2.11, $\mathcal{E}$ is a weakly Ulrich bundle. □

Lemma 5.5.10. *Let $\pi : S \longrightarrow C_0$ be a Weierstrass fibration with section C, general fiber F, $e \geq 1$, and let $H \equiv \alpha C + \beta F$ be a very ample divisor. Let us suppose that either $\alpha \leq 2e + 5$ or $\beta \geq (e+1)\alpha - 2e - 2$. Then, the vector bundles constructed as extensions* (5.41) *are simple.*

Proof. Denote by $c_1 = c_1(\mathcal{E})$. If we tensor (5.41) by $\mathcal{E}^\vee \cong \mathcal{E}(-c_1) = \mathcal{E}(-3H - K_S)$, we get

$$0 \longrightarrow \mathcal{E}(-c_1 + A) \longrightarrow \mathcal{E} \otimes \mathcal{E}^\vee \longrightarrow \mathcal{E}^\vee \otimes \mathcal{I}_Z(B) \longrightarrow 0.$$

Therefore, we will finish if we prove that $H^0(\mathcal{E}(-c_1 + A)) = 0$ and $h^0(\mathcal{E}^\vee \otimes \mathcal{I}_Z(B)) \leq 1$. Observe that $H^0(\mathcal{E}(-c_1 + A)) = 0$ follows from the exact sequence

$$0 \longrightarrow \mathcal{O}_S(-c_1 + 2A) \longrightarrow \mathcal{E}(-c_1 + A) \longrightarrow \mathcal{I}_Z \longrightarrow 0.$$

To prove that $h^0(\mathcal{E}^\vee \otimes \mathcal{I}_Z(B)) \leq 1$, we use the exact sequence

$$0 \longrightarrow \mathcal{O}_S \longrightarrow \mathcal{E}(-A) \longrightarrow \mathcal{I}_Z(B-A) \longrightarrow 0.$$

The same kind of arguments as in the previous propositions show that $h^1(\mathcal{O}_S(B-A)) = h^2(\mathcal{O}_S(B-A)) = 0$. Since $\chi(\mathcal{O}_S(B-A)) = -e\alpha^2/2 + 5/2e\alpha - 2e + \alpha\beta + 2\alpha - 2\beta - 4$, we have

$$h^0(\mathcal{O}_S(B-A)) - |Z| = e\alpha - 2e + \alpha - \beta - 2 \leq 0,$$

for α or β satisfying the hypothesis of the lemma. Since Z is a general set of points, $H^0(\mathcal{I}_Z(B-A)) = 0$. Thus $h^0(\mathcal{E}^\vee \otimes \mathcal{I}_Z(B)) = h^0(\mathcal{E} \otimes \mathcal{I}_Z(-A)) \leq h^0(\mathcal{E}(-A)) = 1$. □

Gluing together Proposition 5.5.9 and Lemma 5.5.10, we get:

Theorem 5.5.11. *Let $\pi : S \longrightarrow C_0$ be a Weierstrass fibration with section C, general fiber F, and $e \geq 1$. Let $H \equiv \alpha C + \beta F$ be a very ample divisor with*

$$\beta \geq \max\{(2e + 2g - 1)(\alpha - 2) + 1, \alpha e + 2g, (e + 1)\alpha - 2e - 2\}.$$

Then, for divisors A, B defined as in (5.40) and a general set Z of z points, there exist rank 2 simple weakly Ulrich bundles $\mathcal{E}$ on $(S, \mathcal{O}_S(H))$ with $c_1(\mathcal{E}) = 3H + K_S$ given by an extension

$$0 \longrightarrow \mathcal{O}_S(A) \longrightarrow \mathcal{E} \longrightarrow \mathcal{I}_Z(B) \longrightarrow 0.$$

Let us study the weakly Ulrich bundles we have constructed for the particular case of regular Weierstrass fibrations $S \longrightarrow C_0 \cong \mathbb{P}^1$, namely $q(S) = g(C_0) = 0$. We have seen that any weakly Ulrich bundle $\mathcal{E}$ constructed in Theorem 5.5.11 verifies $h^0(\mathcal{E}(-H)) \leq g = q(S)$. So, we immediately obtain the following:

Theorem 5.5.12. *Let $\pi : S \longrightarrow \mathbb{P}^1$ be a regular Weierstrass fibration with section C, general fiber F, and $e \geq 1$. Let $H \equiv \alpha C + \beta F$ be a very ample divisor with*

$$\beta \geq \max\{(2e - 1)(\alpha - 2) + 1, (e + 1)\alpha - 2e - 2\}.$$

Then $(S, \mathcal{O}_S(H))$ supports rank 2 simple special Ulrich bundles.

It is proved in [73, Main Theorem] that a very general regular Weierstrass fibration S with $p_g(S) \geq 2$ has Picard number two. Therefore $\mathrm{Pic}(S) \cong \mathbb{Z}[C] \oplus \mathbb{Z}[F]$. So, gathering the previous results, we obtain a stronger statement.

Theorem 5.5.13. *Let $\pi : S \longrightarrow \mathbb{P}^1$ be a very general regular Weierstrass fibration with section C, general fiber F, and $e \geq 1$ (so $p_g(S) \geq 2$). Fix a very ample divisor $H \equiv \alpha C + \beta F$ with*

$$\beta \geq \max\{(2e - 1)(\alpha - 2) + 1, (e + 1)\alpha - 2e - 2\}.$$

Then, there exists a family of rank 2 stable special Ulrich bundles on $(S, \mathcal{O}_S(H))$ of dimension at least $H^2 + 4e + 1$.

Proof. Let $\mathcal{E}$ be a rank 2 special Ulrich bundle on $(S, \mathcal{O}_S(H))$ that exists by Theorem 5.5.12. We know by Proposition 3.3.14 that they are μ-semistable and, moreover, if they are strictly semistable, they will have a presentation as an extension of Ulrich line bundles on $(S, \mathcal{O}_S(H))$ (see [41, Theorem 2.9]). Now, by Proposition 5.5.7, S does not support Ulrich line bundles of the form $aC + bF$. Since, by hypothesis, Pic(S) is generated by C and F, we obtain the stability of $\mathcal{E}$, namely $[\mathcal{E}] \in M_H^s(2; 3H + K_S, c_2(\mathcal{E}))$ where $M_H^s(2; 3H + K_S, c_2(\mathcal{E}))$ is the moduli space of stable rank 2 vector bundles $\mathcal{E}$ on S with Chern classes $c_1(\mathcal{E}) = 3H + K_S$ and $c_2(\mathcal{E})$. Being Ulrich defines an open subscheme of $M_H^s(2; 3H + K_S, c_2(\mathcal{E}))$, and the dimension at a point $[\mathcal{E}]$ is bounded below by $1 - \chi(\mathcal{E} \otimes \mathcal{E}^\vee) = \mathrm{h}^1(\mathcal{E} \otimes \mathcal{E}^\vee) - \mathrm{h}^2(\mathcal{E} \otimes \mathcal{E}^\vee)$. We can compute the Chern classes of $\mathcal{E} \otimes \mathcal{E}^\vee$ and obtain that $c_1(\mathcal{E} \otimes \mathcal{E}^\vee) = 0$ and

$$c_2(\mathcal{E} \otimes \mathcal{E}^\vee) = 4c_2(\mathcal{E}) - c_1^2(\mathcal{E}) = -H^2 - 4e.$$

Therefore, we can conclude applying the Riemann–Roch theorem to $\mathcal{E} \otimes \mathcal{E}^\vee$ to get the lower bound on the dimension of these rank 2 special Ulrich bundles on $(S, \mathcal{O}_S(H))$. □

The last part of this section is concerned with minimal surfaces of general type. Very little is known, and we will just give a few examples.

Example 5.5.14. (1) We first observe that all complete intersection surfaces $S_{d_1,\dots,d_{n-2}} \subset \mathbb{P}^n$ of type $(d_1, \dots, d_{n-2})$ are of general type with the exceptions $S_2, S_3, S_4 \subset \mathbb{P}^3$, $S_{2,2}, S_{2,3} \subset \mathbb{P}^4$, and $S_{2,2,2} \subset \mathbb{P}^5$. By Theorem 4.3.2, any complete intersection surface supports an Ulrich bundle.

(2) Recall that any product $S = C_1 \times C_2$ of smooth curves $(C_i, \mathcal{O}_{C_i}(H_i))$ of genus $g(C_i) \geq 2$ is a surface S of general type. By Theorem 3.4.5, C_i supports an Ulrich line bundle $\mathcal{L}_i$ and applying Lemma 3.3.10 we deduce that $\mathcal{L}_1 \boxtimes \mathcal{L}_2(H_2)$ and $\mathcal{L}_1(H_1) \boxtimes \mathcal{L}_2$ are Ulrich line bundles on $(S, \mathcal{O}_{C_1}(H_1) \boxtimes \mathcal{O}_{C_2}(H_2))$.

The problem of classifying all surfaces S of general type (that is, all surfaces with $\kappa(S) = 2$) seems to be out of reach, except for very favorable choices of the invariants. It is well known that to each minimal surface S of general type one associates a triple of numerical invariants, (p_g, q, K_S^2), which determine all other classical numerical invariants, such as $e_{\mathrm{top}}(S) = 12\chi(\mathcal{O}_S) - K_S^2$ and $P_m(S) := \mathrm{h}^0(S, mK_S) = \chi(\mathcal{O}_S) + \binom{m}{2}K_S^2$. In the remaining part of this section, we will restrict our attention to the most extreme case, that is, the case of $p_g = q = 0$. Surfaces with such invariants are amongst the most famous since they historically represented counterexamples to the famous Max Noether's conjecture, stating that any smooth surface S with $p_q = q = 0$ needs to be rational. The first counterexample to this conjecture was provided by Enriques, and it was a surface with Kodaira dimension 0 (see [90]) and nowadays we call them Enriques surfaces. Currently, a large number of smooth surfaces of general type with $p_g = q = 0$ are known. By Bogomolov–Miyaoka–Yau inequality, we have $K_S^2 \leq 9\chi(\mathcal{O}_S)$. Therefore, for any smooth surface S of general type with $p_g = q = 0$, we have $1 \leq$

$K_S^2 \leq 9$. The lower extreme case is well known. Indeed, Horikawa's results on surfaces S with small K_S^2 contains a fairly explicit treatment of the Godeaux and Campedelli surfaces. For $3 \leq K_S^2 \leq 8$ the situation is, in general, much less understood. Finally, the upper extreme case of surfaces with $p_g = q = 0$ and $K_S^2 = 9$ is that of the famous fake projective planes, i. e., surfaces of general type with Hodge diamond equal to that of $\mathbb{P}^2$. In what follows we will consider smooth surfaces S of general type with $p_g = q = 0$ and $1 \leq K_S^2 \leq 2$.

Definition 5.5.15. A *Godeaux surface* is a minimal surface S of general type with $p_g = q = 0$ and $K_S^2 = 1$. A *Campedelli surface* is a minimal surface S of general type with $p_g = q = 0$ and $K_S^2 = 2$.

We will now give examples of both Godeaux surfaces and Campedelli surfaces carrying an Ulrich bundle.

Example 5.5.16. (1) As an example of Godeaux surface supporting an Ulrich bundle, we have the following one. We consider the Fermat surface $F_5 \subset \mathbb{P}^3$ given by $x_0^5 + x_1^5 + x_2^5 + x_3^5 = 0$ and let $\sigma : F_5 \longrightarrow F_5$ be the automorphism of order 5 of F_5 defined by $\sigma(x_0, x_1, x_2, x_3) = (x_0, \zeta x_1, \zeta^2 x_2, \zeta^3 x_3)$ where ζ is a primitive 5th root of unity. The group $G = \mathbb{Z}/5\mathbb{Z}$ generated by σ acts on F_5 without fixed points and the quotient $S = F_5/G$ is a smooth surface with $\chi(\mathcal{O}_S) = \frac{1}{5}\chi(\mathcal{O}_{F_5}) = 1$, $p_g(S) = q(S) = 0$, and $K_S^2 = \frac{1}{5}K_{F_5}^2 = 1$, so that S is a surface of general type (see [21, Proposition X.1]).

We consider the finite surjective morphism $\pi : F_5 \longrightarrow S = F_5/G$. By Example 4.2.10 (3) and Proposition 4.2.14, F_5 supports an Ulrich line bundle $\mathcal{E}$. Applying Lemma 3.3.12, we conclude that $\pi_*\mathcal{E}$ is an Ulrich bundle of rank 5 on the Godeaux surface S.

(2) As an example of Campedelli surface supporting an Ulrich bundle, we have the following one. We fix coordinates $x_{100}, x_{010}, x_{001}, x_{110}, x_{101}, x_{011}, x_{111}$ on $\mathbb{P}^6$ and the action of $G = (\mathbb{Z}/2\mathbb{Z})^3$ on $\mathbb{P}^6$ given by $g_{ijk}(x_{i'j'k'}) = (-1)^{ii'+jj'+kk'} x_{i'j'k'}$ for $g_{ijk} \in G$. Any diagonal quadric $Q = \sum a_{ijk} x_{ijk}^2$ is invariant under the action of G. A general choice of four diagonal quadrics Q_1, Q_2, Q_3, and Q_4 intersect transversally in a complete intersection smooth surface $S_{2,2,2,2} \subset \mathbb{P}^6$ and the quotient $S := S_{2,2,2,2}/G$ is a smooth Campedelli surface, that is, $p_g(S) = q(S) = 0$ and $K_S^2 = \frac{1}{8}K_{S_{2,2,2,2}}^2 = 2$ (see [21, Proposition X.1]).

We consider the finite surjective morphism $\pi : S_{2,2,2,2} \longrightarrow S = S_{2,2,2,2}/G$. By Theorem 4.3.2, the surface $S_{2,2,2,2} \subset \mathbb{P}^6$ supports an Ulrich bundle $\mathcal{E}$ of rank 2^8. Therefore, applying Lemma 3.3.12 and taking into account that G has order 8, we conclude that $\pi_*\mathcal{E}$ is an Ulrich bundle of rank 2^{11} on the Campedelli surface S.

The existence of Ulrich bundles on surfaces of general type is completely open, and there are very few known examples apart from the case of smooth surfaces $S \subset \mathbb{P}^5$ of degree $d \geq 5$ already discussed in Section 4.4.

5.6 Ulrichness and change of very ample line bundles[1]

As we have seen in previous sections, the existence of Ulrich bundles on a smooth projective surface $(S, \mathcal{O}_S(H))$ strongly depends on the very ample divisor H that we fix. See, for example, Theorem 5.2.7 and Remark 5.2.8 for the case of Ulrich line bundles on ruled surfaces. We have also seen many examples of Ulrich bundles on smooth projective surfaces of Kodaira dimension ≤ 1 and only few examples of Ulrich bundles on smooth projective surfaces of general type are known. In other words, the existence of Ulrich bundles on surfaces of general type is rather open. Nevertheless, in this section we will prove that, for any smooth projective surface $(S, \mathcal{O}_S(H))$, there exists an integer m_0 such that, for any $m \geq m_0$, $(S, \mathcal{O}_S(mH))$ supports an Ulrich bundle. This asymptotic result shows that Ulrichness depends on the choice of the very ample divisor.

Definition 5.6.1. Let $(X, \mathcal{O}_X(H))$ be a smooth, irreducible, projective surface. Denote by $\overline{M}_{X,H}(r; c_1, c_2)$ the moduli space of S-equivalence classes of Gieseker semistable (with respect to H) rank r sheaves on X with Chern classes c_1 and c_2. We say that a moduli space $\overline{M}_{X,H}(r; c_1, c_2)$ satisfies the *Brill–Noether property* (BN property, for short) if the general sheaf in every component of $\overline{M}_{X,H}(r; c_1, c_2)$ has at most one nonzero cohomology group.

Given a rank r vector bundle $\mathcal{E}$ on a smooth projective surface $(X, \mathcal{O}_X(H))$ with Chern classes (c_1, c_2), we define the *discriminant* of $\mathcal{E}$ as follows:

$$\Delta(\mathcal{E}) := \frac{1}{r}\left(c_2 - \frac{r-1}{2r}c_1^2\right).$$

We sometimes denote it by $\Delta(r, c_1, c_2)$. We easily check that for any divisor D on X,

$$\Delta(\mathcal{E}(D)) = \Delta(\mathcal{E}) \quad \text{and} \quad \Delta(\mathcal{E}^\vee) = \Delta(\mathcal{E}).$$

The relation between moduli spaces with the BN property and Ulrich bundles is outlined in the next proposition.

Proposition 5.6.2. *Let $\mathcal{E}$ be a rank r Ulrich bundle on a smooth projective surface $(X, \mathcal{O}_X(H))$ with $c_1(\mathcal{E}) = c_1 + rH$ and discriminant Δ. Then*

$$2c_1H = r(H^2 + HK_X) \quad \textit{and} \quad 2r^2\Delta = c_1^2 - rc_1K_X + 2r^2\chi(\mathcal{O}_X).$$

Conversely, if r, c_1, and Δ satisfy these equalities, the moduli space $\overline{M}_{X,H}(r; c_1, c_2)$ contains vector bundles and the moduli spaces $\overline{M}_{X,H}(r; c_1, c_2)$ and $\overline{M}_{X,H}(r; c_1 - rH, c_2 - (r -$

1 The material in Section 5.6 is based on the paper I. Coskun and J. Huizenga, *Brill–Noether Problems, Ulrich bundles and the Cohomology of Moduli Spaces of Sheaves*, Matemática Contemporânea 47 (2020), 21–72 [63].

$1)c_1H+\binom{r}{2}H^2)$ *satisfy both the BN property, then the general vector bundle in* $\overline{M}_{X,H}(r;c_1+rH,c_2+(r-1)c_1H+\binom{r}{2}H^2)$ *is an Ulrich bundle.*

Furthermore, if $2c_1=r(H+K_X)$, $\overline{M}_{X,H}(r;c_1,c_2)$ *contains vector bundles and satisfies the BN property, then* $(X,\mathcal{O}_X(H))$ *admits Ulrich bundles of every rank divisible by r.*

Proof. The first part of the statement immediately follows from Proposition 5.1.1.

Assume $\overline{M}_{X,H}(r;c_1,c_2)$ contains vector bundles and let $\mathcal{E}$ be a general such vector bundle. Then by twisting $\mathcal{E}$ by $\mathcal{O}_X(H)$ and $\mathcal{O}_X(-H)$, we also obtain vector bundles in the moduli spaces $\overline{M}_{X,H}(r;c_1+rH,c_2+(r-1)c_1H+\binom{r}{2}H^2)$ and $\overline{M}_{X,H}(r;c_1-rH,c_2-(r-1)c_1H+\binom{r}{2}H^2)$, respectively. By the numerical assumptions on c_1 and Δ, we have that

$$\chi(\mathcal{E})=\chi(\mathcal{E}(-H))=0.$$

Hence, if the BN property holds for the general sheaf in every component of $\overline{M}_{X,H}(r;c_1,c_2)$ and $\overline{M}_{X,H}(r;c_1-rH,c_2-(r-1)c_1H+\binom{r}{2}H^2)$, then $\mathcal{E}$ and $\mathcal{E}(-H)$ have no cohomology, and hence $\mathcal{E}(H)$ is an Ulrich bundle on X.

Observe that $2c_1=r(H+K_X)$ always satisfies the equality $2c_1H=r(H^2+HK_X)$. In this case, by Serre's duality (see Theorem 1.3.2), the vanishing of the cohomology groups for the general vector bundle in $\overline{M}_{X,H}(r;c_1,c_2)$ implies the same vanishing for a general vector bundle in $\overline{M}_{X,H}(r;c_1-rH,c_2-(r-1)c_1H+\binom{r}{2}H^2)$. Hence, if $\overline{M}_{X,H}(r;c_1,c_2)$ contains vector bundles and satisfies the BN property, then for a general vector bundle $\mathcal{E}\in\overline{M}_{X,H}(r;c_1,c_2)$, $\mathcal{E}(H)$ is an Ulrich bundle on $(X,\mathcal{O}_X(H))$ which concludes the proof of the proposition. □

The main result of this section concerning the asymptotic existence of Ulrich bundles on any smooth projective surface will follow after a sequence of elementary modifications (see Definition 5.4.1).

Theorem 5.6.3. *Let* $(X,\mathcal{O}_X(H))$ *be a smooth projective surface. There is an integer* $m_0\in\mathbb{Z}$ *such that, for all* $m\geq m_0$, $(X,\mathcal{O}_X(mH))$ *admits an Ulrich bundle of every positive even rank. Furthermore, if* K_X *(respectively,* K_X+H*) is divisible by 2 in the Picard group and* $2m\geq m_0$ *(respectively,* $2m+1\geq m_0$*), then* $(X,\mathcal{O}_X(2mH))$ *(respectively,* $(X,\mathcal{O}_X((2m+1)H))$*) admits an Ulrich bundle of every rank* $r\geq 2$.

Proof. By [168, Theorem D], for any integer $r\geq 2$ and for any first Chern class c_1, there exists a constant Δ_0 such that for $\Delta(r,c_1,c_2)\geq\Delta_0$ the moduli space $\overline{M}_{X,H}(r;c_1,c_2)$ of S equivalence classes of rank r semistable sheaves on X with Chern classes (c_1,c_2) is irreducible, contains a μ_H-stable vector bundle and it has the expected dimension (see Proposition 1.5.15)

$$2rc_2-(r-1)c_1^2-(r^2-1)\chi(\mathcal{O}_X).$$

Let $\mathcal{E}\in M^s_{X,H}(r;c_1,c_2)$ be a general vector bundle. By Serre's theorem (see Theorem 1.3.1) and semicontinuity of cohomology, there exists an integer $m_1(r;c_1)$ such that, for $m \geq m_1(r; c_1)$,

$$\mathrm{H}^i(X, \mathcal{E}(mH)) = 0 \quad \text{for } i = 1, 2.$$

By applying a sequence of $\mathrm{h}^0(X, \mathcal{E}(mH))$ general elementary modifications to $\mathcal{E}(mH)$, we obtain a new μ_H-stable sheaf $\mathcal{E}'$ that satisfies $\mathrm{H}^i(X, \mathcal{E}') = 0$ for all i. Since elementary modifications preserve μ-stability and only increase the discriminant, by [168, Theorem D], the moduli space containing $\mathcal{E}'$ is irreducible and the general sheaf in this moduli space is locally free. By semicontinuity of cohomology, a general deformation $\mathcal{E}_0$ of $\mathcal{E}'$ is locally free, has no cohomology,

$$\mathrm{rank}(\mathcal{E}_0) = r, \quad \text{and} \quad c_1(\mathcal{E}_0) = c_1 + rmH.$$

By Serre's duality (see Theorem 1.3.2), $\mathcal{E}_0^\vee(K_X)$ also does not have any cohomology,

$$\mathrm{rank}(\mathcal{E}_0^\vee(K_X)) = r, \quad \text{and} \quad c_1(\mathcal{E}_0^\vee(K_X)) = -c_1 - rmH + rK_X.$$

Set $c_1 = \frac{rK_X}{2}$, which we can do if r is even or if K_X is divisible by 2 in the Picard group. We claim that $\mathcal{E}_0(2mH)$ is a rank r Ulrich bundle on $(X, \mathcal{O}_X(2mH))$. Observe that $\mathcal{E}_0(-2mH)$ is a μ_H-stable vector bundle,

$$\mathrm{rank}(\mathcal{E}_0(-2mH)) = \mathrm{rank}(\mathcal{E}_0^\vee(K_X)) = r, \quad c_1(\mathcal{E}_0(-2mH)) = c_1(\mathcal{E}_0^\vee(K_X)), \quad \text{and}$$
$$\Delta(\mathcal{E}_0(-2mH)) = \Delta(\mathcal{E}_0^\vee(K_X)).$$

Hence, if $\mathcal{E}_0$ is general, $\mathcal{E}_0(-2mH)$ has no cohomology by semicontinuity. This shows that $\mathcal{E}_0(2mH)$ is an Ulrich bundle of rank r on $(X, \mathcal{O}_X(2mH))$.

Similarly, set $c_1 = \frac{r(K_X+H)}{2}$, which we can do if the rank r is even or if $K_X + H$ is divisible by 2 in the Picard group. We claim that $\mathcal{E}_0((2m+1)H)$ is a rank r Ulrich bundle on $(X, \mathcal{O}_X((2m+1)H))$. Observe that $\mathcal{E}_0(-(2m+1)H)$ is a μ_H-stable bundle with the same rank, first Chern class, and discriminant as $\mathcal{E}_0^\vee(K_X)$, and the same argument as in the previous case applies.

If we let

$$m_0 = \max\{2m_1(2; K_X), 2m_1(2; K_X + H)\}$$

(where m_1 is the bound required to apply Serre's theorem), we have constructed a rank 2 Ulrich bundle on $(X, \mathcal{O}_X(mH))$ for $m \geq m_0$. If we have an Ulrich bundle $\mathcal{F}$ of rank 2, then $\mathcal{F}^{\oplus j}$ is an Ulrich bundle of rank $2j$.

Similarly, if K_X (respectively, $K_X + H$) is divisible by 2, we can construct both rank 2 and rank 3 Ulrich bundles on $(X, \mathcal{O}_X(2mH))$ (respectively, $(X, \mathcal{O}_X((2m+1)H))$) by this construction for

$$m \geq \max\left\{m_1(2; K_X), m_1\left(3; \frac{3K_X}{2}\right)\right\}$$
$$\left(\text{respectively, } m \geq \max\left\{m_1(2; K_X + H), m_1\left(3; \frac{3(K_X + H)}{2}\right)\right\}\right).$$

Once we have Ulrich bundles of ranks 2 and 3 on $(X, \mathcal{O}_X(2mH))$ (respectively, $(X, \mathcal{O}_X((2m+1)H))$, by taking appropriate direct sums, we obtain Ulrich bundles of every rank at least 2, which concludes the proof of the theorem. □

5.7 Final comments and additional reading

As one can expect, since Ulrich bundles on smooth curves were from the beginning completely understood, a lot of the research on this field has been focused on the case of smooth projective surfaces. However, the study of Ulrich bundles on projective surfaces is still far from being completed. Nevertheless, the result explained in Sections 5.2–5.5 strongly support Conjecture 3.4.8.

Section 5.1 presents the main tools to deal with Ulrich bundles on projective surfaces. The classification of surfaces according to their Kodaira dimension supposes a cornerstone in Algebraic Geometry. The reader can use the books [21] and [20] to learn about these achievements. Proposition 5.1.1 can be found in [42, Proposition 2.1] (see also [43]). Reference [83] offers a complete account concerning the Cayley–Bacharach property. We have followed [140] for the statement and proof of Theorem 5.1.6 and later material. Lazarsfeld–Mukai bundles were introduced during the 1980s to study two independent kinds of problem. On the one hand, the existence of smooth curves which are generic in the sense of Brill–Noether–Petri theory and, on the other, the classification of prime Fano threefolds of coindex three. The reader can consider [6] for an introduction. Proposition 5.1.11 connecting Lazarsfeld–Mukai and Ulrich bundles is taken from [85, Proposition 6.2].

A quite complete understanding of Ulrich bundles on minimal projective surfaces with Kodaira dimension $-\infty$ is by now reached. The results in Section 5.2 concerning Ulrich bundles on geometrically ruled surfaces are taken from [8] and [7]. For the particular case of Hirzebruch surfaces, we have used [5]. For the Veronese surface $(\mathbb{P}^2, \mathcal{O}_{\mathbb{P}^2}(d))$, the complete classification is taken from [70]. Notice that the case of Ulrich bundles on Veronese embeddings of projective spaces of dimension greater than or equal to three is not yet fully understood. The reader can consult [70] for the state-of-the-art on this topic.

Del Pezzo surfaces are the main topic of Section 5.3. The reader can use [148] for a classical account and [175] for the classification of Ulrich line bundles on del Pezzo surfaces. By the results from [45], as well as those from [126] and [194], explained in Chapter 3, describing the behavior of Ulrich bundles under pullback and pushforward for blow-up morphisms, it is clear that the case of cubic surfaces plays a central role. They are the main topic of [40] and [41]. A different approach to the construction of Ulrich bundles on del Pezzo surfaces and their relation with the minimal resolution conjecture as stated by Mustata can be studied in [158].

The central result in Section 5.4 shows that any minimal smooth surface of Kodaira dimension zero supports rank 2 special Ulrich bundles. The case of K3 surfaces, our Theorem 5.4.3, is the main result from [92]. The rest of surfaces of Kodaira dimension zero are treated in [23] and [24].

As it has already been mentioned, much less is known for Ulrich bundles on surfaces with positive Kodaira dimension. Section 5.5 gives an approach to this ongoing research. For the case of Kodaira dimension one, Theorem 5.5.13 about stable and special Ulrich bundles on Weierstrass fibrations, the reader is addressed to [159]. The examples of general surfaces worked out in this section are taken from [21]. More results on this direction can be found in [144] and references therein.

Even if Conjecture 3.4.8 is far from being settled, the results from Section 5.6, taken from the paper [63], show that it is true at least asymptotically for smooth surfaces.

6 Intersection of two quadrics: an example

The goal of this chapter is to work out a complete classification of Ulrich bundles in a particularly nice case, the complete intersection $X \subset \mathbb{P}^5$ of two general quadrics Q_0 and Q_∞ of $\mathbb{P}^5$.

The reason for this choice is twofold. Firstly, since the case of Ulrich bundles on hypersurfaces, as well as their relation with matrix factorizations, has been thoroughly studied in Chapter 4, it is natural to move forward to the next case, namely those complete intersections of codimension two and lowest degree. Secondly, since we devoted Chapter 5 to projective surfaces, it is meaningful to increase the dimension just by one. In particular, since in Section 5.3 we performed a careful classification of Ulrich bundles on projective surfaces with ample anticanonical bundle, we are especially interested in those threefolds whose anticanonical line bundles share the same property, the so-called *Fano varieties*. It turns out that the complete intersection $X \subset \mathbb{P}^5$ of two general quadrics Q_0 and Q_∞ of $\mathbb{P}^5$ lies at the intersection of these two approaches. Indeed, as it will be explained in Section 6.1, X is an archetype of the class of Fano varieties (in fact, X is a Fano threefold of index two, which means that $\omega_X \cong \mathcal{O}_X(-2)$ with $\mathrm{Pic}(X) \cong \mathbb{Z}\langle\mathcal{O}_X(1)\rangle$). This large family of varieties has been largely studied in many different frameworks. This fact makes automatically available a large toolkit of techniques and results on Fano varieties that can be applied to the construction of Ulrich bundles on $X \subset \mathbb{P}^5$. Indeed, in Section 6.4, the existence of rank-two Ulrich bundles will be proved for an arbitrary Fano threefold of index two.

On the other hand, we are going to use a particular connection of X with a smooth genus two curve C, via a Fourier–Mukai transform, as discovered in [31]. This connection, exclusive for this particular Fano threefold, will be the key point to understand the structure of the moduli space of Ulrich bundles of higher rank. Indeed, a certain class of vector bundles on C, the so-called Raynaud bundles to whom we devote Section 6.2, will play a central role on their classification. Raynaud bundles on C will allow us to show that the moduli space of stable Ulrich bundles of rank $r \geq 2$ on $X \subset \mathbb{P}^5$ is isomorphic to a nonempty subscheme of $M_C^s(r; 2r)$, the moduli space of stable vector bundles of rank r and degree $2r$ on the hyperelliptic curve C naturally associated to X (see Theorem 6.3.10 for the precise statement). This highlights the deep connections that Ulrich bundles have with some a priori unrelated geometric objects and becomes the main reason to thoroughly work out this example in this chapter. We devote Section 6.3 to proving the existence of Ulrich bundles of arbitrary rank greater than or equal to two on the complete intersection of two quadrics in $\mathbb{P}^5$. Indeed, the case of rank-two Ulrich bundles can be worked out in a unified way for most of the Fano threefolds of index one and two; this approach will be explained in Section 6.4.

https://doi.org/10.1515/9783110647686-006

6.1 Complete intersections of two quadrics and curves of genus two

The complete intersection of two quadrics is a particular case of Fano variety. We are going to start this section recalling the definition of Fano variety and basic facts on its classification.

Definition 6.1.1. A *Fano variety* X is a smooth projective variety with ample anticanonical line bundle $\omega_X^\vee$. The *index* i_X of X is the maximal integer dividing the anticanonical class $-K_X$ in $\mathrm{Pic}(X)$. When $\mathrm{Pic}(X) \cong \mathbb{Z}$ is generated by the ample line bundle $\mathcal{O}_X(1)$, we have that $\omega_X \cong \mathcal{O}_X(-i_X)$.

Example 6.1.2. (1) Let $X \subset \mathbb{P}^{n+c}$ be a complete intersection of c hypersurfaces of respective degree $d_1, \dots, d_c$. By the adjunction formula, we know that the canonical line bundle of X is of the form $\mathcal{O}_X(-n-c-1+\sum_{i=1}^{c} d_i)$. Therefore, if $\sum_{i=1}^{c} d_i < n+c+1$, X will be an n-dimensional Fano variety of index $n+c+1-\sum_{i=1}^{c} d_i$.

(2) Let $\pi : X := \mathbb{P}(T_{\mathbb{P}^n}) \longrightarrow \mathbb{P}^n$ be the projective tangent bundle over $\mathbb{P}^n$. Then the canonical divisor of X is of the form

$$K_X \equiv -n\xi + \pi^*(K_{\mathbb{P}^n} + \det(T_{\mathbb{P}^n})) \equiv -n\xi$$

where ξ is the relative ample bundle. Therefore X is a Fano variety of dimension $2n-1$ and index n. Mori proved (see [160]) Hartshorne's conjecture asserting that $\mathbb{P}^n$ is the only variety whose associated projective tangent bundle is a Fano variety.

It is classically known that the index of an m-dimensional Fano variety is less than or equal to $m+1$ (see [121]). Moreover, $i_X = m+1$ (respectively $i_X = m$) if and only if $X \cong \mathbb{P}^m$ (respectively the smooth quadric $Q \subset \mathbb{P}^{m+1}$). The rest of Fano threefolds, namely those of index one and two, are classified in [121]. Among Fano threefolds with index two and Picard group generated by a very ample line bundle, those with lowest degree correspond to cubic hypersurfaces $Y \subset \mathbb{P}^4$. The case of Ulrich bundles on hypersurfaces, as well as their relation with matrix factorizations, has been thoroughly studied in Chapter 4. The next case, Fano threefolds of degree four correspond to the complete intersection $X = Q_0 \cap Q_\infty$ of two general quadrics Q_0 and Q_∞ of $\mathbb{P}^5$. This is the object of study of this chapter.

Let X be a smooth Fano threefold of degree four defined by the complete intersection of two quadrics of $\mathbb{P}^5$. By [183, Proposition 2.1], we can assume that X is defined by the quadrics

$$Q_0 := V(x_0^2 + \cdots + x_5^2) \quad \text{and} \quad Q_\infty := V(\lambda_0 x_0^2 + \cdots + \lambda_5 x_5^2)$$

for $0 \neq \lambda_i \in \mathbf{k}$ with $\lambda_i \neq \lambda_j$ if $i \neq j$. Associated to these two quadrics we have the quadratic pencil $\Lambda := |sQ_0 + tQ_\infty|_{[s:t]\in\mathbb{P}^1}$ on $\mathbb{P}^5$ with base locus X. By the hypothesis on the λ_i, each of the six singular quadrics $Q_{\lambda_i} := \lambda_i Q_0 - Q_\infty$, $i = 0, \dots, 5$ in the pencil is isomorphic to the cone of a smooth quadric $Q \subset \mathbb{P}^4$ over a point $p \in \mathbb{P}^5$. In the terminology of [183], the pencil Λ is said to be *nonsingular*.

Let $\sigma : C \longrightarrow \mathbb{P}^1$ be the double cover of Λ, ramified at the six points $[1 : \lambda_0], \dots, [1 : \lambda_5]$. By the Riemann–Hurwitz formula,

$$2g(C) - 2 = \deg(\sigma)(2g(\mathbb{P}^1) - 2) + \deg(R)$$

where R is the divisor of branch points. Hence C is a smooth curve of genus two. Let us denote by $\tau : C \longrightarrow C$ its hyperelliptic involution.

Bondal and Orlov in [31] showed that there exists a strict relationship between the derived categories $D^b(X)$ of bounded complexes of coherent sheaves of $\mathcal{O}_X$-modules and $D^b(C)$ of bounded complexes of coherent sheaves of $\mathcal{O}_C$-modules. In order to see this, first of all notice that, as for any Fano threefold of index two, the derived category $D^b(X)$ of X has a semiorthogonal decomposition of the form

$$D^b(X) = \langle \mathcal{O}_X(-1), \mathcal{O}_X, \mathcal{B}_X \rangle$$

where $\mathcal{B}_X$ is the *nontrivial component* of $D^b(X)$,

$$\mathcal{B}_X := \{\mathcal{F} \in D^b(X) \mid \operatorname{Ext}^\bullet(\mathcal{F}, \mathcal{O}_X) = \operatorname{Ext}^\bullet(\mathcal{F}, \mathcal{O}_X(-1)) = 0\}.$$

On $X \times C$ it is possible to define a rank-two universal vector bundle $\mathcal{S}$ as follows. Let $c, c' \in C$ be nonramified points such that $\sigma(c) = \sigma(c') \in \mathbb{P}^1$ correspond to a smooth four-dimensional quadric $Q_{\sigma(c)} \cong Q \subset \mathbb{P}^5$ of the pencil Λ. Then, on Q there are two spinor bundles corresponding to the universal quotient bundle and to the dual of the universal subbundle under the identification $Q \cong \operatorname{Gr}(1,3) \subset \mathbb{P}^5$. Then $\mathcal{S}_{|X\times\{c\}}$ and $\mathcal{S}_{|X\times\{c'\}}$ correspond to the restriction of the two spinor bundles on Q to X.

On the other hand, let $c \in C$ be a ramification point. Namely, $\sigma(c) \in \mathbb{P}^1$ corresponds to a cone $Q_{\sigma(c)}$ over a three-dimensional smooth quadric Q with vertex $p \in \mathbb{P}^5$. Then $\mathcal{S}_{|X\times\{c\}}$ is the pullback of the unique spinor bundle on Q under the composition of maps,

$$X \hookrightarrow Q_{\sigma(c)} \setminus \{p\} \twoheadrightarrow Q^3.$$

Denote by p (respectively, by q) the projection from $X \times C$ to X (respectively, to C). According to this notation, we have the following

Theorem 6.1.3. *The Fourier–Mukai functor*

$$\begin{array}{rccc} \Phi_{\mathcal{S}}: & D^b(C) & \longrightarrow & D^b(X), \\ & \mathcal{F} & \mapsto & Rp_*(Lq^*(\mathcal{F}) \otimes \mathcal{S}) \end{array}$$

is fully faithful and induces a semiorthogonal decomposition

$$D^b(X) = \langle \mathcal{O}_X(-1), \mathcal{O}_X, \Phi_{\mathcal{S}}(D^b(C)) \rangle$$

with $\Phi_{\mathcal{S}}(D^b(C)) \cong \mathcal{B}_X$.

Proof. See Definition A.0.29 for the notion of Fourier–Mukai functor and [31] for the proof of the statement. □

There is another approach for understanding the relation between X and the smooth curve C. Let us fix first some notation. For a line bundle ξ of degree $d \in \mathbb{Z}$, we denote by $M_C(r;\xi)$ (respectively, by $M_C(r;d)$) the moduli space of semistable rank r vector bundles on C with determinant ξ (respectively, the moduli space of semistable rank r vector bundles on C of degree d). Then $M_C(r;\xi)$ is a normal projective variety of dimension $r^2 - 1$ (see [165]). Analogously, $M_C(r;d)$ is also a normal projective variety of dimension $r^2 + 1$. We are going to denote by $M_C^s(r;\xi)$ (respectively, by $M_C^s(r;d)$) their open subvarieties parameterizing stable objects.

With this notation, let us fix a line bundle ξ of degree one on the curve C and consider the fine moduli space $M_C^s(2;\xi^\vee)$ that parameterizes rank-two stable vector bundles with determinant $\xi^\vee$ on C. It was proved in [166] that $M_C^s(2;\xi^\vee) \cong X$ and that $\mathcal{S}$ is the universal bundle of this moduli space. In particular, we have

$$\det(\mathcal{S}) = \mathcal{O}_X(1) \boxtimes \xi^\vee.$$

The Chow ring of X has a particularly simple structure. Denote by $[H_X]$, $[L_X]$, and $[P_X]$ the classes of a hyperplane section, a line, and a point in X, respectively. Then it turns out that

$$\mathrm{H}^2(X) \cong \mathbb{Z}[H_X], \quad \mathrm{H}^4(X) \cong \mathbb{Z}[L_X], \quad \text{and} \quad \mathrm{H}^6(X) \cong \mathbb{Z}[P_X].$$

Moreover, the intersection product is as follows: $H_X^2 = 4L_X$ and $H_X L_X = P_X$.

Let $\mathcal{E}(1)$ be a rank r Ulrich bundle on $(X, \mathcal{O}_X(1))$. We will see that $\mathcal{E}$ dwells inside the nontrivial component $\mathcal{B}_X$ of $D^b(X)$.

Proposition 6.1.4. *Let $\mathcal{E}(1)$ be a rank r Ulrich bundle on $(X, \mathcal{O}_X(1))$. Then*

$$\mathrm{Hom}_{D^b(X)}(\mathcal{E}, \mathcal{O}_X(-i)[p]) = 0$$

for all p and $i = 0, 1, 2$. In particular, $\mathcal{E} \in \Phi_{\mathcal{S}}(D^b(C)) \cong \mathcal{B}_X$.

Proof. By Lemma 6.3.2, we have that $\mathcal{E}^\vee(1)$ is also an Ulrich bundle (it is the dual Ulrich bundle). Therefore, applying Serre's duality, we have

$$\mathrm{Hom}_{D^b(X)}(\mathcal{E}, \mathcal{O}_X(-i)[p]) = \mathrm{H}^p(\mathcal{E}^\vee(-i)) = 0$$

for all p and $i = 0, 1, 2$. The second part of the statement follows from the semiorthogonal decomposition of $D^b(X)$ described in Theorem 6.1.3. □

Therefore, we should study the right adjoint functor $\Phi_{\mathcal{S}}^!$ of $\Phi_{\mathcal{S}}$ (see Definition A.0.30):

$$\begin{array}{rccl} \Phi_{\mathcal{S}}^!: & D^b(X) & \longrightarrow & D^b(C), \\ & \mathcal{E} & \mapsto & Rq_*(Lp^*(\mathcal{E}) \otimes \mathcal{S}^\vee) \otimes \omega_C[1]. \end{array}$$

Note that in order to see that $\mathcal{E} \in \Phi_{\mathcal{S}}(D^b(C)) \cong \mathcal{B}_X$ for an Ulrich bundle $\mathcal{E}(1)$ on X, we are using just a part of the characterization of Ulrich bundles, namely $\mathrm{H}^p(\mathcal{E}(1-i)) = 0$ for $i = 1, 2$. The remaining cohomology vanishing is exploited in the following lemma.

Lemma 6.1.5. *Let $\mathcal{E}(1)$ be an Ulrich bundle on $(X, \mathcal{O}_X(1))$. Then, the vector bundle $\mathcal{F} := \Phi^!_{\mathcal{S}}(\mathcal{E}) \in D^b(C)$ satisfies*

$$\mathrm{Hom}_{D^b(C)}(\mathcal{F}, \Phi^!_{\mathcal{S}}(\mathcal{O}_X(-2))[p]) = 0.$$

Proof. It is an immediate consequence of the following chain of equivalences, using the definition of Ulrich and the adjunction formula:

$$\begin{aligned}
&\mathrm{Hom}_{D^b(X)}(\mathcal{E}, \mathcal{O}_X(-2)[p]) = 0 \iff \\
&\mathrm{Hom}_{D^b(X)}(\Phi_{\mathcal{S}}\Phi^!_{\mathcal{S}}(\mathcal{E}), \mathcal{O}_X(-2)[p]) = 0 \iff \\
&\mathrm{Hom}_{D^b(C)}(\mathcal{F}, \Phi^!_{\mathcal{S}}(\mathcal{O}_X(-2))[p]) = 0.
\end{aligned}$$

□

Therefore, we will be interested in understanding the object $\Phi^!_{\mathcal{S}}(\mathcal{O}_X(-2)) \in D^b(C)$. It turns out that it is a vector bundle that had already been singled out in the theory of algebraic curves, $\Phi^!_{\mathcal{S}}(\mathcal{O}_X(-2)) \cong \omega_C^2 \otimes \mathcal{R}^\vee$ where $\mathcal{R}$ is a second Raynaud bundle, as defined in [182]. We are going to devote the next section to introduce this family of bundles and give their basic properties.

6.2 Introduction to Raynaud bundles

Raynaud bundles on the smooth curve C of genus two are fundamental objects to understand Ulrich bundles on a complete intersection $X \subset \mathbb{P}^5$ of two general quadrics. In this section we offer a short introduction to them, as well as present their main properties that will be used later on.

Let D be a smooth curve of genus $g \geq 2$. Let us fix a point $c \in D$. Associated to this choice there is the Abel–Jacobi embedding

$$\begin{aligned}
i : D &\longrightarrow J(D) := \mathrm{Pic}^0(D),\\
x &\mapsto \mathcal{O}_D(x - c)
\end{aligned}$$

of D in the Jacobian $A := J(D)$ of D, the set of line bundles on D of degree zero. This A has the structure of an abelian variety of dimension g. Moreover, A comes equipped with a canonical *principal polarization* $\mathcal{L}$. It can be characterized (modulo translation) as an ample line bundle $\mathcal{L}$ such that $\mathrm{h}^0(\mathcal{L}) = 1$. The unique (up to scalar) divisor Θ on the associated linear class is called a *theta divisor* of A. So $\mathcal{L} \cong \mathcal{O}_D(\Theta)$. Also $\mathcal{L}$ defines a canonical isomorphism between A and its *dual variety* $\hat{A} := \mathrm{Pic}^0(A)$. Let $\mathcal{P}$ be the *Poincaré bundle* on $A \times \hat{A}$. Then $\mathcal{P}$ is the universal line bundle parameterizing line bundles of degree zero on A.

For any $m \geq 1$, let us define the *Fourier–Mukai transform* of $\mathcal{O}_{\hat{A}}(-m\Theta)$ by

$$\mathcal{G}_m := R^g p_{A*}(Rp_{\hat{A}}^* \mathcal{O}(-m\Theta) \otimes \mathcal{P})$$

where p_A (respectively $p_{\hat{A}}$) is the natural projection from $A \times \hat{A}$ to A (respectively $\hat{A}$).

Proposition 6.2.1. *$\mathcal{G}_m$ is a vector bundle on A of rank m^g and degree gm^{g-1}.*

Proof. By base change theorem, $\mathcal{G}_m$ is a vector bundle on A with fiber over $x \in A$ isomorphic to

$$\mathrm{H}^g(\hat{A}, \mathcal{O}_{\hat{A}}(-m\Theta) \otimes \mathcal{P}_{|\{x\}\times\hat{A}}) \cong \mathrm{H}^0(J(D), \mathcal{O}_{J(D)}(m\Theta) \otimes \mathcal{P}_x)^\vee$$

where $\mathcal{P}_x$ is the line bundle in $\mathrm{Pic}^0(J(D))$ corresponding to $x \in J(D)$. Therefore this bundle has m^g global sections. □

Definition 6.2.2. With the previous notation, the vector bundle $\mathcal{R}_m := i^*(\mathcal{G}_m)$ on D is called the *m-Raynaud bundle*.

Remark 6.2.3. Note that the choice of a principal polarization $\mathcal{L}$ is fixed up to the translation. Consequently, the Raynaud bundles are fixed up to tensor by a line bundle of degree zero on D.

Proposition 6.2.4. *Let D be a smooth curve of genus $g \geq 2$. Then the Raynaud bundles $\mathcal{R}_m$ on D are rank m^g stable vector bundles of slope $\mu = g/m$.*

Proof. The fact that Raynaud bundles are semistable with slope $\mu = g/m$ can be found in [112, Lemma 2.3] following the original Raynaud's argument in [182, Section 3.1]. In [191, Corollary 2.2], using an argument based on theta group actions, it is proved that Raynaud bundles are indeed stable. □

Now we fix our attention to the case relevant in this chapter. Let C be a smooth curve of genus two and denote by $\mathcal{R} := \mathcal{R}_2$ the second Raynaud bundle on C. The next proposition gives a neat characterization of them.

Proposition 6.2.5. *Let $\mathcal{G}$ be a rank-four semistable vector bundle of degree four on a smooth curve C of genus two. If for any line bundle $\mathcal{L} \in \mathrm{Pic}^0(C)$, it holds that $\mathrm{Hom}(\mathcal{L}, \mathcal{G}) \neq 0$, then $\mathcal{G} \cong \mathcal{R}$ is a second Raynaud bundle.*

Proof. See [137, Lemma 5.8]. □

In the next lemma we can see, using universal properties of Raynaud bundles, that $\Phi_S^!(\mathcal{O}_X(-2))[2]$ is closely related to the second Raynaud bundle. In fact, we have the following result.

Proposition 6.2.6. *Let*

$$\begin{array}{rccc} \Phi_S^!: & D^b(X) & \longrightarrow & D^b(C), \\ & \mathcal{E} & \mapsto & Rq_*(Lp^*(\mathcal{E}) \otimes \mathcal{S}^\vee) \otimes \omega_C[1] \end{array}$$

be the right adjoint functor of the Fourier–Mukai functor $\Phi_{\mathcal{S}}$*. Then* $\Phi^!_{\mathcal{S}}(\mathcal{O}_X) \cong \mathcal{R}[1]$*, where* $\mathcal{R}$ *is the second Raynaud bundle.*

Proof. See [137, Lemma 5.9]. □

6.3 Ulrich bundles on complete intersections of two quadrics

In this section we are going to prove the existence of Ulrich bundles of arbitrary rank $r \geq 2$ on the complete intersection of two quadrics $Q_0 \cap Q_\infty \subset \mathbb{P}^5$. In particular, we are going to prove Theorem 6.3.10. Whereas the existence for rank $r \geq 3$ will rely on the previous results on derived categories of bounded complexes of coherent sheaves and the theory of Raynaud bundles, the case $r = 2$ admits a straight presentation in terms of Serre's correspondence as can be seen in the following result.

Proposition 6.3.1. *Let* $X := Q_0 \cap Q_\infty \subset \mathbb{P}^5$ *be the smooth complete intersection of two quadrics. Then* $(X, \mathcal{O}_X(1))$ *supports a rank-two Ulrich bundle* $\mathcal{E}$ *associated, through Serre's correspondence, to an elliptic normal curve* $D \subset X \subset \mathbb{P}^5$ *of degree* 6.

Proof. The existence of rank-two Ulrich bundles will be proved in full generality for an arbitrary Fano threefold of index two in Proposition 6.4.3. □

The next three statements are ancillary results to prove our main result. In what follows, X stands for the smooth complete intersection of the two quadrics Q_0 and Q_∞.

Proposition 6.3.2. *Let* $\mathcal{E}$ *be a rank r Ulrich bundle on* $(X, \mathcal{O}_X(H))$*. Then:*
(1) $\mu(\mathcal{E}) = H^3 = 4$.
(2) $\mathcal{E}^\vee(2)$ *is its Ulrich dual, as in Definition* 3.3.4.

Proof. (1) This follow from Proposition 3.2.5.
(2) It is a consequence of Definition 3.3.4 and the fact that $\omega_X \cong \mathcal{O}_X(-2)$. □

Lemma 6.3.3. *Let D be a smooth curve of genus g. Let* $\mathcal{F}$ *be a stable (respectively semistable) rank* $r \geq 2$ *vector bundle on D such that* $\mu(\mathcal{F}) \geq 2g - 2$ *(respectively* $\mu(\mathcal{F}) > 2g - 2$*). Then* $\mathrm{h}^1(\mathcal{F}) = 0$.

Proof. If $\mathrm{h}^1(\mathcal{F}) \neq 0$, by Serre's duality, it also holds that $\mathrm{h}^0(\mathcal{F}^\vee \otimes \omega_D) \neq 0$. This implies that there is a nontrivial map from $\mathcal{F}$ to ω_D. Since $\mathcal{F}$ is stable (respectively semistable) and $\mu(\mathcal{F}) \geq \mu(\omega_D) = 2g - 2$ (respectively $\mu(\mathcal{F}) > \mu(\omega_D)$), the only possibility is $\mathcal{F} \cong \omega_D$, in contradiction with the fact that $\mathrm{rank}(\mathcal{F}) \geq 2$. □

Lemma 6.3.4. *Let* $\mathcal{F}$ *and* $\mathcal{G}$ *be vector bundles on a smooth curve D such that* $\mathrm{H}^i(\mathcal{F} \otimes \mathcal{G}) = 0$ *for* $i = 0, 1$*. Then* $\mathcal{F}$ *and* $\mathcal{G}$ *are semistable.*

Proof. By Proposition 3.2.7, the tensor product $\mathcal{F}\otimes\mathcal{G}$ is an Ulrich bundle on the smooth curve D. Therefore, by Proposition 3.3.14, $\mathcal{F}\otimes\mathcal{G}$ is semistable. This forces $\mathcal{F}$ (and $\mathcal{G}$) to be semistable as well. □

Lemma 6.3.5. *Let*

$$\begin{array}{rccl} \Phi^!_{\mathcal{S}}\colon & D^b(X) & \longrightarrow & D^b(C), \\ & \mathcal{E} & \mapsto & Rq_*(Lp^*(\mathcal{E})\otimes\mathcal{S}^\vee)\otimes\omega_C[1] \end{array}$$

be the right adjoint functor of the Fourier–Mukai functor $\Phi_{\mathcal{S}}$. Then, $\Phi^!_{\mathcal{S}}(\mathcal{O}_X(-2))[2]\cong\mathcal{R}^\vee\otimes\omega_C^2$, where $\mathcal{R}$ is the second Raynaud bundle.

Proof. It follows from Proposition 6.2.6, using Grothendieck–Verdier duality and the projection formula. □

The next proposition is the key link between Ulrich bundles on X and certain semistable vector bundles on the curve C.

Proposition 6.3.6. *Let $\mathcal{E}(1)$ be a rank r Ulrich bundle on $(X,\mathcal{O}_X(1))$. Then, $\mathcal{F}:=\Phi^!_S(\mathcal{E})\in D^b(C)$ is a rank r semistable vector bundle of degree $2r$ satisfying:*
(1) $\mathrm{Ext}^i_C(\mathcal{R},\mathcal{F}^\vee\otimes\omega_C^{\otimes 2})=0$ *for* $i=0,1$.
(2) $\mathrm{H}^1(\mathcal{F}\otimes\mathcal{S}_x)=0$ *for all* $x\in X$.

Conversely, if $\mathcal{F}$ is a semistable rank r vector bundle on C of degree $2r$ satisfying the conditions (1) and (2), then $\Phi_S(\mathcal{F})\cong\mathcal{E}$ for some rank r Ulrich bundle $\mathcal{E}(1)$ on X.

Proof. Since it is given by a Fourier–Mukai functor, $\mathcal{F}$ is a complex of coherent sheaves with cohomology stalks over a point $c\in C$ isomorphic to

$$\mathrm{H}^{i+1}(X,\mathcal{E}\otimes\mathcal{S}_c^\vee)\cong\mathrm{Ext}^{i+1}(\mathcal{E}^\vee,\mathcal{S}_c^\vee).$$

By Proposition 6.3.2 (2), $\mathcal{E}^\vee$ is the twist of an Ulrich bundle of degree 0. Since $\mu(\mathcal{S}_c^\vee)=-1/2$ (independently whether $c\in C$ is a ramification point or not), it turns out that $\mathrm{Hom}(\mathcal{E}^\vee,\mathcal{S}_c^\vee)=0$.

Now consider the structural short exact sequence

$$0\longrightarrow\mathcal{S}^\vee_{\tau(c)}\longrightarrow\mathcal{O}_X^4\longrightarrow\mathcal{S}_c^\vee(1)\longrightarrow 0, \tag{6.1}$$

where τ denotes the hyperelliptic involution of the smooth curve C. Recursively, we can tensor (6.1) by $\mathcal{E}(j)$ for $j=0,-1,-2$, and use that $\mathcal{E}(1)$ is an Ulrich bundle to see that

$$\mathrm{H}^{i+1}(\mathcal{E}\otimes\mathcal{S}_c^\vee)\cong\mathrm{H}^{i+2}(\mathcal{E}(-1)\otimes\mathcal{S}^\vee_{\tau(c)})\cong\mathrm{H}^{i+3}(\mathcal{E}(-2)\otimes\mathcal{S}_c^\vee)=0$$

for $i\geq 1$. Therefore, $\mathcal{F}$ is concentrated on degree 0. Moreover, by flatness of the projection map, $\chi(\mathcal{E}\otimes\mathcal{S}_c)$ is independent on $c\in C$, which means that $\mathcal{F}$ is a rank r vector bundle of degree $2r$.

The fact that $\mathcal{F}$ satisfies (1) follows from Lemmas 6.1.5 and 6.3.5. Indeed, notice that since $\mathcal{F}$ and $\mathcal{R}$ are vector bundles on C, the vanishing conditions to be fulfilled from Lemma 6.1.5 are reduced to the degrees $p = 0, 1$.

Moreover, $\mathcal{F}$ is semistable by Lemma 6.3.4. Finally, to see that $\mathcal{F}$ satisfies (2), notice that $\Phi_{\mathcal{S}}(\mathcal{F}) = \mathcal{E}$ is a vector bundle on X and therefore $R^1p_*(Lq^*(\mathcal{F}) \otimes \mathcal{S}) = 0$, which implies that $\mathrm{H}^1(\mathcal{F} \otimes \mathcal{S}_x) = 0$ for all $x \in X$.

Conversely, if $\mathcal{F}$ is a vector bundle on C satisfying all the conditions from the statement, we get that $\mathcal{E} := \Phi_{\mathcal{S}}(\mathcal{F})$ is a vector bundle on X by (2). Moreover, $\mathcal{E} \in \langle \mathcal{O}_X, \mathcal{O}_X(-1)\rangle^{\perp}$. By (1), it also holds that $\mathrm{H}^i(\mathcal{E}^{\vee}(-2)) = 0$, which implies that $\mathcal{E}^{\vee}(1)$ is a rank r Ulrich bundle on $(X, \mathcal{O}_X(1))$. Therefore, by Lemma 6.3.2, $\mathcal{E}(1)$ is also an Ulrich bundle. □

By the previous proposition, the existence of rank r Ulrich bundles on $(X, \mathcal{O}_X(1))$ is therefore equivalent to the existence of certain semistable rank r vector bundles of degree $2r$ on C verifying some extra properties. Next results will be devoted to show their existence.

Proposition 6.3.7. *Let $X \subset \mathbb{P}^5$ be the complete intersection of two smooth quadrics and C its associated hyperelliptic smooth curve of genus two. Let $r \geq 2$ and let $M_C^s(r; 2r)$ be the moduli space of stable rank r vector bundles on C of degree $2r$. The subset*

$$V := \{\mathcal{F} \in M_C^s(r; 2r) \mid \mathrm{H}^1(\mathcal{F} \otimes \mathcal{S}_x) = 0 \textit{ for all } x \in X\}$$

is a nonempty open subset.

Proof. In order to see that V is open, we consider the subscheme

$$\Delta := \{(x, \mathcal{F}) \mid \mathrm{h}^1(\mathcal{F} \otimes \mathcal{S}_x) \geq 1\} \subset X \times M_C^s(r; 2r).$$

The set Δ is closed by upper-semicontinuity. Now, the projection map

$$p : X \times M_C^s(r; 2r) \longrightarrow M_C^s(r; 2r)$$

is proper which implies that $V = M_C^s(r; 2r)\backslash p(\Delta)$ is open.

It remains to see that V is nonempty. For $r = 2$, since we know that X carries a rank-two Ulrich bundle $\mathcal{E}(1)$, Proposition 6.3.6 shows that $\Phi^!_{\mathcal{S}}(\mathcal{E}) \in V$. For $r \geq 3$, let $\mathcal{F} \in M_C^s(r; 2r)\backslash V$. By Serre's duality, there exists a nonzero map $\mathcal{F} \longrightarrow \mathcal{S}_x^{\vee} \otimes \omega_C$ for some $x \in X$. Recall that $\mathcal{S}_x$ is a stable rank-two vector bundle of degree -1 so $\mu(\mathcal{S}_x^{\vee} \otimes \omega_C) = \frac{5}{2}$.

Since $\mathcal{F}$ is stable, such a map should be surjective, so there exists a short exact sequence

$$0 \longrightarrow \mathcal{F}' \longrightarrow \mathcal{F} \longrightarrow \mathcal{S}_x^{\vee} \otimes \omega_C \longrightarrow 0, \tag{6.2}$$

where $\mathcal{F}'$ is a semistable rank $r-2$ bundle of degree $2r-5$. Let us compute the dimension of vector bundles $\mathcal{F}$ fitting in such exact sequence for some $x \in X$ and $\mathcal{F}' \in M_C(r-2; 2r-5)$.

By Riemann–Roch theorem, $\operatorname{ext}^1(\mathcal{S}_x^\vee \otimes \omega_C, \mathcal{F}') = 3r-4$. Hence, for a fixed $x \in X$, the dimension of such vector bundles $\mathcal{F}$ is at most $(r-2)^2+1+(3r-5) = r^2-r$. Sweeping the variety X, the locus of vector bundles $\mathcal{F}$ that can be written as an extension (6.2) has dimension at most $r^2-r+3 < r^2+1$. Therefore, we can conclude that a general $\mathcal{F} \in M_C^s(r;2r)$ satisfies $\mathrm{H}^1(\mathcal{F} \otimes \mathcal{S}_x) = 0$ for every $x \in X$. In particular, this proves that V is nonempty. □

Proposition 6.3.8. *Let C be a genus-two smooth curve and $\mathcal{R}$ the second Raynaud bundle on C. Then, for a generic rank 3 stable vector bundle $\mathcal{G}$ of degree 6, it holds*

$$\mathrm{H}^i(\mathcal{R}^\vee \otimes \mathcal{G}) = 0 \quad \textit{for } i = 0,1.$$

Proof. Since, by Riemann–Roch theorem, $\chi(\mathcal{R}^\vee \otimes \mathcal{G}) = 0$, it is enough to see that $\mathrm{H}^0(\mathcal{R}^\vee \otimes \mathcal{G}) = 0$. Suppose on the contrary that there exists a nonzero map $\phi : \mathcal{R} \longrightarrow \mathcal{G}$. Since $\mathcal{R}$ is stable, ϕ is surjective or

$$(\operatorname{rank}(\operatorname{im}\phi), \deg(\operatorname{im}\phi)) \in \{(2,3),(3,4),(3,5)\}.$$

We will see that none of these possibilities is accomplished by a generic $\mathcal{G}$. For instance, let us suppose that ϕ is surjective. Then there exists a short exact sequence

$$0 \longrightarrow \mathcal{L} \longrightarrow \mathcal{R} \longrightarrow \mathcal{G} \longrightarrow 0, \tag{6.3}$$

where $\mathcal{L}$ is a line bundle of degree -2. In this situation, $\mathcal{R} \otimes \mathcal{L}^\vee$ will be a stable rank-four vector bundle of degree 12. Namely, $\mu(\mathcal{R} \otimes \mathcal{L}^\vee) = 3 > 2g(C)-2 = 2$. Therefore, by Lemma 6.3.3 and Riemann–Roch theorem, it holds that $\hom(\mathcal{L},\mathcal{R}) = 8$. Hence, the dimension of the family of stable bundles fitting into the short exact sequence (6.3) is at most $\dim \operatorname{Pic}^{-2}(C) + \dim \mathbb{P}(\operatorname{Hom}(\mathcal{L},\mathcal{R})) = 9 < 10 = \dim M_C(3;6)$.

A similar argument rules out the rest of cases. The reader can consult [56, Proposition 3.12] for the details. □

Proposition 6.3.9. *Let C be a smooth genus-two curve and let $\mathcal{R}$ be the second Raynaud bundle on C. For each $r \geq 2$, a generic stable vector bundle $\mathcal{G} \in M_C^s(r;2r)$ satisfies*

$$\operatorname{Ext}^i(\mathcal{R},\mathcal{G}) = 0$$

for $i = 0,1$.

Proof. By semicontinuity, it is enough to show the existence of a single vector bundle $\mathcal{G} \in M_C^s(r;2r)$ satisfying those vanishings. Propositions 6.3.1 and 6.3.6 deal with the case of rank 2. Proposition 6.3.8 shows us that the statement is true for $r = 3$. For the rest of ranks, we argue by induction on r. So let $r \geq 4$ and take $r_1, r_2 \geq 2$ such that $r = r_1 + r_2$. Consider two vector bundles $\mathcal{G}_j \in M_C^s(r_j;2r_j)$, $\mathcal{G}_1 \not\cong \mathcal{G}_2$, satisfying $\operatorname{Ext}^i(\mathcal{R},\mathcal{G}_j) = 0$. Since they are stable vector bundles with the same slope, we have $\operatorname{Hom}(\mathcal{G}_2,\mathcal{G}_1) = 0$.

Therefore, using Riemann–Roch theorem, we see that $\mathrm{Ext}^1(\mathcal{G}_2, \mathcal{G}_1) \neq 0$. We consider a nonsplitting extension

$$0 \longrightarrow \mathcal{G}_1 \longrightarrow \mathcal{G}' \longrightarrow \mathcal{G}_2 \longrightarrow 0.$$

Then $\mathcal{G}'$ is a rank r simple semistable vector bundle of degree $2r$ satisfying the vanishing conditions. The moduli space $\mathrm{Spl}_C(r; 2r)$ of simple sheaves of rank r and degree $2r$ on C is a well-defined smooth variety of dimension r^2+1 (see [3]). Therefore a generic element around $\mathcal{G}'$ is semistable. We are going to see that a generic element in an open neighborhood of $\mathcal{G}' \in \mathrm{Spl}_C(r; 2r)$ should be stable by a dimension counting argument. Otherwise, for a generic simple bundle $\mathcal{G}$ around $\mathcal{G}'$ in $\mathrm{Spl}_C(r; 2r)$ there would exist a destabilizing short exact sequence

$$0 \longrightarrow \mathcal{F}_1 \longrightarrow \mathcal{G} \longrightarrow \mathcal{F}_2 \longrightarrow 0$$

where $\mathcal{F}_1$ is a rank r' semistable vector bundle of degree $2r'$ (and therefore $\mathcal{F}_2$ is a rank $r - r'$ semistable vector bundle of degree $2(r - r'))$ for some $1 \le r' \le r - 1$. But the dimension of this family of extensions is bounded by

$$\dim M_C(r'; 2r') + \dim M_C(r - r'; 2(r - r')) + \mathrm{ext}^1(\mathcal{F}_2, \mathcal{F}_1) - 1 = r^2 - r'(r - r') + 1 < r^2 + 1,$$

for any $1 \le r' \le r - 1$. □

We can summarize the previous results in the following theorem, which is the central theorem of this chapter as it was sketched at its introduction.

Theorem 6.3.10. *For $r \ge 2$, let $M_X^U(r)$ be the moduli space of rank r Ulrich bundles on the complete intersection $(X, \mathcal{O}_X(1))$ of two quadrics and let $M_C(r; 2r)$ be the moduli space of semistable rank r vector bundles of degree $2r$ on C. Then the functor $\Phi^!_S : D^b(X) \longrightarrow D^b(C)$ induces a map*

$$\begin{array}{rcl} \phi : M_X^U(r) & \longrightarrow & M_C(r; 2r), \\ \mathcal{E}(1) & \mapsto & \phi(\mathcal{E}(1)) := [\Phi^!_S(\mathcal{E})] \end{array}$$

satisfying the following properties:
1. *ϕ is set-theoretically an injection.*
2. *ϕ maps exactly stable objects to stable objects.*
3. *ϕ induces an isomorphism of the stable locus $M_X^{U,s}(r)$ onto a nonempty open subscheme of $M_C^s(r; 2r)$. More precisely, we have*

$$\phi(M_X^{U,s}(r)) = \{\mathcal{F} \in M_C^s(r; 2r) \mid \mathrm{Ext}^i_C(\mathcal{R}, \mathcal{F}^\vee \otimes \omega_C^{\otimes 2}) = 0 \text{ for } i = 0, 1,$$
$$\text{and } \mathrm{H}^1(\mathcal{F} \otimes \mathcal{S}_x) = 0 \text{ for all } x \in X\},$$

Proof. First of all, a standard argument shows that ϕ is well-defined. Claim (1) follows from the fact that $\Phi_{\mathcal{S}} : D^b(C) \longrightarrow \Phi_{\mathcal{S}}(D^b(C))$ is an equivalence of categories and that $\mathcal{E} \in \Phi_{\mathcal{S}}(D^b(C))$ for any Ulrich bundle $\mathcal{E}(1)$ on X.

(2) Assume that $\mathcal{F} = \Phi^!_{\mathcal{S}}(\mathcal{E})$ is strictly semistable. Then there exists a destabilizing sequence

$$0 \longrightarrow \mathcal{F}'' \longrightarrow \mathcal{F} \longrightarrow \mathcal{F}' \longrightarrow 0.$$

Constructing the long exact sequences of cohomology associated to this short exact sequence, we see that $\mathcal{F}'$ satisfies the conditions of Proposition 6.3.6. Therefore $\mathcal{E}' := \Phi_{\mathcal{S}}(\mathcal{F}')$ is an Ulrich bundle twisted by -1 on X. Moreover, there exists a nonzero map between $\mathcal{E} \longrightarrow \mathcal{E}'$ showing that $\mathcal{E}$ should also be strictly semistable. On the other hand, if we start assuming that $\mathcal{E}$ is strictly semistable, Proposition 3.3.1 and the exactness of the functor $\Phi^!_{\mathcal{S}}$ shows that $\mathcal{F}$ is also strictly semistable.

In order to see (3), take any Ulrich bundle $\mathcal{E}(1)$ on X. Then the functor $\Phi^!_{\mathcal{S}}$ induces the following isomorphisms:

$$\begin{aligned} T_{[\mathcal{E}]}M_X^s(r) &\cong \operatorname{Ext}^1_X(\mathcal{E}, \mathcal{E}) \\ &\cong \operatorname{Ext}^1_C(\phi(\mathcal{E}), \phi(\mathcal{E})) \\ &\cong T_{[\phi(\mathcal{E})]}M_C(r; 2r), \end{aligned}$$

showing that ϕ is an isomorphism around $[\mathcal{E}]$. To conclude, we use Propositions 6.3.7 and 6.3.9 to observe that $\phi(M_X^{U,s}(r))$ is open and nonempty. □

6.4 Rank-two Ulrich bundles on Fano threefolds

As stated in Proposition 6.3.1, through Serre's correspondence it is possible to show the existence of rank-two Ulrich bundles on the complete intersection of two quadrics $Q_0 \cap Q_\infty \subset \mathbb{P}^5$. Indeed, this is a common feature of most of the Fano threefolds. In this section we are going to construct rank-two Ulrich bundles on them. Recall, that the index i_X of a Fano threefold X is bounded by four. When $i_X = 4$, we are dealing with the projective space $(\mathbb{P}^3, \mathcal{O}_{\mathbb{P}^3}(1))$. In this case, by Horrocks' theorem (see Theorem 2.2.2), the only rank-two Ulrich bundle is $\mathcal{O}^2_{\mathbb{P}^3}$. The next case, namely when $i_X = 3$, corresponds to the quadric threefold $X = Q_3 \subset \mathbb{P}^4$. In this case, by Knörrer's theorem (see Theorem 2.3.4), the only Ulrich bundle is the spinor bundle of rank two.

The most interesting cases correspond therefore to index two and one. The method to construct rank-two Ulrich bundles on them differs drastically. Indeed, for Fano threefolds of index two, or more generally of even index, we are going to construct them directly by Serre's correspondence. On the other hand, the method for constructing rank-two Ulrich bundles on (general) Fano threefolds of index one will rely on an induction procedure and a deformation argument on a certain moduli space. Let us start first with the index-two case.

Recall, as it was stated in Definition 6.1.1, that a Fano threefold has even index if $\omega_X^\vee \cong \mathcal{L}^{\otimes 2}$ for some ample line bundle $\mathcal{L}$ in $\mathrm{Pic}(X)$. In this situation, the degree d of X with respect to $\mathcal{L}$ is bounded by $3 \le d \le 8$. Besides, for $d = 8$, we recover $\mathbb{P}^3$ of index four (since $\omega_{\mathbb{P}^3} \cong \mathcal{O}_{\mathbb{P}^3}(-4)$). Otherwise $\mathcal{L}$ is not divisible in $\mathrm{Pic}(X)$, namely X has index two. This is the complete list of Fano threefolds of even index. Moreover, $\mathcal{L}$ gives an embedding $X \subset \mathbb{P}^{d+1}$.

Definition 6.4.1. A *normal elliptic curve* $C \subset \mathbb{P}^{d+1}$ is a smooth projective curve of genus one that it is projectively normal, namely $\mathrm{H}^1(\mathcal{I}_{C|\mathbb{P}^{d+1}}(k)) = 0$ for all $k \in \mathbb{Z}$. In other words, C is projectively normal if any divisor from the linear system $|\mathcal{O}_C(k)|$ is defined by the intersection of C with a hypersurface of $\mathbb{P}^{d+1}$ of degree k. The degree of C in $\mathbb{P}^{d+1}$ turns out to be $d + 2$.

Proposition 6.4.2. *Let $X \subset \mathbb{P}^{d+1}$ be a Fano threefold of even index. Then X contains a normal elliptic curve $C \hookrightarrow X$ of degree $d + 2$.*

Proof. Let us first deal with the case $d \le 7$. Let $S := X \cap H \subset H \cong \mathbb{P}^d$ be the intersection of X with a general hyperplane $H \subset \mathbb{P}^{d+1}$. Notice that S is a del Pezzo surface of degree d, namely, S is isomorphic to the blow-up of $\mathbb{P}^2$ at $9 - d$ points. Using the notation for the divisors of del Pezzo surfaces from Theorem 5.3.8, let us consider the linear system $|4l - 2e_1 - 2e_2 - e_3 - \cdots - e_{9-d}|$ of quartic curves of $\mathbb{P}^2$ with double points at p_1 and p_2 and simple points at the rest. This is a base point free linear system and therefore we can take a smooth curve D from this linear system. Also $D \subset H$ is a normal elliptic curve of degree $d + 1$ inside $H \cong \mathbb{P}^d$.

Now let us consider $C' := D \cup L$ a reducible curve with $L \subset X$ a line not contained in H and such that $D \cap L = \{p\}$ is a simple point. Notice that it is possible to find such a line L by [121, Proposition 3.3.5]. If we manage to see that C' can be deformed inside X to a smooth curve C, then we conclude since C should be a normal elliptic curve of degree $d+2$. In order to see that C' can be deformed, it is enough, by [111, Theorem 4.1], to see that $\mathrm{H}^1(L, \mathcal{N}_{L|X}) = 0$ and $\mathrm{H}^1(D, \mathcal{N}_{D|X}(-p)) = 0$.

It is easy to see that $\mathcal{N}_{L|X}$ is isomorphic to either $\mathcal{O}_L^2$ or $\mathcal{O}_L(-1) \oplus \mathcal{O}(1)$. So in both cases we have the first requested vanishing. In order to see the second one, notice that $\mathcal{N}_{D|X}$ lies in the exact sequence

$$0 \longrightarrow \mathcal{N}_{D|S} \longrightarrow \mathcal{N}_{D|X} \longrightarrow \mathcal{N}_{S|X\,|D} \longrightarrow 0.$$

Since both bundles $\mathcal{N}_{D|S}$ and $\mathcal{N}_{S|X\,|D} \cong \mathcal{O}_D(1)$ have degree $d + 1 \ge 4$, we get $\mathrm{H}^1(D, \mathcal{N}_{D|X}(-p)) = 0$.

Finally, we deal with the Fano threefold X of degree $d = 8$, namely

$$v : \mathbb{P}^3 \xrightarrow{\cong} X \subset \mathbb{P}^9,$$

which is the Veronese embedding of $\mathbb{P}^3$ by $\mathcal{O}_{\mathbb{P}^3}(2)$. Let us consider an elliptic curve $C \subset \mathbb{P}^3$ of degree five. Since C is not contained in any quadratic form, we have an isomorphism

$$\nu^* : \mathrm{H}^0(\mathcal{O}_{\mathbb{P}^3}(2)) \xrightarrow{\cong} \mathrm{H}^0(\mathcal{O}_C(2)),$$

and therefore $\nu(C)$ is a normal elliptic curve of degree ten. □

Proposition 6.4.3. *Let $(X, \mathcal{L})$ be a Fano threefold of even index with $\mathcal{L}$ the very ample line bundle such that $\omega_X^\vee \cong \mathcal{L}^{\otimes 2}$. Then $(X, \mathcal{L})$ supports a rank-two special Ulrich bundle.*

Proof. Let $C \subset X \subset \mathbb{P}^{d+1}$ be a normal elliptic curve whose existence has been proven in Proposition 6.4.2. Since $\omega_C \cong \mathcal{O}_C$, it holds that $\det(\mathcal{N}_{C|X}) \cong \mathcal{O}_C(2)$. Therefore, since $\mathrm{Ext}^1(\mathcal{I}_{C|X}(2), \mathcal{O}_X) \cong \mathrm{H}^2(\mathcal{O}_C) \cong \mathbf{k}$, by Serre's correspondence, there exists a (unique) nontrivial extension

$$0 \longrightarrow \mathcal{O}_X \longrightarrow \mathcal{E} \longrightarrow \mathcal{I}_{C|X}(2) \longrightarrow 0 \tag{6.4}$$

where $\mathcal{E}$ is a rank-two vector bundle. Let us prove that $\mathrm{H}^i(\mathcal{E}(-j)) = 0$ for $j = 1, 2, 3$ and $i = 0, \ldots, 3$. This, by Theorem 3.2.9, implies that $\mathcal{E}$ is an Ulrich bundle. Indeed, since $c_1(\mathcal{E}) = -K_X$, by Serre's duality it is enough to show that $\mathrm{H}^i(\mathcal{E}(-1)) = 0$ for $i = 0, \ldots, 3$ and $\mathrm{H}^i(\mathcal{E}(-2)) = 0$ for $i = 0, 1$.

Now, since $C \subset X \subset \mathbb{P}^{d+1}$ is a nondegenerate normal elliptic curve, it holds that $\mathrm{H}^i(\mathcal{I}_{C|\mathbb{P}^{d+1}}(1)) = 0$ for $i = 0, \ldots, 3$ and $\mathrm{H}^i(\mathcal{I}_{C|\mathbb{P}^{d+1}}) = 0$ for $i = 0, 1$. On the other hand, it also holds that $\mathrm{H}^i(\mathcal{O}_X(-1)) = 0$ for $i = 0, \ldots, 3$ and $\mathrm{H}^i(\mathcal{O}_X(-2)) = 0$ for $i = 0, 1$. Then applying the functor of global sections to the exact sequence (6.4), we obtain that $\mathcal{E}$ is an Ulrich bundle. □

Finally, we are going to deal with the case of Fano threefold of index one and $\mathrm{Pic}(X) \cong \mathbb{Z}$. They are known in the literature as *prime Fano threefolds*. In this setting, we see that the Picard group $\mathrm{Pic}(X)$ of X is generated by the ample divisor $H := -K_X$. They have been classified in terms of its *genus* $g := H^3/2 + 1$. Notice that it corresponds to the arithmetic genus of a smooth curve defined by the intersection of two general hyperplanes H. The genus g of a prime Fano threefold X belongs to the set $g \in \{2, 3, \ldots, 10, 12\}$. Moreover, there is a single deformation class for each prime Fano threefold of fixed genus.

In order to be able to construct rank-two Ulrich bundles, we need to impose two further restrictions to our prime Fano threefold X. First, X should be *nonhyperelliptic*, meaning that $-K_X$ is indeed very ample. Notice this is a minor restriction. In fact, any prime Fano threefold of genus $g \geq 4$ is nonhyperelliptic while there is one family of prime hyperelliptic Fano threefolds of genus two and also one family of prime hyperelliptic Fano threefolds of genus three.

The other restriction concerns the Hilbert scheme $\mathrm{Hilb}^{t+1}(X)$ of lines contained in X. It is a projective curve. The prime Fano threefold X will be called *exotic* if $\mathrm{Hilb}^{t+1}(X)$ has a component nonreduced at any point. This is equivalent to the fact that

$\mathcal{N}_{L|X} \cong \mathcal{O}_L(1) \oplus \mathcal{O}_L(-2)$ for any line L from that component. Otherwise, X is called *ordinary*. If X is ordinary, $\mathrm{Hilb}^{t+1}(X)$ has a generically smooth component or, equivalently, the splitting type of the normal bundle $\mathcal{N}_{L|X}$ of a general line L from that component is isomorphic to $\mathcal{O}_L \oplus \mathcal{O}_L(-1)$. We are going to work only with ordinary prime Fano threefolds. Again it is a minor restriction since a general prime Fano threefold is ordinary. Moreover, for $g_X \geq 9$, the only exotic prime Fano threefold is the *Mukai–Umemura threefold.*

Theorem 6.4.4. *Let $(X, \mathcal{O}_X(-K_X))$ be a smooth ordinary nonhyperelliptic prime Fano threefold of genus g. Then X supports stable special rank-two Ulrich bundles $\mathcal{E}$. Their Chern classes are $c_1(\mathcal{E}) = -3K_X$ and $c_2(\mathcal{E}) = 5g - 1$.*

Proof. By [32, Lemma 3.8], X supports a rank-two aCM bundle $\mathcal{F}$ with Chern classes $c_1(\mathcal{F}) = H := -K_X$ and $c_2(\mathcal{F}) = g + 2$. In addition, $\mathrm{h}^0(\mathcal{F}) = 1$ and $\mathcal{F}_{|L} \cong \mathcal{O}_L \oplus \mathcal{O}_L(1)$ for any line $L \subset X$ with $\mathcal{N}_{L|X} \cong \mathcal{O}_L \oplus \mathcal{O}_L(-1)$.

The splitting of $\mathcal{F}_{|L}$ provides a short exact sequence

$$0 \longrightarrow \mathcal{K} \longrightarrow \mathcal{F} \longrightarrow \mathcal{O}_L \longrightarrow 0 \tag{6.5}$$

with $\mathcal{K}$ a stable rank-two torsion-free sheaf with Chern classes $c_1(\mathcal{K}) = H$ and $c_2(\mathcal{K}) = g + 3$, namely an element of the moduli space of stable torsion-free sheaves $\overline{M}^s_{X,H}(2; H, g + 3)$. Let $\mathcal{E}$ be a general locally free deformation of $\mathcal{K}$. We are going to show that $\mathcal{E}(1)$ is Ulrich. First of all, notice that $\mathrm{H}^0(\mathcal{E}(-1)) = \mathrm{H}^1(\mathcal{E}(-1)) = 0$ since the same vanishings are true for $\mathcal{F}$. So by Serre duality (taking into account that $\mathcal{E}^\vee \cong \mathcal{E}(-1)$), we see that $\mathrm{H}^i(\mathcal{E}(-1)) = 0$ for all $i = 0, \ldots, 3$. On the other hand, from the exact sequence (6.5) and semicontinuity, we obtain $\mathrm{H}^2(\mathcal{E}) = 0$. Moreover, $\mathrm{H}^3(\mathcal{E})$ is also trivial by Serre duality and stability. Suppose we prove that $\mathrm{H}^0(\mathcal{E}) = 0$. Then, since, by Riemann–Roch formula, $\chi(\mathcal{E}) = 0$, we obtain that $\mathrm{H}^1(\mathcal{E}) = 0$. But therefore $\mathrm{H}^i(\mathcal{E}(-2)) \cong \mathrm{H}^{3-i}(\mathcal{E}^\vee(1)) \cong \mathrm{H}^{3-i}(\mathcal{E}) = 0$ for all $i = 0, \ldots, 3$. We can conclude by Theorem 3.2.9 that $\mathcal{E}(1)$ is a rank-two Ulrich bundle on $(X, \mathcal{O}_X(-K_X))$ with the aforementioned Chern classes.

So, it only remains to prove that for a general locally free deformation $\mathcal{E}$ of $\mathcal{K}$ it holds that $\mathrm{H}^0(\mathcal{E}) = 0$. Notice that, by the exact sequence (6.5) and the hypothesis on $\mathcal{F}$, in any case $\mathrm{H}^0(\mathcal{E}) \leq 1$. If we had equality in general, we could define a map

$$\phi : U \subset \overline{M}^s_{X,H}(2; H, g + 3) \longrightarrow \mathrm{Hilb}^{(g+3)t}(X)$$

from an open neighborhood U of $\mathcal{K}$ to the Hilbert scheme of curves of arithmetic genus one and degree $g + 3$, by sending the unique nonzero section of the stable sheaf $\mathcal{E}$ to its zero locus D. Notice that this map ϕ is injective and that the local dimension of U is $g+4$ again by [32, Lemma 3.8]. Now, using (6.5), it can be seen that the zero locus of the nonzero section s of $\mathcal{K}$ is $C \cup L$ where C is the zero locus of the section of $\mathcal{F}$. Moreover, along the same lines, the following cohomology group can be computed:

$$\mathrm{ext}^2_X(\mathcal{I}_{C\cup L|X}, \mathcal{I}_{C\cup L|X}) = 0.$$

Since D is a flat deformation of $C \cup L$, we also have $\mathrm{ext}^2_X(\mathcal{I}_{D|X}, \mathcal{I}_{D|X}) = 0$. Again, by Riemann–Roch formula, we get that $\mathrm{ext}^1_X(\mathcal{I}_{D|X}, \mathcal{I}_{D|X}) = g + 3$ so $\mathrm{Hilb}^{(g+3)t}(X)$ is smooth and locally of dimension $g + 3$ around D. But this gives a contradiction with ϕ being an injection. This contradiction forces us to conclude that $\mathrm{H}^0(\mathcal{E}) = 0$ and the proof is completed. □

6.5 Final comments and additional reading

As it was mentioned at the introduction, the goal of this chapter was to perform a complete study of Ulrich bundles on the particular case of the smooth intersection $X = Q_0 \cap Q_\infty \subset \mathbb{P}^5$ of two four-dimensional quadrics. The study of a semiorthogonal decomposition of the derived category of coherent sheaves on this variety and, in particular, its connection with its associated hyperelliptic curve, was one of the central subjects of the groundbreaking paper [31]. The structure and most of the results of this chapter are taken from [56], as well as from [137]. The natural link between these two papers is the fact that rank-two Ulrich bundles on X are the so-called *instanton bundles* studied in [137]. A key role in the constructions done in this chapter has been played by the *Raynaud bundles* that were introduced in [182]. The reader can also check [166]. An introduction to these vector bundles is given in [177]. The reader can consult [121] for a complete introduction to the study and classification of Fano threefolds. It is worthwhile to point out that in the series of papers [32, 33], and [34] the authors had exploited similar techniques on semiorthogonal decomposition and homological projective duality to study moduli spaces of rank 2 aCM sheaves on prime Fano threefolds.

The existence of elliptic normal curves on Fano threefolds of index two associated to Ulrich rank-two bundles from Proposition 6.4.2 was proven in [24]. The reader can also see [12]. Finally, the construction of Ulrich bundles on Fano threefolds of index one has been learnt from [32]. The results on abelian varieties mentioned throughout this chapter can be found in [25]. In the next chapters we will move forward to study Ulrich bundles on higher dimensional varieties. The reader particularly interested in the theory of Ulrich bundles on threefolds and the sparse results obtained in this area can read [138, 141], and [94], among other references.

7 Ulrich bundles on higher-dimensional varieties

In this chapter, we focus our attention on the existence of Ulrich bundles on higher-dimensional varieties. In particular, we will give new evidence to support one of the central conjectures of this book, Conjecture 3.4.8. Nevertheless, we will see that the picture is much more unclear if we compare it with the scenario that we have when the underlying variety is a curve or a surface. In fact, in general it is a difficult problem to deal with vector bundles on higher-dimensional varieties, and the difficulty increases if one tries to deal with vector bundles of small rank comparing with the dimension of the variety. It is a very difficult problem to construct low rank vector bundles on higher-dimensional varieties. Another difficulty that one faces is the fact that if the underlying variety has dimension greater than or equal to three, there is no characterization of Ulrich bundles in terms of numerical invariants of the bundle and a few cohomological vanishing as it exists in the case of curves (see Lemma 3.4.2) and in the case of surfaces (see Proposition 5.1.1). Another obstruction can be found in the fact that, in general, the construction and, hence, existence of Ulrich bundles on a variety strongly depends on the geometry of the variety and the classification and characterization of higher-dimensional varieties is out of reach in the sense that we do not have a classification result in the flavor of Enriques–Kodaira theorem for surfaces. As a consequence, there are few general results concerning the existence of Ulrich bundles on higher-dimensional varieties.

In Chapter 4, we have seen the existence of Ulrich bundles on complete intersections and their relation with the matrix factorization. But, if one delves into the idea that the existence of Ulrich bundles on higher-dimensional varieties is closely related with the geometry of the variety, one can also start to look for Ulrich bundles inside a set of bundles that share another property closely linked with the geometry of the variety. For instance, homogeneous bundles on homogeneous varieties. This is exactly what we have done in the last two sections of this chapter. In Section 7.2 we completely classify all irreducible homogeneous Ulrich bundles on any Grassmann variety $\mathrm{Gr}(k,n)$. In this case we give a criterion to check if an irreducible homogeneous bundle is Ulrich or not. In particular, we prove the existence of Ulrich bundles on any Grassmann variety $\mathrm{Gr}(k,n)$. It should be mention that the rank of this bundle is in general very large, and in this context the existence of Ulrich bundles (not homogeneous) of lower rank (see Problem 7.2.12) is not clear. The situation changes completely when we deal with homogeneous bundles on flag varieties $\mathrm{F}(k_1,\dots,k_r;n)$. In this case we will see that if the flag has more than two steps then there are no irreducible homogeneous Ulrich bundles supported on it. If the flag has two steps then we will see the existence of irreducible homogeneous Ulrich bundles in many cases and we will state a conjecture (see Conjecture 7.3.70) which exactly describes the existence of irreducible

The original version of this chapter was revised: the footnote on p. 197, line 1 has been added. An Erratum is available at DOI: https://doi.org/10.1515/9783110647686-013
https://doi.org/10.1515/9783110647686-007

homogeneous Ulrich bundles on such flag varieties. All the results are obtained as a consequence of the fact that in this case the existence of Ulrich bundles is translated into a combinatorial problem.

Segre varieties are another example of higher-dimensional varieties that have very specific properties coming from the fact that they are immersions of products of projective spaces. This rich geometry allows constructing low rank Ulrich bundles in some cases, in fact, Ulrich line bundles, and in other cases higher rank Ulrich bundles as a tensor product of bundles on the factors that are not Ulrich on the corresponding factor. The existence of Ulrich bundles on Segre varieties is the content of Section 7.1.

7.1 Ulrich bundles on Segre varieties

In this section, we are going to explain how to construct large families of Ulrich bundles on most of the Segre varieties. Let us start recalling their definition. Given integers $1 \leq n_1, \ldots, n_s$, we consider the product of projective spaces of the form $\mathbb{P}^{n_1} \times \cdots \times \mathbb{P}^{n_s}$. Let p_i denote the ith projection

$$\mathbb{P}^{n_1} \times \cdots \times \mathbb{P}^{n_s} \longrightarrow \mathbb{P}^{n_i}.$$

There is a canonical isomorphism $\mathbb{Z}^s \longrightarrow \operatorname{Pic}(\mathbb{P}^{n_1} \times \cdots \times \mathbb{P}^{n_s})$, given by

$$(a_1, \ldots, a_s) \mapsto \mathcal{O}_{\mathbb{P}^{n_1} \times \cdots \times \mathbb{P}^{n_s}}(a_1, \ldots, a_s) := p_1^*(\mathcal{O}_{\mathbb{P}^{n_1}}(a_1)) \otimes \cdots \otimes p_s^*(\mathcal{O}_{\mathbb{P}^{n_s}}(a_s)).$$

Definition 7.1.1. Given integers $1 \leq n_1, \ldots, n_s$, we denote by

$$\sigma_{n_1,\ldots,n_s} : \mathbb{P}^{n_1} \times \cdots \times \mathbb{P}^{n_s} \longrightarrow \mathbb{P}^N, \quad N = \prod_{i=1}^{s}(n_i + 1) - 1$$

the *Segre embedding* of $\mathbb{P}^{n_1} \times \cdots \times \mathbb{P}^{n_s}$ through the very ample line bundle $\mathcal{O}_{\mathbb{P}^{n_1} \times \cdots \times \mathbb{P}^{n_s}}(1, \ldots, 1)$ on $\mathbb{P}^{n_1} \times \cdots \times \mathbb{P}^{n_s}$.

The image of $\sigma_{n_1,\ldots,n_s}$ is the *Segre variety* $\Sigma_{n_1,\ldots,n_s} := \sigma_{n_1,\ldots,n_s}(\mathbb{P}^{n_1} \times \cdots \times \mathbb{P}^{n_s}) \subseteq \mathbb{P}^N$, $N = \prod_{i=1}^{s}(n_i + 1) - 1$.

In the next proposition we gather some basic properties of Segre varieties.

Proposition 7.1.2. *Fix integer $1 \leq n_1, \ldots, n_s$ and denote by $\Sigma_{n_1,\ldots,n_s} \subseteq \mathbb{P}^N$, $N = \prod_{i=1}^{s}(n_i + 1) - 1$, the Segre variety. Then:*
1. $\dim \Sigma_{n_1,\ldots,n_s} = \sum_{i=1}^{s} n_i$ *and* $\deg(\Sigma_{n_1,\ldots,n_s}) = \frac{(\sum_{i=1}^{s} n_i)!}{\prod_{i=1}^{s}(n_i)!}$.
2. $\Sigma_{n_1,\ldots,n_s}$ *is an aCM variety and the ideal* $I(\Sigma_{n_1,\ldots,n_s})$ *of* $\Sigma_{n_1,\ldots,n_s}$ *is generated by* $\binom{N+2}{2} - \prod_{i=1}^{s}\binom{n_i+2}{2}$ *hyperquadrics.*

Proof. (1) The dimension of $\Sigma_{n_1,\ldots,n_s}$ is clearly $\sum_{i=1}^{s} n_i$. On the other hand, notice that the degree of $\Sigma_{n_1,\ldots,n_s}$ with respect to $\mathcal{O}_{\Sigma_{n_1,\ldots,n_s}}(1, \ldots, 1)$ can be computed as the coefficient of $H_1^{n_1} \cdots H_s^{n_s}$ in the expansion of $(H_1 + \cdots + H_s)^{\sum_{i=1}^{s} n_i}$ where H_i is the divisor corresponding to the line bundle $p_i^*(\mathcal{O}_{\mathbb{P}^{n_i}}(1))$. This number is the multinomial coefficient $\frac{(\sum_{i=1}^{s} n_i)!}{\prod_{i=1}^{s}(n_i)!}$.

(2) See [72, Proposition 2.2]. □

We are going to fix the notation that we will use for the rest of the section. As mentioned, for any coherent sheaf $\mathcal{E}_i$ on $\mathbb{P}^{n_i}$, we set

$$\mathcal{E}_1 \boxtimes \cdots \boxtimes \mathcal{E}_s := p_1^*(\mathcal{E}_1) \otimes \cdots \otimes p_s^*(\mathcal{E}_s).$$

We will denote by

$$\pi_i : \mathbb{P}^{n_1} \times \cdots \times \mathbb{P}^{n_s} \longrightarrow X_i := \mathbb{P}^{n_1} \times \cdots \times \widehat{\mathbb{P}^{n_i}} \times \cdots \times \mathbb{P}^{n_s}$$

the natural projection and, given sheaves $\mathcal{E}$ and $\mathcal{F}$ on X_i and $\mathbb{P}^{n_i}$, respectively, $\mathcal{E} \boxtimes \mathcal{F}$ stands for $\pi_i^*(\mathcal{E}) \otimes p_i^*(\mathcal{F})$.

By the Künneth formula (see Example 2.2 (3)), we have

$$\mathrm{H}^l(\Sigma_{n_1,\dots,n_s}, \mathcal{E} \boxtimes \mathcal{F}) = \bigoplus_{p+q=l} \mathrm{H}^p(X_i, \mathcal{E}) \otimes \mathrm{H}^q(\mathbb{P}^{n_i}, \mathcal{F}).$$

To simplify the notation, we are going to denote by $\mathcal{O}_{\Sigma_{n_1,\dots,n_s}}(1)$ the very ample line bundle $\mathcal{O}_{\Sigma_{n_1,\dots,n_s}}(1, \dots, 1)$ while given a coherent sheaf $\mathcal{H}$ on $\Sigma_{n_1,\dots,n_s}$, $\mathcal{H}(t)$ stands for $\mathcal{H} \otimes \mathcal{O}_{\Sigma_{n_1,\dots,n_s}}(t, \dots, t)$.

Recall that we saw in Example 3.3.11 (2) that the line bundles of the form $\mathcal{O}_{\Sigma_{n_1,\dots,n_s}}(a_1, \dots, a_s)$ where, if we order the coefficients $0 = a_{i_1} \leq \cdots \leq a_{i_k} \leq \cdots \leq a_{i_s}$ then $a_{i_k} = \sum_{1 \leq j < k} n_{i_j}$, are Ulrich line bundles on $\Sigma_{n_1,\dots,n_s}$. Here we can show that this list is exhaustive.

Proposition 7.1.3. *Let $(\Sigma_{n_1,\dots,n_s}, \mathcal{O}_{\Sigma_{n_1,\dots,n_s}}(1))$ be a Segre variety. Then there exist s! Ulrich line bundles on $\Sigma_{n_1,\dots,n_s}$. They are of the form*

$$\mathcal{L}_{X_i} \boxtimes \mathcal{O}_{\mathbb{P}^{n_i}}\left(\sum_{k \neq i} n_k\right),$$

where $\mathcal{L}_{X_i}$ is a rank-one Ulrich bundle on the Segre variety $X_i := \Sigma_{n_1,\dots,\widehat{n_i},\dots,n_s} \subseteq \mathbb{P}^{N'}$, $N' = \prod_{\substack{1 \leq j \leq d \\ j \neq i}} (n_j + 1) - 1$. More explicitly, the Ulrich line bundles on $\Sigma_{n_1,\dots,n_s}$ are of the form $\mathcal{O}_{\Sigma_{n_1,\dots,n_s}}(a_1, \dots, a_s)$ where, if we order the coefficients $0 = a_{i_1} \leq \cdots \leq a_{i_k} \leq \cdots \leq a_{i_s}$ then $a_{i_k} = \sum_{1 \leq j < k} n_{i_j}$.

Proof. According to Example 3.3.11 (2), we only have to check that the above list is exhaustive. To this end, let us consider an Ulrich line bundle $\mathcal{L} := \mathcal{O}_{\Sigma_{n_1,\dots,n_s}}(a_1, \dots, a_s)$ with $a_{i_1} \leq \cdots \leq a_{i_k} \leq \cdots \leq a_{i_s}$. Given that $\mathcal{L}$ is initialized, it holds that $a_{i_1} = 0$. Since $\mathcal{L}$ is aCM, we have

$$\mathrm{H}^{\sum_{j=1}^k n_{i_j}}(\Sigma_{n_1,\dots,n_s}, \mathcal{L}(-\Sigma_{j=1}^k n_{i_j} - 1)) = 0$$

for $k = 1, \dots, s-1$. In particular, using Künneth formula, it holds

$$\prod_{l=1}^{k} \mathrm{h}^{n_{i_l}}\left(\mathbb{P}^{n_{i_l}}, \mathcal{O}_{\mathbb{P}^{n_{i_l}}}\left(a_{i_l} - \sum_{j=1}^{k} n_{i_j} - 1\right)\right) \cdot \prod_{l=k+1}^{s} \mathrm{h}^0(\mathbb{P}^{n_{i_l}}, \mathcal{O}_{\mathbb{P}^{n_{i_l}}}(a_{i_l} - \Sigma_{j=1}^k n_{i_j} - 1)) = 0,$$

from where it follows that, by induction, $a_{i_{k+1}} \leq b_{i_{k+1}} := \Sigma_{1\leq j\leq k} n_{i_j}$ for $k = 1, \ldots, s-1$ (and $b_{i_1} := 0$). But, by Künneth formula and Proposition 7.1.2,

$$\mathrm{h}^0(\Sigma_{n_1,\ldots,n_s}, \mathcal{O}_{\Sigma_{n_1,\ldots,n_s}}(b_1,\ldots,b_s)) = \frac{(\sum_{i=1}^{s} n_i)!}{\prod_{i=1}^{s}(n_i)!} = \deg(\Sigma_{n_1,\ldots,n_s}).$$

Hence, we are forced to have $a_{i_j} = b_{i_j}$ for $j = 1, \ldots, s$. □

On the other hand, we got the complete list of Ulrich bundles on the case of the Segre variety $\mathbb{P}^1 \times \mathbb{P}^1 \subset \mathbb{P}^3$ or, in other words, on the quadric surface. They are $\mathcal{O}_{\mathbb{P}^1\times\mathbb{P}^1}(1,0)$ and $\mathcal{O}_{\mathbb{P}^1\times\mathbb{P}^1}(0,1)$. The goal of the remainder of this section is to show that for the rest of Segre varieties, we can construct large families of indecomposable Ulrich bundles of arbitrarily large rank. The proof will be different depending on whether $n_1 \geq 2$ or not.

First, we are going to focus on the case $n_1 \geq 2$ and start constructing Ulrich bundles on the Segre variety with only two factors (see Theorem 7.1.10). Then, in Theorem 7.1.11 we will extend this result to Segre varieties with an arbitrary number of factors (always with $n_1 \geq 2$).

On the last part of this section we deal with the case $n_1 = 1$. The results will be slightly different depending on whether $n_2 \geq 3$ or $s \geq 3$. This is the content of Theorem 7.1.14. Notice that the remaining case $\mathbb{P}^1\times\mathbb{P}^1$ is the quadric surface already studied in Chapter 2, Section 2.3.

Let us start then with the former case, that is, with the Segre variety with only two factors which we will denote by $\Sigma_{n,m} \subset \mathbb{P}^{nm+n+m}$ with $n \geq 2$. The first step will be to construct some ancillary vector bundles on $\mathbb{P}^m$, which are far from being aCM but will allow us to construct Ulrich bundles on the Segre variety. To this end, we start fixing some notation.

Let us consider $\mathbf{k}$-vector spaces A and B of respective dimension a and b. Set $V = \mathrm{H}^0(\mathbb{P}^m, \mathcal{O}_{\mathbb{P}^m}(1))$ and let $M = \mathrm{Hom}(B, A \otimes V)$ be the space of $(a \times b)$-matrices of linear forms. It is well-known that there exists a bijection between the elements $\phi \in M$ and the morphisms $\phi : B \otimes \mathcal{O}_{\mathbb{P}^m} \longrightarrow A \otimes \mathcal{O}_{\mathbb{P}^m}(1)$. Taking the tensor with $\mathcal{O}_{\mathbb{P}^m}(1)$ and considering global sections, we have morphisms

$$\mathrm{H}^0(\phi(1)) : \mathrm{H}^0(\mathbb{P}^m, \mathcal{O}_{\mathbb{P}^m}(1)^b) \longrightarrow \mathrm{H}^0(\mathbb{P}^m, \mathcal{O}_{\mathbb{P}^m}(2)^a).$$

The following result tells us under which conditions the aforementioned morphisms ϕ and $\mathrm{H}^0(\phi(1))$ are surjective.

Proposition 7.1.4. *For $a \geq 1$, $b \geq a + m$, and $2b \geq (m+2)a$, the set of elements $\phi \in M$ such that*

$$\phi : B \otimes \mathcal{O}_{\mathbb{P}^m} \longrightarrow A \otimes \mathcal{O}_{\mathbb{P}^m}(1)$$

and

$$H^0(\phi(1)) : H^0(\mathbb{P}^m, \mathcal{O}_{\mathbb{P}^m}(1)^b) \longrightarrow H^0(\mathbb{P}^m, \mathcal{O}_{\mathbb{P}^m}(2)^a)$$

are surjective forms a nonempty open dense subset that we will denote by V_m.

Proof. See [89, Proposition 4.1]. □

Definition 7.1.5. For any $2 \le m$ and any $1 \le a$, we denote by $\mathcal{E}_{m,a}$ a vector bundle on $\mathbb{P}^m$ given by the exact sequence

$$0 \longrightarrow \mathcal{E}_{m,a} \longrightarrow \mathcal{O}_{\mathbb{P}^m}(1)^{(m+2)a} \overset{\phi(1)}{\longrightarrow} \mathcal{O}_{\mathbb{P}^m}(2)^{2a} \longrightarrow 0 \tag{7.1}$$

where $\phi \in V_m$. Note that $\mathcal{E}_{m,a}$ has rank ma.

We need to know the value of the cohomology groups of these vector bundles.

Lemma 7.1.6. *With the above notation we have:*

(1)

$$h^0(\mathbb{P}^m, \mathcal{E}_{m,a}(t)) = \begin{cases} 0 & \text{for } t \le 0, \\ a((m+2)\binom{m+t+1}{m} - 2\binom{m+t+2}{m}) & \text{for } t > 0; \end{cases}$$

(2)

$$h^1(\mathbb{P}^m, \mathcal{E}_{m,a}(t)) = \begin{cases} 0 & \text{for } t < -2 \text{ or } t \ge 0, \\ am & \text{for } t = -1, \\ 2a & \text{for } t = -2; \end{cases}$$

(3) $h^i(\mathbb{P}^m, \mathcal{E}_{m,a}(t)) = 0$ *for all* $t \in \mathbb{Z}$ *and* $2 \le i \le m-1$;
(4) $h^m(\mathbb{P}^m, \mathcal{E}_{m,a}(t)) = 0$ *for* $t \ge -m-1$.

Proof. Since $\phi \in V_m$, by Proposition 7.1.4, $H^0(\phi(1))$ is surjective. But, since the $\mathbf{k}$-vector spaces $H^0(\mathbb{P}^m, \mathcal{O}_{\mathbb{P}^m}(1)^{(m+2)a})$ and $H^0(\mathbb{P}^m, \mathcal{O}_{\mathbb{P}^m}(2)^{2a})$ have the same dimension, $H^0(\phi(1))$ is an isomorphism, and therefore $H^0(\mathcal{E}_{m,a}) = 0$. A fortiori, $H^0(\mathcal{E}_{m,a}(t)) = 0$ for $t \le 0$. On the other hand, again by the surjectivity of $H^0(\phi(1))$, $H^1(\mathcal{E}_{m,a}) = 0$. Since it is obvious that $H^i(\mathcal{E}_{m,a}(1-i)) = 0$ for $i \ge 2$, it turns out that $\mathcal{E}_{m,a}$ is 1-regular and, in particular, $H^1(\mathcal{E}_{m,a}(t)) = 0$ for $t \ge 0$. The rest of cohomology groups can be deduced from the long exact cohomology sequence associated to the exact sequence (7.1). □

The next step will be show the simplicity of the vector bundles $\mathcal{E}_{m,a}$. Indeed, it will turn out to be more convenient to work with the dual vector bundles given by the short exact sequence

$$0 \longrightarrow \mathcal{O}_{\mathbb{P}^m}(-2)^{2a} \longrightarrow \mathcal{O}_{\mathbb{P}^m}(-1)^{(m+2)a} \longrightarrow \mathcal{E}_{m,a}^{\vee} \longrightarrow 0. \tag{7.2}$$

Notice that $M \cong \mathrm{Hom}(\mathcal{O}_{\mathbb{P}^m}(-2)^{2a}, \mathcal{O}_{\mathbb{P}^m}(-1)^{(m+2)a})$. Moreover, since $\mathrm{Aut}(\mathcal{O}_{\mathbb{P}^m}(-1)^{(m+2)a}) \cong \mathrm{GL}((m+2)a)$ and $\mathrm{Aut}(\mathcal{O}_{\mathbb{P}^m}(-2)^{2a}) \cong \mathrm{GL}(2a)$, the group $\mathrm{GL}((m+2)a) \times \mathrm{GL}(2a)$ acts naturally on $\mathrm{Hom}(\mathcal{O}_{\mathbb{P}^m}(-2)^{2a}, \mathcal{O}_{\mathbb{P}^m}(-1)^{(m+2)a})$ by

$$\begin{array}{ccc} \mathrm{GL}((m+2)a) \times \mathrm{GL}(2a) \times M & \longrightarrow & M, \\ (g_1, g_2, \phi) & \mapsto & g_1^{-1}\phi g_2. \end{array}$$

Moreover, for all $\phi \in M$ and $\lambda \in \mathbf{k}^*$, $(\lambda \,\mathrm{Id}_b, \lambda \,\mathrm{Id}_a)$ belongs to the stabilizer of ϕ. Hence $\dim \mathrm{Stab}(\phi) \geq 1$. We have the following result about the dimension of the stabilizer of such representations.

Proposition 7.1.7. *Let $M = \mathbf{k}^{m+1} \otimes \mathbf{k}^{\alpha} \otimes \mathbf{k}^{\beta}$ be endowed with the natural action of $\mathrm{GL}(\alpha) \times \mathrm{GL}(\beta)$. If $\alpha^2 + \beta^2 - (m+1)\alpha\beta \leq 1$ then, for a general element $\phi \in M$,*

$$\dim \mathrm{Stab}(\phi) = 1.$$

Proof. See [122, Theorem 4]. □

Proposition 7.1.8. *Let $\mathcal{E}_{m,a}$ be the vector bundle obtained by the short exact sequence (7.1). Then:*

(1) *The vector bundles $\mathcal{E}_{m,a}$ are simple.*
(2) *The vector bundles $\mathcal{E}_{m,1}$ are μ-stable.*

Proof. (1) We are going to work with the dual exact sequence (7.2) since simplicity is respected under taking duals. First of all, notice that the values of $2a$ and $(m+2)a$ appearing in the statement of this proposition verify the inequality of Proposition 7.1.7.

Let ϕ be the element from M associated to the short exact sequence

$$0 \longrightarrow \mathcal{O}_{\mathbb{P}^m}(-2)^{2a} \xrightarrow{\phi} \mathcal{O}_{\mathbb{P}^m}(-1)^{(m+2)a} \xrightarrow{g} \mathcal{E}_{m,a}^{\vee} \longrightarrow 0$$

whose cokernel is $\mathcal{E}_{m,a}^{\vee}$. We will now check that $\mathcal{E}_{m,a}^{\vee}$ is simple. Otherwise, there exists a nontrivial morphism $\psi : \mathcal{E}_{m,a}^{\vee} \longrightarrow \mathcal{E}_{m,a}^{\vee}$. Then we get a morphism $\overline{\psi} = \psi \circ g : \mathcal{O}_{\mathbb{P}^m}(-1)^{(m+2)a} \longrightarrow \mathcal{E}_{m,a}^{\vee}$. Applying $\mathrm{Hom}(\mathcal{O}_{\mathbb{P}^m}(-1)^{(m+2)a}, -)$ to the exact sequence (7.2) and taking into account that $\mathrm{Hom}(\mathcal{O}_{\mathbb{P}^m}(-1), \mathcal{O}_{\mathbb{P}^m}(-2)) = \mathrm{Ext}^1(\mathcal{O}_{\mathbb{P}^m}(-1), \mathcal{O}_{\mathbb{P}^m}(-2)) = 0$, we get

$$\mathrm{Hom}(\mathcal{O}_{\mathbb{P}^m}(-1)^{(m+2)a}, \mathcal{O}_{\mathbb{P}^m}(-1)^{(m+2)a}) \cong \mathrm{Hom}(\mathcal{O}_{\mathbb{P}^m}(-1)^{(m+2)a}, \mathcal{E}_{m,a}^{\vee}).$$

Hence there is a nontrivial morphism $\widetilde{\psi} \in \mathrm{Hom}(\mathcal{O}_{\mathbb{P}^m}(-1)^{(m+2)a}, \mathcal{O}_{\mathbb{P}^m}(-1)^{(m+2)a})$ induced by $\overline{\psi}$ and represented by a matrix $B \neq \mu \,\mathrm{Id} \in \mathrm{Mat}_{(m+2)a \times (m+2)a}(\mathbf{k})$ such that the following diagram commutes:

$$\begin{array}{ccccccccc} 0 & \longrightarrow & \mathcal{O}_{\mathbb{P}^m}(-2)^{2a} & \xrightarrow{\phi} & \mathcal{O}_{\mathbb{P}^m}(-1)^{(m+2)a} & \longrightarrow & \mathcal{E}_{m,a}^{\vee} & \longrightarrow & 0 \\ & & \downarrow C & & \downarrow B & \searrow^{\overline{\psi}} & \downarrow \psi & & \\ 0 & \longrightarrow & \mathcal{O}_{\mathbb{P}^m}(-2)^{2a} & \xrightarrow{\phi} & \mathcal{O}_{\mathbb{P}^m}(-1)^{(m+2)a} & \longrightarrow & \mathcal{E}_{m,a}^{\vee} & \longrightarrow & 0 \end{array}$$

where $C \in \mathrm{Mat}_{2a\times 2a}(\mathbf{k})$ is the matrix associated to $\widetilde{\psi}_{|\mathcal{O}_{\mathbb{P}^m}(-2)^{2a}}$. Then the pair $(C, B) \neq (\mu\,\mathrm{Id}, \mu\,\mathrm{Id})$ verifies $AC = BA$. Let us consider an element $\alpha \in \mathbf{k}$ that does not belong to the set of eigenvalues of B and C. Then the pair

$$(B - \alpha\,\mathrm{Id}, C - \alpha\,\mathrm{Id}) \in \mathrm{GL}((m+2)a) \times \mathrm{GL}(2a)$$

belongs to $\mathrm{Stab}(\phi)$ and therefore $\dim \mathrm{Stab}(\phi) > 1$ which produces the desired contradiction with Proposition 7.1.7.

(2) Since μ-stability is preserved under taking duals, it is enough to check that $\mathcal{E}_{m,1}^{\vee}$ is μ-stable. Now, since $\mathcal{E}_{m,1}^{\vee}$ is a rank m vector bundle on $\mathbb{P}^m$ sitting in a exact sequence of the form

$$0 \longrightarrow \mathcal{O}_{\mathbb{P}^m}(-2)^2 \longrightarrow \mathcal{O}_{\mathbb{P}^m}(-1)^{m+2} \longrightarrow \mathcal{E}_{m,1}^{\vee} \longrightarrow 0,$$

its μ-stability follows from [28, Theorem 2.7]. □

The main ingredient on the construction of simple Ulrich bundles on $\Sigma_{n,m} \subseteq \mathbb{P}^{nm+n+m}$, $2 \le n \le m$, will be the family of simple vector bundles $\mathcal{E}_{m,a}$ on $\mathbb{P}^m$ given by the exact sequence (7.1), as well as the vector bundles of p-holomorphic forms of $\mathbb{P}^n$, $\Omega^p_{\mathbb{P}^n} := \wedge^p \Omega^1_{\mathbb{P}^n}$, where $\Omega^1_{\mathbb{P}^n}$ is the cotangent bundle. The values of $\mathrm{h}^i(\Omega^p_{\mathbb{P}^n}(t))$ are given by *Bott formula* (see (1.2)).

Theorem 7.1.9. *Fix integers $2 \le n \le m$ and let $(\Sigma_{n,m}, \mathcal{O}_{\Sigma_{n,m}}(1,1))$ be the Segre variety. For any integer $a \ge 1$, there exists a family of dimension $a^2(m^2 + 2m - 4) + 1$ of simple Ulrich bundles of rank $am\binom{n}{2}$.*

Proof. Let us define the vector bundle

$$\mathcal{F} := \Omega^{n-2}_{\mathbb{P}^n}(n-1) \boxtimes \mathcal{E}_{m,a}(n-1)$$

where $\mathcal{E}_{m,a}$ is a general vector bundle obtained on $\mathbb{P}^m$ from the exact sequence (7.1). The first goal is to prove that $\mathcal{F}$ is aCM, namely that $\mathrm{H}^i(\Sigma_{n,m}, \mathcal{F} \otimes \mathcal{O}_{\Sigma_{n,m}}(t,t)) = 0$ for $1 \le i \le n+m-1$ and $t \in \mathbb{Z}$. By Künneth formula,

$$\mathrm{H}^i(\Sigma_{n,m}, \mathcal{F} \otimes \mathcal{O}_{\Sigma_{n,m}}(t,t)) = \bigoplus_{p+q=i} \mathrm{H}^p(\mathbb{P}^n, \Omega^{n-2}_{\mathbb{P}^n}(n-1+t)) \otimes \mathrm{H}^q(\mathbb{P}^m, \mathcal{E}_{m,a}(n-1+t)). \quad (7.3)$$

According to Bott formula (1.2), the only nonzero cohomology groups of $\Omega^{n-2}_{\mathbb{P}^n}(n-1+t)$ are:

$$\begin{array}{ll} \mathrm{H}^0(\mathbb{P}^n, \Omega^{n-2}_{\mathbb{P}^n}(n-1+t)) & \text{for } t \ge 0 \text{ and } n \ge 3 \text{ or } t \ge -1 \text{ and } n = 2, \\ \mathrm{H}^{n-2}(\mathbb{P}^n, \Omega^{n-2}_{\mathbb{P}^n}(n-1+t)) & \text{for } t = -n+1, \\ \mathrm{H}^n(\mathbb{P}^n, \Omega^{n-2}_{\mathbb{P}^n}(n-1+t)) & \text{for } t \le -n-2. \end{array}$$

On the other hand, by Lemma 7.1.6, the only nonzero cohomology groups of $\mathcal{E}_{m,a}(n-1+t)$ are:

$$\begin{aligned} &\mathrm{H}^0(\mathbb{P}^m, \mathcal{E}_{m,a}(n-1+t)) && \text{for } t \geq -n+2,\\ &\mathrm{H}^1(\mathbb{P}^m, \mathcal{E}_{m,a}(n-1+t)) && \text{for } -n-1 \leq t \leq -n,\\ &\mathrm{H}^m(\mathbb{P}^m, \mathcal{E}_{m,a}(n-1+t)) && \text{for } t \leq -n-m-1. \end{aligned}$$

Hence, by (7.3), $\mathrm{H}^i(\Sigma_{n,m}, \mathcal{F} \otimes \mathcal{O}_{\Sigma_{n,m}}(t,t)) = 0$ for $1 \leq i \leq n+m-1$ and $t \in \mathbb{Z}$. Since for $n \geq 3$, $\mathrm{H}^0(\mathbb{P}^n, \Omega^{n-2}_{\mathbb{P}^n}(n-2)) = 0$ and, by Lemma 7.1.6, for $n = 2$ we have that $\mathrm{H}^0(\mathbb{P}^m, \mathcal{E}_{m,a}) = 0$, $\mathcal{F}$ is an initialized aCM bundle on $\Sigma_{n,m}$. Let us compute the number of global sections. Recall that, by Bott formula, $\mathrm{h}^0(\mathbb{P}^n, \Omega^{n-2}_{\mathbb{P}^n}(n-1)) = \binom{n+1}{2}$. Hence,

$$\begin{aligned} \mathrm{h}^0(\mathcal{F}) &= \mathrm{h}^0(\Sigma_{n,m}, \Omega^{n-2}_{\mathbb{P}^n}(n-1) \boxtimes \mathcal{E}_{m,a}(n-1))\\ &= \mathrm{h}^0(\mathbb{P}^n, \Omega^{n-2}_{\mathbb{P}^n}(n-1))\, \mathrm{h}^0(\mathbb{P}^m, \mathcal{E}_{m,a}(n-1))\\ &= \binom{n+1}{2} a\left((m+2)\binom{m+n}{m} - 2\binom{m+n+1}{m}\right)\\ &= a\left(\frac{(m+2)(m+n)!(n+1)!}{m!n!(n-1)!2!} - \frac{2(m+n+1)!(n+1)!}{m!(n+1)!(n-1)!2!}\right)\\ &= a\left(\frac{n!(m+n)!}{2!(n-2)!m!n!} \cdot \frac{(n+1)(m+2)-2(m+n+1)}{n-1}\right)\\ &= a\binom{n}{2}\binom{m+n}{m}\frac{m(n-1)}{n-1}\\ &= a\binom{n}{2}\binom{m+n}{m} m\\ &= \operatorname{rank}(\mathcal{F}) \deg(\Sigma_{n,m}) \end{aligned}$$

where the last equality follows from the fact that $\deg(\Sigma_{n,m}) = \binom{m+n}{m}$ and

$$\operatorname{rank}(\mathcal{F}) = \operatorname{rank}(\mathcal{E}_{m,a}) \operatorname{rank}(\Omega^{n-2}_{\mathbb{P}^n}) = am\binom{n}{2}.$$

Therefore, $\mathcal{F}$ is an Ulrich bundle on $\Sigma_{n,m}$. With respect to simplicity, we need only to observe that $\mathrm{Hom}(\mathcal{F},\mathcal{F}) \cong \mathrm{H}^0(\Sigma_{n,m}, \mathcal{F}^\vee \otimes \mathcal{F})$ and the latter is isomorphic to

$$\mathrm{H}^0(\mathbb{P}^n, \Omega^{n-2}_{\mathbb{P}^n}(n-1)^\vee \otimes \Omega^{n-2}_{\mathbb{P}^n}(n-1)) \otimes \mathrm{H}^0(\mathbb{P}^m, \mathcal{E}_{m,a}(n-1)^\vee \otimes \mathcal{E}_{m,a}(n-1)).$$

Then, using the fact that $\Omega^{n-2}_{\mathbb{P}^n}$ and $\mathcal{E}_{m,a}$ are both simple, by Proposition 7.1.8 (1), we have that $\mathcal{F}$ is also a simple vector bundle.

It only remains to compute the dimension of the family of simple Ulrich bundles $\mathcal{F} := \Omega^{n-2}_{\mathbb{P}^n}(n-1) \boxtimes \mathcal{E}_{m,a}(n-1)$ on $\Sigma_{n,m}$. Since they are completely determined by a general morphism $\phi \in M := \mathrm{Hom}(\mathcal{O}^{(m+2)a}_{\mathbb{P}^m}, \mathcal{O}_{\mathbb{P}^m}(1)^{2a})$, this dimension turns out to be:

$$\begin{aligned} &\dim M - \dim \operatorname{Aut}(\mathcal{O}^{(m+2)a}_{\mathbb{P}^m}) - \dim \operatorname{Aut}(\mathcal{O}_{\mathbb{P}^m}(1)^{2a}) + 1\\ &= 2a^2(m+2)(m+1) - a^2(m+2)^2 - 4a^2 + 1 = a^2(m^2+2m-4) + 1, \end{aligned}$$

which proves what we want. □

Notice that in Theorem 7.1.9 we were able to construct simple Ulrich bundles on $\Sigma_{n,m} \subset \mathbb{P}^N$ for some scattered ranks, namely for ranks of the form $am\binom{n}{2}$, $a \geq 1$. The next goal will be to construct simple Ulrich bundles on $\Sigma_{n,m} \subset \mathbb{P}^{nm+n+m}$, $2 \leq n \leq m$, of the remaining ranks $r \geq m\binom{n}{2}$.

Theorem 7.1.10. *Fix integers $2 \leq n \leq m$ and let $(\Sigma_{n,m}, \mathcal{O}_{\Sigma_{n,m}}(1,1))$ be the Segre variety. For any integer $r \geq m\binom{n}{2}$, set $r = qm\binom{n}{2} + l$ with $q \geq 1$ and $0 \leq l \leq m\binom{n}{2} - 1$. Then, there exists a family of dimension $q^2(m^2+2m-4)+1+l(qm\binom{n+1}{2}-l)$ of simple Ulrich bundles $\mathcal{G}$ on $\Sigma_{n,m}$ of rank r.*

Proof. Note that for any $r \geq m\binom{n}{2}$, there exist an unique $q \geq 1$ and $m\binom{n}{2} - 1 \geq l \geq 0$, such that $r = qm\binom{n}{2}+l$. For such q, consider the family P_q of rank $qm\binom{n}{2}$ Ulrich bundles given by Theorem 7.1.9. Notice that

$$\dim P_q = q^2(m^2+2m-4)+1.$$

Hence it is enough to consider the case $l > 0$. To this end, for any $l > 0$, we construct the family $P_{q,l}$ of vector bundles $\mathcal{G}$ given by a nontrivial extension

$$e : 0 \longrightarrow \mathcal{F} \longrightarrow \mathcal{G} \longrightarrow \mathcal{O}_{\Sigma_{n,m}}(0,n)^l \longrightarrow 0 \tag{7.4}$$

where $\mathcal{F} \in P_q$ and $e := (e_1,\dots,e_l) \in \operatorname{Ext}^1(\mathcal{O}_{\Sigma_{n,m}}(0,n)^l, \mathcal{F}) \cong \operatorname{Ext}^1(\mathcal{O}_{\Sigma_{n,m}}(0,n), \mathcal{F})^l$ with $e_1,\dots,e_l$ linearly independent.

Since

$$\begin{aligned}
\operatorname{ext}^1(\mathcal{O}_{\Sigma_{n,m}}(0,n), \mathcal{F}) &= \mathrm{h}^1(\Sigma_{n,m}, \Omega^{n-2}_{\mathbb{P}^n}(n-1) \boxtimes \mathcal{E}(-1)) \\
&= \mathrm{h}^0(\mathbb{P}^n, \Omega^{n-2}_{\mathbb{P}^n}(n-1))\, \mathrm{h}^1(\mathbb{P}^m, \mathcal{E}(-1)) \\
&= qm\binom{n+1}{2} \\
&> m\binom{n}{2},
\end{aligned} \tag{7.5}$$

such an extension exists.

Being an extension of Ulrich bundles, $\mathcal{G}$ is also an Ulrich bundle by Proposition 3.3.1. Let us see that $\mathcal{G}$ is simple, which means that $\operatorname{Hom}(\mathcal{G},\mathcal{G}) \cong \mathbf{k}$. If we apply the functor $\operatorname{Hom}(-,\mathcal{G})$ to the exact sequence (7.4), we obtain

$$0 \longrightarrow \operatorname{Hom}(\mathcal{O}_{\Sigma_{n,m}}(0,n)^l, \mathcal{G}) \longrightarrow \operatorname{Hom}(\mathcal{G},\mathcal{G}) \longrightarrow \operatorname{Hom}(\mathcal{F},\mathcal{G}).$$

On the other hand, if we apply $\operatorname{Hom}(\mathcal{F},-)$ to the same exact sequence, we have

$$0 \longrightarrow \mathbf{k} \cong \operatorname{Hom}(\mathcal{F},\mathcal{F}) \longrightarrow \operatorname{Hom}(\mathcal{F},\mathcal{G}) \longrightarrow \operatorname{Hom}(\mathcal{F}, \mathcal{O}_{\Sigma_{n,m}}(0,n)^l).$$

But

$$\begin{aligned}\operatorname{Hom}(\mathcal{F},\mathcal{O}_{\Sigma_{n,m}}(0,n)) &\cong \operatorname{Ext}^{n+m}(\mathcal{O}_{\Sigma_{n,m}}(0,n),\mathcal{F}(-n-1,-m-1))\\ &\cong \mathrm{H}^{n+m}(\Sigma_{n,m},\mathcal{F}(-n-1,-m-n-1))\\ &= \mathrm{H}^{n}(\mathbb{P}^n,\Omega^{n-2}_{\mathbb{P}^n}(-2))\otimes \mathrm{H}^{m}(\mathbb{P}^m,\mathcal{E}(-m-2)) = 0,\end{aligned} \tag{7.6}$$

by Serre's duality and Bott formula. This implies that $\operatorname{Hom}(\mathcal{F},\mathcal{G})\cong\mathbf{k}$.

Finally, using the fact that $\operatorname{Hom}(\mathcal{O}_{\Sigma_{n,m}}(0,n),\mathcal{F})\cong \mathrm{H}^0(\mathcal{F}(0,-n)) = 0$ and applying the functor $\operatorname{Hom}(\mathcal{O}_{\Sigma_{n,m}}(0,n),-)$ to the short exact sequence (7.4), we obtain

$$\begin{aligned}0 &\longrightarrow \operatorname{Hom}(\mathcal{O}_{\Sigma_{n,m}}(0,n),\mathcal{G}) \longrightarrow \operatorname{Hom}(\mathcal{O}_{\Sigma_{n,m}}(0,n),\mathcal{O}_{\Sigma_{n,m}}(0,n)^l)\\ &\cong \mathbf{k}^l \xrightarrow{\ \phi\ } \operatorname{Ext}^1(\mathcal{O}_{\Sigma_{n,m}}(0,n),\mathcal{F}) \longrightarrow \operatorname{Ext}^1(\mathcal{O}_{\Sigma_{n,m}}(0,n),\mathcal{G}).\end{aligned}$$

Since, by construction, the image of ϕ is the subvector space generated by $e_1,\dots,e_l$, it turns out that ϕ is injective and, in particular, $\operatorname{Hom}(\mathcal{O}_{\Sigma_{n,m}}(0,n),\mathcal{G}) = 0$. Summing up, $\operatorname{Hom}(\mathcal{G},\mathcal{G})\cong\mathbf{k}$, namely, $\mathcal{G}$ is simple.

It only remains to compute the dimension of $P_{q,l}$. Assume that there exist vector bundles $\mathcal{F},\mathcal{F}'\in P_q$ giving rise to isomorphic vector bundles. In this case we have the following diagram:

$$\begin{array}{ccccccccc} 0 & \longrightarrow & \mathcal{F} & \xrightarrow{j_1} & \mathcal{G} & \xrightarrow{\alpha} & \mathcal{O}_{\Sigma_{n,m}}(0,n)^l & \longrightarrow & 0\\ & & & & i\|\wr & & & & \\ 0 & \longrightarrow & \mathcal{F}' & \xrightarrow{j_2} & \mathcal{G}' & \xrightarrow{\beta} & \mathcal{O}_{\Sigma_{n,m}}(0,n)^l & \longrightarrow & 0.\end{array}$$

Since by (7.6) $\operatorname{Hom}(\mathcal{F},\mathcal{O}_{\Sigma_{n,m}}(0,n)) = 0$, the isomorphism i between $\mathcal{G}$ and $\mathcal{G}'$ lifts to an automorphism f of $\mathcal{O}_{\Sigma_{n,m}}(0,n)^l$ such that $f\alpha=\beta i$, which allows us to conclude that the morphism $ij_1:\mathcal{F}\longrightarrow\mathcal{G}'$ factorizes through $\mathcal{F}'$ and produces an isomorphism from $\mathcal{F}$ to $\mathcal{F}'$.

Therefore, since $\dim\operatorname{Hom}(\mathcal{F},\mathcal{G}) = 1$ and by (7.1.11) $\dim\operatorname{Ext}^1(\mathcal{O}_{\Sigma_{n,m}}(0,n),\mathcal{F}) = qm\binom{n+1}{2}$, we have

$$\begin{aligned}\dim P_{q,l} &= \dim P_q + \dim \mathrm{Gr}^a\left(l, qm\binom{n+1}{2}\right)\\ &= \dim P_q + lqm\binom{n+1}{2} - l^2\\ &= q^2(m^2+2m-4)+1+l\left(qm\binom{n+1}{2}-l\right).\end{aligned}$$

□

As a by-product of the previous results, we can extend the construction of simple Ulrich bundles on $\Sigma_{n,m}$, $n\ge 2$, to the case of Segre embeddings of more than two factors.

Theorem 7.1.11. *Fix integers $2\le n_1\le\cdots\le n_s$ and let $(\Sigma_{n_1,\dots,n_s},\mathcal{O}_{\Sigma_{n_1,\dots,n_s}}(1))$ be a Segre variety. For any integer $r\ge n_2\binom{n_1}{2}$, set $r = qn_2\binom{n_1}{2}+l$ with $q\ge 1$ and $0\le l\le n_2\binom{n_1}{2}-1$.*

Then there exists a family of dimension $q^2(n_2^2 + 2n_2 - 4) + 1 + l(qn_2\binom{n_1+1}{2} - l)$ of simple rank r Ulrich bundles on $(\Sigma_{n_1,\dots,n_s}, \mathcal{O}_{\Sigma_{n_1,\dots,n_s}}(1))$.

Proof. By Theorem 7.1.9, we can suppose that $s \geq 3$. Therefore, by Lemma 3.3.10, the vector bundle of the form $\mathcal{H} := \mathcal{G} \boxtimes \mathcal{L}(n_1 + n_2)$, for $\mathcal{G}$ belonging to the family constructed in Theorem 7.1.10 and $\mathcal{L}$ an Ulrich line bundle on $\mathbb{P}^{n_3} \times \cdots \times \mathbb{P}^{n_s}$ as constructed in Proposition 7.1.3, is a simple Ulrich bundle. In order to show that in this way we obtain a family of the aforementioned dimension, it only remains to show that whenever $\mathcal{G} \not\cong \mathcal{G}'$ then $\mathcal{H} \not\cong \mathcal{H}'$ or, equivalently, $\mathcal{G} \boxtimes \mathcal{O}_{\mathbb{P}^{n_3}\times\cdots\times\mathbb{P}^{n_s}} \not\cong \mathcal{G}' \boxtimes \mathcal{O}_{\mathbb{P}^{n_3}\times\cdots\times\mathbb{P}^{n_s}}$. But if there were an isomorphism

$$\phi : \mathcal{G} \boxtimes \mathcal{O}_{\mathbb{P}^{n_3}\times\cdots\times\mathbb{P}^{n_s}} \xrightarrow{\cong} \mathcal{G}' \boxtimes \mathcal{O}_{\mathbb{P}^{n_3}\times\cdots\times\mathbb{P}^{n_s}}$$

through the projection

$$\pi : \mathbb{P}^{n_1} \times \cdots \times \mathbb{P}^{n_s} \longrightarrow \mathbb{P}^{n_1} \times \mathbb{P}^{n_2},$$

$\pi_*\phi$ would also be an isomorphism between $\pi_*(\mathcal{G} \boxtimes \mathcal{O}_{\mathbb{P}^{n_3}\times\cdots\times\mathbb{P}^{n_s}}) \cong \mathcal{G}$ and $\pi_*(\mathcal{G}' \boxtimes \mathcal{O}_{\mathbb{P}^{n_3}\times\cdots\times\mathbb{P}^{n_s}}) \cong \mathcal{G}'$, in contradiction with the hypothesis. □

Now we move to prove the existence of Ulrich bundles on Segre varieties of the form $\Sigma_{1,n_2,\dots,n_s}$ for $s \geq 3$. In this case, the technique used to construct them will be quite different. On $(\Sigma_{1,n_2\dots,n_s}, \mathcal{O}_{\Sigma_{1,n_2\dots,n_s}}(1))$, Ulrich bundles will not be obtained as products of vector bundles constructed on each factor, but they will be obtained directly as iterated extensions. In the next theorem we will use the following notation:

$$\begin{aligned}
\mathcal{A} &:= \mathcal{O}_{\Sigma_{1,\dots,n_s}}\left(0, 1, 1 + n_2, \dots, 1 + \sum_{i=2}^{s-1} n_i\right),\\
\mathcal{B} &:= \mathcal{O}_{\Sigma_{1,\dots,n_s}}\left(\sum_{i=2}^{s} n_i, 0, n_2, \dots, \sum_{i=2}^{s-1} n_i\right),\\
\mathcal{C} &:= \mathcal{O}_{\Sigma_{1,\dots,n_s}}\left(0, 1 + n_3, 1, 1 + n_2 + n_3, \dots, 1 + \sum_{i=2}^{s-1} n_i\right), \quad \text{and}\\
\mathcal{D} &:= \mathcal{O}_{\Sigma_{1,\dots,n_s}}\left(\sum_{i=2}^{s} n_i, n_3, 0, n_2 + n_3, \dots, \sum_{i=2}^{s-1} n_i\right).
\end{aligned} \tag{7.7}$$

Theorem 7.1.12. *Let $(\Sigma_{1,\dots,n_s}, \mathcal{O}_{\Sigma_{1,\dots,n_s}}(1))$ be a Segre variety with either $s \geq 3$ or $n_2 \geq 2$. Let r be an integer, $2 \leq r \leq (\sum_{i=2}^{s} n_i - 1)\prod_{i=2}^{s}(n_i + 1)$. Then, with the notation (7.7):*

(1) There exists a family Λ_r of rank r simple Ulrich bundles $\mathcal{E}$ on $(\Sigma_{1,\dots,n_s}, \mathcal{O}_{\Sigma_{1,\dots,n_s}}(1))$ given by nontrivial extensions

$$0 \longrightarrow \mathcal{A} \longrightarrow \mathcal{E} \longrightarrow \mathcal{B}^{r-1} \longrightarrow 0 \tag{7.8}$$

with the first Chern class $c_1(\mathcal{E}) = ((r-1)\sum_{i=2}^{s} n_i, 1, 1 + rn_2, \dots, 1 + r(\sum_{i=2}^{s-1} n_i))$.

(2) There exists a family Γ_r of rank r simple Ulrich bundles $\mathcal{F}$ on $(\Sigma_{1,\dots,n_s}, \mathcal{O}_{\Sigma_{1,\dots,n_s}}(1))$ given by nontrivial extensions

$$0 \longrightarrow \mathcal{C} \longrightarrow \mathcal{F} \longrightarrow \mathcal{D}^{r-1} \longrightarrow 0 \tag{7.9}$$

with the first Chern class $c_1(\mathcal{F}) = ((r-1)\sum_{i=2}^{s} n_i, 1 + rn_3, 1, \dots, 1 + r(\sum_{i=2}^{s-1} n_i))$.

Proof. We are going to give the details of the proof of statement (1) since statement (2) is proved analogously. Recall that by Proposition 7.1.3, $\mathcal{A}$ and $\mathcal{B}$ are Ulrich line bundles on $\Sigma_{1,\dots,n_s}$. On the other hand, the dimension of $\mathrm{Ext}^1(\mathcal{B}, \mathcal{A})$ can be computed as:

$$\begin{aligned} \dim \mathrm{Ext}^1(\mathcal{B}, \mathcal{A}) &= \mathrm{h}^1\left(\Sigma_{1,\dots,n_s}, \mathcal{O}_{\Sigma_{1,\dots,n_s}}\left(-\sum_{i=2}^{s} n_i, 1, \dots, 1\right)\right) \\ &= \mathrm{h}^1\left(\mathbb{P}^1, \mathcal{O}_{\mathbb{P}^1}\left(-\sum_{i=2}^{s} n_i\right)\right) \prod_{i=2}^{s} \mathrm{h}^0(\mathbb{P}^{n_i}, \mathcal{O}_{\mathbb{P}^{n_i}}(1)) \\ &= \left(\sum_{i=2}^{s} n_i - 1\right) \prod_{i=2}^{s} (n_i + 1). \end{aligned}$$

So, exactly as in the proof of Theorem 7.1.10, if we take l linearly independent elements $e_1, \dots, e_l$ in $\mathrm{Ext}^1(\mathcal{B}, \mathcal{A})$, $1 \le l \le (\sum_{i=2}^{s} n_i - 1)\prod_{i=2}^{s}(n_i + 1) - 1$, these elements provide with an element $e := (e_1, \dots, e_l)$ of $\mathrm{Ext}^1(\mathcal{B}^l, \mathcal{A}) \cong \mathrm{Ext}^1(\mathcal{B}, \mathcal{A})^l$. Then the associated extension

$$0 \longrightarrow \mathcal{A} \longrightarrow \mathcal{E} \longrightarrow \mathcal{B}^l \longrightarrow 0 \tag{7.10}$$

defines a rank $l + 1$ simple Ulrich bundle. □

Lemma 7.1.13. *Consider the Segre variety $(\Sigma_{1,\dots,n_s}, \mathcal{O}_{\Sigma_{1,\dots,n_s}}(1))$ with either $s \ge 3$ or $n_2 \ge 2$ and keep the notation introduced in Theorem 7.1.12. We have:*

(1) For any two nonisomorphic rank 2 Ulrich bundles $\mathcal{E}$ and $\mathcal{E}'$ from the family Λ_2 obtained from the exact sequence (7.8), it holds that $\mathrm{Hom}(\mathcal{E}, \mathcal{E}') = 0$. Moreover, the set of nonisomorphic classes of elements of Λ_2 is parameterized by

$$\mathbb{P}(\mathrm{Ext}^1(\mathcal{B}, \mathcal{A})) \cong \mathbb{P}\left(\mathrm{H}^1\left(\Sigma_{1,n_2\dots,n_s}, \mathcal{O}_{\Sigma_{1,n_2\dots,n_s}}\left(-\sum_{i=2}^{s} n_i, 1, \dots, 1\right)\right)\right)$$

and, in particular, it has dimension $(\sum_{i=2}^{s} n_i - 1)\prod_{i=2}^{s}(n_i + 1) - 1$.

(2) For any pair of bundles $\mathcal{E} \in \Lambda_2$ and $\mathcal{F} \in \Gamma_3$ obtained from the exact sequences (7.8) and (7.9), it holds that $\mathrm{Hom}(\mathcal{E}, \mathcal{F}) = 0$ and $\mathrm{Hom}(\mathcal{F}, \mathcal{E}) = 0$.

Proof. (1) It is a direct instance of Proposition 3.3.2 (1).

(2) It is a straightforward computation applying the functor $\mathrm{Hom}(\mathcal{F}, -)$ (respectively, $\mathrm{Hom}(-, \mathcal{E})$) to the short exact sequence (7.8) (respectively, the short exact sequence (7.9)) and taking into account that there are no nontrivial morphisms among the line bundles $\mathcal{A}$, $\mathcal{B}$, $\mathcal{C}$, and $\mathcal{D}$. □

In the next theorem we are going to construct families of increasing dimension of simple Ulrich bundles for arbitrary large rank on the Segre variety $\Sigma_{1,n_2,\dots,n_s}$. In the case $s \geq 3$, we can use the two distinct families of rank 2 and rank 3 Ulrich bundles obtained in Theorem 7.1.12 to cover all the possible ranks. However, when $s = 2$, since there exists just a unique family, we will have to restraint ourselves to constructing Ulrich bundles of arbitrary even rank.

Theorem 7.1.14. *Consider the Segre variety $(\Sigma_{1,\dots,n_s}, \mathcal{O}_{\Sigma_{1,\dots,n_s}}(1))$ with either $s \geq 3$ or $n_2 \geq 2$.*

(1) Then, for any $r = 2t$, $t \geq 2$, there exists a family of dimension

$$(2t-1)\left(\sum_{i=2}^{s} n_i - 1\right)\prod_{i=2}^{s}(n_i+1) - 3(t-1)$$

of simple Ulrich bundles of rank r.

(2) Let us assume that $s \geq 3$ and $n_2 = 1$. Then, for any $r = 2t+1$, $t \geq 2$, there exists a family of dimension greater or equal than $(t-1)((\sum_{i=2}^{s} n_i - 1)(n_3+2)\prod_{i=4}^{s}(n_i+1) - 1)$ of simple Ulrich bundles of rank r.

(3) Let us assume that $s \geq 3$ and $n_2 > 1$. For any integer $r = an_3\binom{n_2}{2} + l \geq n_3\binom{n_2}{2}$ with $a \geq 1$ and $0 \leq l \leq n_3\binom{n_2}{2} - 1$, there exists a family of dimension $a^2(n_3^2 + 2n_3 - 4) + 1 + l(an_3\binom{n_2+1}{2} - l)$ of simple rank r Ulrich bundles.

Proof. (1) Let $r = 2t$ be an even integer and set $a := \operatorname{ext}^1(\mathcal{B}, \mathcal{A}) = (\sum_{i=2}^{s} n_i - 1)\prod_{i=2}^{s}(n_i+1)$ with $\mathcal{A}$ and $\mathcal{B}$ defined in (7.7). Denote by U the open subset of $\mathbb{P}^a \times \overset{t)}{\cdots} \times \mathbb{P}^a$, where $\mathbb{P}^a \cong \mathbb{P}(\operatorname{Ext}^1(\mathcal{B}, \mathcal{A})) \cong \Lambda_2$, parameterizing closed points $[\mathcal{E}_1, \dots, \mathcal{E}_t] \in \mathbb{P}^a \times \overset{t)}{\cdots} \times \mathbb{P}^a$ such that $\mathcal{E}_i \ncong \mathcal{E}_j$ for $i \neq j$ (i. e., U is $\mathbb{P}^a \times \overset{t)}{\cdots} \times \mathbb{P}^a$ minus the small diagonals). Given $[\mathcal{E}_1, \dots, \mathcal{E}_t] \in U$, by Lemma 7.1.13, the set of vector bundles $\mathcal{E}_1, \dots, \mathcal{E}_t$ satisfy the hypothesis of Proposition 3.3.2 (1), and therefore there exists a family of rank r simple Ulrich bundles $\mathcal{E}$ parameterized by

$$\mathbb{P}(\operatorname{Ext}^1(\mathcal{E}_t, \mathcal{E}_1)) \times \cdots \times \mathbb{P}(\operatorname{Ext}^1(\mathcal{E}_t, \mathcal{E}_{t-1}))$$

and given as extensions of the form

$$0 \longrightarrow \bigoplus_{i=1}^{t-1} \mathcal{E}_i \longrightarrow \mathcal{E} \longrightarrow \mathcal{E}_t \longrightarrow 0.$$

Next we observe that if we consider $[\mathcal{E}_1, \dots, \mathcal{E}_t] \neq [\mathcal{E}'_1, \dots, \mathcal{E}'_t] \in U$ and the corresponding extensions

$$0 \longrightarrow \bigoplus_{i=1}^{t-1} \mathcal{E}_i \longrightarrow \mathcal{E} \longrightarrow \mathcal{E}_t \longrightarrow 0$$

and

$$0 \longrightarrow \bigoplus_{i=1}^{t-1} \mathcal{E}'_i \longrightarrow \mathcal{E}' \longrightarrow \mathcal{E}'_t \longrightarrow 0$$

then $\operatorname{Hom}(\mathcal{E},\mathcal{E}') = 0$ and, in particular, $\mathcal{E} \not\cong \mathcal{E}'$. Therefore, we have a family of nonisomorphic rank r simple Ulrich bundles $\mathcal{E}$ on $\Sigma_{1,n_2,\dots,n_s}$ parameterized by a projective bundle $\mathbb{P}$ over U of dimension

$$\dim \mathbb{P} = (t-1)\dim \mathbb{P}(\operatorname{Ext}^1(\mathcal{E}_t,\mathcal{E}_1)) + \dim U.$$

Applying the functor $\operatorname{Hom}(-,\mathcal{E}_1)$ to the short exact sequence (7.8), we obtain

$$0 \longrightarrow \operatorname{Hom}(\mathcal{A},\mathcal{E}_1) \cong \mathbf{k} \longrightarrow \operatorname{Ext}^1(\mathcal{B},\mathcal{E}_1) \longrightarrow \operatorname{Ext}^1(\mathcal{E}_t,\mathcal{E}_1) \longrightarrow \operatorname{Ext}^1(\mathcal{A},\mathcal{E}_1) = 0.$$

On the other hand, applying $\operatorname{Hom}(\mathcal{B},-)$ to the same exact sequence we have

$$\begin{aligned} 0 = \operatorname{Hom}(\mathcal{B},\mathcal{E}_1) \longrightarrow \operatorname{Hom}(\mathcal{B},\mathcal{B}) \cong \mathbf{k} \longrightarrow \operatorname{Ext}^1(\mathcal{B},\mathcal{A}) \cong \mathbf{k}^a \longrightarrow \\ \longrightarrow \operatorname{Ext}^1(\mathcal{B},\mathcal{E}_1) \longrightarrow \operatorname{Ext}^1(\mathcal{B},\mathcal{B}) = 0. \end{aligned}$$

Summing up, we obtain $\operatorname{ext}^1(\mathcal{E}_t,\mathcal{E}_1) = a-2$, and so

$$\dim \mathbb{P} = (t-1)(a-3) + ta = (2t-1)a - 3(t-1).$$

(2) Now, let us suppose that $s \geq 3$ and $n_2 = 1$ and take $r = 2t+1$, $t \geq 2$. Let $\mathcal{E}_1,\dots,\mathcal{E}_{t-1}$ be $t-1$ nonisomorphic rank 2 Ulrich bundles from the exact sequence (7.8) and let $\mathcal{F}$ be a rank 3 Ulrich bundle from the exact sequence (7.9). Again, by Lemma 7.1.13, this set of vector bundles satisfy the hypothesis of Proposition 3.3.2 (1) and therefore, there exists a family $\mathbb{G}$ of rank r simple Ulrich bundles $\mathcal{E}$ parameterized by

$$\mathbb{P}(\operatorname{Ext}^1(\mathcal{E}_1,\mathcal{F})) \times \cdots \times \mathbb{P}(\operatorname{Ext}^1(\mathcal{E}_{t-1},\mathcal{F}))$$

and given as extensions of the form

$$0 \longrightarrow \mathcal{F} \longrightarrow \mathcal{E} \longrightarrow \bigoplus_{i=1}^{t-1} \mathcal{E}_i \longrightarrow 0.$$

It only remains to compute the dimension of the family

$$\dim \mathbb{G} = (t-1)\dim \mathbb{P}(\operatorname{Ext}^1(\mathcal{E}_1,\mathcal{F})).$$

Let us fix the notation

$$\begin{aligned} b := \operatorname{ext}^1(\mathcal{B},\mathcal{C}) &= \mathrm{h}^1\left(\mathbb{P}^1,\mathcal{O}_{\mathbb{P}^1}\left(-\sum_{i=2}^{s} n_i\right)\right)\mathrm{h}^0(\mathbb{P}^1,\mathcal{O}_{\mathbb{P}^1}(1+n_3))\prod_{i=4}^{s}\mathrm{h}^0(\mathbb{P}^{n_i},\mathcal{O}_{\mathbb{P}^{n_i}}(1)) \\ &= \left(\sum_{i=2}^{s} n_i - 1\right)(n_3+2)\prod_{i=4}^{s}(n_i+1). \end{aligned}$$

Applying the functor $\operatorname{Hom}(-,\mathcal{F})$ to the short exact sequence (7.8), we obtain

$$0 = \operatorname{Hom}(\mathcal{A},\mathcal{F}) \longrightarrow \operatorname{Ext}^1(\mathcal{B},\mathcal{F}) \longrightarrow \operatorname{Ext}^1(\mathcal{E}_1,\mathcal{F}) \longrightarrow \operatorname{Ext}^1(\mathcal{A},\mathcal{F}).$$

On the other hand, applying $\mathrm{Hom}(\mathcal{B}, -)$ to the short exact sequence (7.9), we have

$$0 = \mathrm{Hom}(\mathcal{B}, \mathcal{D}) \longrightarrow \mathrm{Ext}^1(\mathcal{B}, \mathcal{C}) \cong \mathbf{k}^b \longrightarrow \mathrm{Ext}^1(\mathcal{B}, \mathcal{F}) \longrightarrow \mathrm{Ext}^1(\mathcal{B}, \mathcal{D}) = 0.$$

Summing up, we obtain $\mathrm{ext}^1(\mathcal{E}_1, \mathcal{F}) \geq b$, and therefore $\dim \mathbb{G} \geq (t-1)(b-1)$.

(3) It follows from Theorems 7.1.10 and 3.3.10. □

Corollary 7.1.15. *All Segre varieties $(\Sigma_{n_1,\dots,n_s}, \mathcal{O}_{\Sigma_{n_1,\dots,n_s}}(1))$ are of wild representation type except for the quadric $\mathbb{P}^1 \times \mathbb{P}^1 \subset \mathbb{P}^3$ which is of finite representation type.*

Proof. It follows from Theorems 7.1.9, 7.1.10, 7.1.11, and 7.1.14. □

Remark 7.1.16. (1) The Segre varieties $(\Sigma_{n_1,\dots,n_s}, \mathcal{O}_{\Sigma_{n_1,\dots,n_s}}(1))$ described in Corollary 7.1.15 were the first family of examples of varieties of arbitrary dimension for which wild representation type was witnessed by means of Ulrich vector bundles.

(2) Notice that the previous Proposition 7.1.3 and Theorems 7.1.11 and 7.1.14 give a full answer to Question 3.1.2 for the case of Segre varieties. On the other hand, regarding the existence of Ulrich bundles of lower rank on the Segre varieties $\Sigma_{n,m} \subseteq \mathbb{P}^N$, $N := nm + n + m$, for $2 \leq n, m$, notice that we have shown their existence in the cases of rank one and rank higher than $m\binom{n}{2}$. So, it remains to prove their existence for the intermediate ranks.

Problem 7.1.17. *Construct indecomposable Ulrich bundles of rank r, for $1 < r < m\binom{n}{2}$, on the Segre variety $(\Sigma_{n,m}, \mathcal{O}_{\Sigma_{n,m}}(1,1))$, for $2 \leq n, m$.*

For more details on these issues, the reader can consult [72].

Remark 7.1.18. The Segre variety $\mathbb{P}^1 \times \mathbb{P}^d$ is a particular case of a *rational normal scroll.* These varieties are defined as the projective bundles

$$X := X(a_0, \dots, a_d) = \mathbb{P}(\mathcal{O}_{\mathbb{P}^1}(a_0) \oplus \cdots \oplus \mathcal{O}_{\mathbb{P}^1}(a_d)) \xrightarrow{\pi} \mathbb{P}^1$$

for $0 < a_0 \leq \cdots \leq a_d$. The Picard group of X is freely generated by the tautological line bundle $\mathcal{O}_X(H)$ and by the line bundle $\mathcal{O}_X(F)$ associated to a fiber F of the map π. Also $\mathcal{O}_X(H)$ gives an embedding $i : X \hookrightarrow \mathbb{P}^{\sum_{i=0}^d a_i + d}$ of degree $\sum_{i=0}^d a_i$. Therefore $i(X) \subset \mathbb{P}^{\sum_{i=0}^d a_i + d}$ is a *variety of minimal degree.* The construction of Ulrich bundles of arbitrary large rank on rational normal scrolls was carried out in [156]. In particular, it was proved that all rational normal scrolls, except for $X(1,1)$ (namely, the smooth quadric $\mathbb{P}^1 \times \mathbb{P}^1$) and $X(1,2) \subset \mathbb{P}^4$, support families of arbitrarily large dimension of indecomposable Ulrich bundles (i. e., all rational normal scrolls $X(m,n)$ with $(m,n) \neq (1,1), (1,2)$ are of wild representation type). On the other hand, it was shown in [10] that any Ulrich bundle on $X(a_0, \dots, a_d)$ has a filtration with quotients being a direct sum of vector bundles of the form $\wedge^i \Omega_{X(a_0,\dots,a_d)|\mathbb{P}^1}((i+1)H - F)$.

7.2 Ulrich bundles on Grassmann varieties

In Section 2.4, we characterized all irreducible homogeneous aCM bundles on a Grassmann variety $\mathrm{Gr}(k,n)$ in terms of the so-called step matrices. In this section, we will determine which of them are Ulrich. More precisely, we will give a full classification of all irreducible homogeneous Ulrich bundles on a Grassmann variety $\mathrm{Gr}(k,n)$ of k-planes on $\mathbb{P}^n$. In particular, we have that any Grassmann variety $\mathrm{Gr}(k,n) \hookrightarrow \mathbb{P}(\bigwedge^{k+1} V)$ embedded by the Plücker embedding supports an Ulrich bundle. In that way, for the case of Grassmann varieties, we have an affirmative answer to Question 3.1.2 (see also Conjecture 3.4.8). As a main tool we will use Schur functors and the Borel–Bott–Weil theorem (see Theorem 2.4.4) which computes the cohomology of any irreducible homogeneous bundle $\Sigma^{\alpha}\mathcal{Q}\otimes\Sigma^{\gamma}\mathcal{S}^{\vee}$ on $\mathrm{Gr}(k,n)$.

We keep the notation introduced in Section 2.4. We denote by $(\mathrm{Gr}(k,n), \mathcal{O}_G(1))$ the Grassmann variety embedded by the Plücker embedding, where for short we set $\mathcal{O}_G(1) = \mathcal{O}_{\mathrm{Gr}(k,n)}(1)$. On the other hand, since we are dealing with initialized irreducible homogeneous bundles $\mathcal{E} = \Sigma^{(b_1,\dots,b_{k+1})}\mathcal{Q}\otimes\Sigma^{(a_1,\dots,a_s)}\mathcal{S}^{\vee}$ on the Grassmann variety $\mathrm{Gr}(k,n)$, by Remark 2.4.5 they have natural cohomology (see Definition 2.1.16) and thus we have $\mathrm{H}^0(\mathrm{Gr}(k,n),\mathcal{E}) \neq 0$ and $\mathrm{H}^i(\mathrm{Gr}(k,n),\mathcal{E}) = 0$ for any $i > 0$. Therefore, applying Theorem 2.4.4, we have

$$b_1 \geq \cdots \geq b_{k+1} = a_1 \geq \cdots \geq a_s > 0.$$

As we know, the slope of any Ulrich bundle is fixed (Proposition 3.2.5). So, let us start computing the slope of any Ulrich bundle on a $\mathrm{Gr}(k,n)$. To compute it we will need the following result.

Lemma 7.2.1. *Let $\beta = (\beta_1,\dots,\beta_{k+1}) \in \mathbb{Z}^{k+1}$ and $\gamma = (\gamma_1,\dots,\gamma_{n-k}) \in \mathbb{Z}^{n-k}$ be two nonincreasing sequences. Then,*

$$\mu(\Sigma^{\beta}\mathcal{Q}\otimes\Sigma^{\gamma}\mathcal{S}^{\vee}) = \left(\frac{|\beta_1+\cdots+\beta_{k+1}|}{k+1} - \frac{|\gamma_1+\cdots+\gamma_{n-k}|}{n-k}\right)d$$

with $d := \deg(\mathrm{Gr}(k,n)) = \frac{((k+1)(n-k))!k!(k-1)!\cdots 2!}{n!(n-1)!\cdots(n-k)!}$.

Proof. By [185, p. 4], for any rank r vector bundle $\mathcal{E}$ on $\mathrm{Gr}(k,n)$ and any nonincreasing sequence $\lambda \in \mathbb{Z}^r$, we have

$$\mu(\Sigma^{\lambda}\mathcal{E}) = |\lambda_1+\cdots+\lambda_r|\mu(\mathcal{E}).$$

By the multiplicative character of the Chern classes (see (1.5.4)),

$$\begin{aligned}\mu(\Sigma^{\beta}\mathcal{Q}\otimes\Sigma^{\gamma}\mathcal{S}^{\vee}) &= \frac{\deg(c_1(\Sigma^{\beta}\mathcal{Q}\otimes\Sigma^{\gamma}\mathcal{S}^{\vee}))}{\mathrm{rank}(\Sigma^{\beta}\mathcal{Q}\otimes\Sigma^{\gamma}\mathcal{S}^{\vee})}\\ &= \frac{\deg(c_1(\Sigma^{\beta}\mathcal{Q})\,\mathrm{rank}(\Sigma^{\gamma}\mathcal{S}^{\vee}) + c_1(\Sigma^{\gamma}\mathcal{S}^{\vee})\,\mathrm{rank}(\Sigma^{\beta}\mathcal{Q}))}{\mathrm{rank}(\Sigma^{\beta}\mathcal{Q})\,\mathrm{rank}(\Sigma^{\gamma}\mathcal{S}^{\vee})}\end{aligned}$$

$$= \mu(\Sigma^\beta \mathcal{Q}) + \mu(\Sigma^\gamma \mathcal{S}^\vee).$$

Now, the result follows from the fact that $\mu(\mathcal{Q}) = \frac{d}{k+1}$ and $\mu(\mathcal{S}^\vee) = -\frac{d}{n-k}$. □

Let us start giving a necessary condition for an irreducible homogeneous bundle $\mathcal{E} = \Sigma^{(b_1,\dots,b_{k+1})}\mathcal{Q} \otimes \Sigma^{(a_1,\dots,a_s)}\mathcal{S}^\vee$ on $\mathrm{Gr}(k,n)$ to be an Ulrich bundle.

Proposition 7.2.2. *Let $\mathcal{E} = \Sigma^{(b_1,\dots,b_{k+1})}\mathcal{Q} \otimes \Sigma^{(a_1,\dots,a_s)}\mathcal{S}^\vee$ be an irreducible initialized homogeneous bundle on the Grassmann variety $(\mathrm{Gr}(k,n), \mathcal{O}_G(1))$. If $\mathcal{E}$ is an Ulrich bundle then*

$$b_1 = k(n-k-1).$$

Proof. Recall that for any integer $t \in \mathbb{Z}$,

$$\mathcal{E}(t) = \Sigma^{(b_1+t,\dots,b_{k+1}+t)}\mathcal{Q} \otimes \Sigma^{(a_1,\dots,a_s)}\mathcal{S}^\vee$$

and, by Theorem 3.2.9 (2), we have

$$\mathrm{H}^d(\mathrm{Gr}(k,n), \mathcal{E}(-(k+1)(n-k))) = 0 \text{ and } \mathrm{H}^d(\mathrm{Gr}(k,n), \mathcal{E}(-(k+1)(n-k)-1)) \neq 0$$

where $d = \dim \mathrm{Gr}(k,n) = (k+1)(n-k)$. Hence, it follows from Theorem 2.4.4 that for an Ulrich bundle $\mathcal{E}$ on $\mathrm{Gr}(k,n)$, the first vanishing forces $b_1 \geq k(n-k-1)$ while the fact that $\mathrm{H}^d(\mathrm{Gr}(k,n), \mathcal{E}(-(k+1)(n-k)-1)) \neq 0$ implies $b_1 \leq k(n-k-1)$. Putting together both inequalities, we get $b_1 = k(n-k-1)$. □

In the next example we will illustrate the idea behind the main result.

Example 7.2.3. We consider the Grassmann variety $\mathrm{Gr}(7,19)$, which means that $k+1 = 8$ and $n-k = 12$, together with two ordered sequences of integers of the same length

$$(k_1, k_2) = (2,4) \quad \text{and} \quad (n_1, n_2) = (3,4)$$

such that $k+1 = k_1 \cdot k_2$ and $n-k = n_1 \cdot n_2$. We are going to associate univocally to them an irreducible homogenous Ulrich bundle $\mathcal{E} = \Sigma^{(b_1,\dots,b_8)}\mathcal{Q} \otimes \Sigma^{(a_1,\dots,a_{12})}\mathcal{S}^\vee$ on $\mathrm{Gr}(7,19)$. Given the pair $(n_1, k_1) = (3,2)$, we consider the 3×2 block B_1 with 3 columns and 2 rows filled by integers in the following way:

4	5	6
1	2	3

Block of type B_1

Now, we consider the next pair $(n_2, k_2) = (4,4)$ and build the block B_2 with 4 rows of 4 blocks of type B_1 on each row. Place first the 4 blocks of the first row, then the

4 blocks of the second and third rows, and, finally, the 4 blocks of the last row, each block filled by integers as shown in the picture:

76	77	78	82	83	84	88	89	90	94	95	96
73	74	75	79	80	81	85	86	87	91	92	93
52	53	54	58	59	60	64	65	66	70	71	72
49	50	51	55	56	56	61	62	63	67	68	69
28	29	30	34	35	36	40	41	42	46	47	48
25	26	27	31	32	33	37	38	39	43	44	45
4	5	6	10	11	12	16	17	18	22	23	24
1	2	3	7	8	9	13	14	15	19	20	21

Block of type B_2

We define two nonincreasing sequences of integers

$$\begin{aligned}(a_1, a_2, a_3, a_5, a_5, a_6, a_7, a_8, a_9, a_{10}, a_{11}, a_{12}) &:= (21, 20, 19, 15, 14, 13, 9, 8, 7, 3, 2, 1)\\ &\quad - (12, 11, 10, 9, 8, 7, 6, 5, 4, 3, 2, 1)\\ &= (9, 9, 9, 6, 6, 6, 3, 3, 3, 0, 0, 0),\\ (b_1, b_2, b_3, b_4, b_5, b_6, b_7, b_8) &:= (76, 73, 52, 49, 28, 25, 4, 1)\\ &\quad - (20, 19, 18, 17, 16, 15, 14, 13)\\ &\quad + 21(1, 1, 1, 1, 1, 1, 1, 1)\\ &= (77, 75, 55, 53, 33, 31, 11, 9).\end{aligned}$$

Let us see that $\mathcal{E} = \Sigma^{(b_1,\dots,b_8)}\mathcal{Q} \otimes \Sigma^{(a_1,\dots,a_{12})}\mathcal{S}^\vee$ is an Ulrich bundle on the 96-dimensional Grassmann variety Gr(7, 19). First of all, notice that $\mathcal{E}$ is an initialized vector bundle. Indeed, since $b_8 = a_1$ and $(b_1, b_2, \dots, b_8, a_1, \dots, a_{12})$ is a nonincreasing sequence by Theorem 2.4.4, $H^0(\mathcal{E}(-1)) = 0$ and $H^0(\mathcal{E}) \neq 0$. According to the characterization of Ulrich bundles given in Theorem 3.2.9, we only need to check that for any integer $t \in [1, 96]$,

$$H^i(\mathcal{E}(-t)) = 0, \quad 0 \le i \le 96.$$

To this end, we observe that any integer $t \in [1, 96]$ corresponds to a unique point of position $(i, 9 - j)$ for some $1 \le i \le 12$, $1 \le j \le 8$ inside the block B_2. Moreover, if t is in the position $(i, 9 - j)$ then

$$b_j + 19 - j + 2 - t = a_i + 12 - (i - 1).$$

This means that for any integer $t \in [1, 96]$ at least two entries of

$$\begin{gathered}(b_1 - t, b_2 - t, b_3 - t, b_4 - t, b_5 - t, b_6 - t, b_7 - t, b_8 - t, a_1, a_2, a_3, a_5, a_5, a_6, a_7, a_8, a_9,\\ a_{10}, a_{11}, a_{12})\\ + (20, 19, 18, 17, 16, 15, 14, 13, 12, 11, 10, 9, 8, 7, 6, 5, 4, 3, 2, 1)\end{gathered}$$

coincide. Therefore, by Theorem 2.4.4, this means that for any integer $t \in [1, 96]$,

$$\mathrm{H}^i(\mathcal{E}(-t)) = 0, \quad 0 \le i \le 96.$$

Thus, $\mathcal{E}$ is an Ulrich bundle on $\mathrm{Gr}(7, 19)$.

Let us fix the definition that we will use in the statement of the main result concerning Ulrich bundles on Grassmann varieties.

Definition 7.2.4. For any $k, n \in \mathbb{N}$ and for any pair of ordered sequences of integers of the same length $(k_1, \dots, k_s)$ and $(n_1, \dots, n_s)$ such that

$$k + 1 = k_1 \cdot k_2 \cdots k_s,$$
$$n - k = n_1 \cdot n_2 \cdots n_s,$$

with $n_l \cdot k_l \neq 1$ for $1 \le l \le s$, we define

$$\alpha = n_1 + \sum_{i=1}^{s-1} (n_{s-i+1} - 1) k_1 \cdots k_{s-i} \cdot n_1 \cdots n_{s-i}.$$

For any integer j, $1 \le j \le k + 1$, consider the unique expression

$$j = j_1 \cdot k_1 \cdots k_{s-1} + j_2 \cdot k_1 \cdots k_{s-2} + \cdots + j_{s-2} k_1 k_2 + j_{s-1} k_1 + \rho_j$$

with $1 \le \rho_j \le k_1$ and $0 \le j_l < k_{s-l+1}$ for $1 \le l \le s - 1$ and define

$$b_{k+2-j} = \alpha + (\rho_j - 1) n_1 + 1 + j_{s-1} k_1 n_1 n_2 + j_{s-2} k_1 k_2 n_1 n_2 n_3 + \cdots + j_1 k_1 k_2 \cdots k_{s-1} n_1 \cdots n_s.$$

For any integer i, $1 \le i \le n - k$, if

$$i = i_1 \cdot n_1 \cdots n_{s-1} + i_2 \cdot n_2 \cdot n_1 \cdots n_{s-2} + \cdots + i_{s-2} n_1 n_2 + i_{s-1} n_1 + \rho_i,$$

with $1 \le \rho_i \le n_1$ and $0 \le i_l \le n_{s-l+1}$ for $1 \le l \le s - 1$, we define

$$a_{n-k+1-j} = \rho_i + 1 - i + i_{s-1} k_1 n_1 + i_{s-2} k_1 k_2 n_1 n_2 + \cdots + i_1 k_1 k_2 \cdots k_{s-1} n_1 \cdots n_{s-1}.$$

Finally, we define

$$\mathcal{E}^{(k_1,\dots,k_s)}_{(n_1,\dots,n_s)} := \Sigma^{(b_1,\dots,b_{k+1})} \mathcal{Q} \otimes \Sigma^{(a_1,\dots,a_{n-k})} \mathcal{S}^\vee$$

the corresponding irreducible homogenous bundle on $\mathrm{Gr}(k, n)$ with the a_i, $1 \le i \le n-k$, and b_j, $1 \le j \le k + 1$, given as above.

Theorem 7.2.5. *Let $(\mathrm{Gr}(k, n), \mathcal{O}_G(1))$ be the Grassmann variety. Then, for any pair of ordered sequences of integers of the same length $(k_1, \dots, k_s)$ and $(n_1, \dots, n_s)$ such that*

$$k + 1 = k_1 \cdot k_2 \cdots k_s, \quad n - k = n_1 \cdot n_2 \cdots n_s,$$

with $n_l \cdot k_l \neq 1$ for $1 \leq l \leq s$, the irreducible homogeneous bundle

$$\mathcal{E} = \mathcal{E}^{(k_1,\ldots,k_s)}_{(n_1,\ldots,n_s)}$$

is an Ulrich bundle on $\mathrm{Gr}(k,n)$. Conversely, any irreducible homogeneous Ulrich bundle on $\mathrm{Gr}(k,n)$ is of this form for some ordered sequences of integers of the same length $(k_1,\ldots,k_s)$ and $(n_1,\ldots,n_s)$ such that

$$k+1 = k_1 \cdot k_2 \cdots k_s,$$
$$n-k = n_1 \cdot n_2 \cdots n_s,$$

with $n_l \cdot k_l \neq 1$ for $1 \leq l \leq s$.

Proof. First of all, we observe that $\mathcal{E}$ is an initialized vector bundle. Indeed, since $b_{k+1} = a_1$, by Theorem 2.4.4, we have $\mathrm{H}^0(\mathcal{E}(-1)) = 0$ and, since

$$(b_1+n+1,\ldots,b_{k+1}+n-k+1, a_1+n-k,\ldots,a_{n-k}+1)$$

is a strictly decreasing sequence, we have $\mathrm{H}^0(\mathcal{E}) \neq 0$. Therefore, by the characterization of Ulrich bundles given in Theorem 3.2.9, $\mathcal{E}$ is an Ulrich bundle if and only if for any integer $t \in [1,(k+1)(n-k)]$,

$$\mathrm{H}^i(\mathcal{E}(-t)) = 0, \quad 0 \leq i \leq (k+1)(n-k).$$

In order to prove all these vanishings, we will proceed as in Example 7.2.3. Given the pair of integers (n_1,k_1), we consider the $n_1 \times k_1$ block B_1 with n_1 columns and k_1 rows filled by integers in the following way:

$(k_1-1)n_1+1$	$(k_1-1)n_1+2$	...	...	...	k_1n_1
$\vdots$	$\vdots$				$\vdots$
$\vdots$	$\vdots$				$\vdots$
$2n_1+1$	$2n_1+2$	...	...	...	$3n_1$
n_1+1	n_1+2	...	...	...	$2n_1$
1	2	...	...	...	n_1

Block of type B_1

For any integer i, $2 \leq i \leq s$, we consider the pair (n_i,k_i) and we build the block B_i with k_i rows of n_i blocks of type B_{i-1} on each row. We first place the n_i blocks of the first row, then the n_i blocks of the second row, and so on, until the k_ith row.

Now we observe that any integer $t \in [1,(k+1)(n-k)]$ corresponds to a unique point of position $(i,k+2-j)$ for some $1 \leq i \leq n-k$, $1 \leq j \leq k+1$, inside the block B_s. Moreover, if t is in the position $(i,k+2-j)$ then

$$b_j + n - j + 2 - t = a_i + (n-k) - (i-1).$$

That means that for any integer $t \in [1,(k+1)(n-k)]$ at least two entries of

$$(b_1 - t, b_2 - t, \dots, b_{k+1} - t, a_1, a_2, \dots, a_{n-k}) + (n+1, n, \dots, 2, 1)$$

coincide. Hence, by Theorem 2.4.4, for any integer $t \in [1,(k+1)(n-k)]$,

$$\mathrm{H}^i(\mathcal{E}(-t)) = 0, \quad 0 \le i \le (k+1)(n-k),$$

and we conclude that $\mathcal{E}$ is an Ulrich bundle on $\mathrm{Gr}(k,n)$.

Let us prove the converse. Let $\mathcal{E} = \Sigma^{(b_1,\dots,b_{k+1})}\mathcal{Q} \otimes \Sigma^{(a_1,\dots,a_{n-k})}\mathcal{S}^\vee$ be an initialized irreducible homogeneous Ulrich bundle on $\mathrm{Gr}(k,n)$. For any integer $t \in [1,(k+1)(n-k)]$, we have

$$\mathrm{H}^i(\mathcal{E}(-t)) = 0, \quad 0 \le i \le (k+1)(n-k).$$

Thus, by Theorem 2.4.4, for any integer $t \in [1,(k+1)(n-k)]$, there exists a pair (i,j) such that

$$b_j + n - j + 2 - t = a_i + (n-k) - (i-1). \tag{7.11}$$

Consider the $(n-k) \times (k+1)$ block B with $(n-k)$ columns and $(k+1)$ rows, put at the position $(i, k+2-j)$ the value of t such that (7.11) holds and denote by $t_{i,j}$ that value. That is,

$$b_j + n - j + 2 - t_{i,j} = a_i + (n-k) - (i-1). \tag{7.12}$$

Notice that for any integer i, $1 \le i \le n-k$, and any integer j, $1 \le j \le k+1$, we have

$$t_{i,j+1} - t_{i,j} = b_{j+1} - b_j - 1 \quad \text{and} \quad t_{i+1,j} - t_{i,j} = a_i - a_{i+1} + 1, \tag{7.13}$$

and that there is a bijection between the integers $t \in [1,(k+1)(n-k)]$ and the values inside the block B.

It follows from the equality (7.12) that $t_{1,k+1} = 1$. Denote by n_1, $1 \le n_1 \le n-k$, the integer such that $t_{i,k+1} = i$ for $1 \le i \le n_1$ and $t_{n_1+1,k+1} > n_1 + 1$. If $n_1 = n-k$, take $s = 1$, $k_1 = k+1$, and then relations (7.12) give us $\mathcal{E} = \mathcal{E}^{(k+1)}_{(n-k)}$. Assume $n_1 < n-k$. Then, by (7.12) and (7.13), we must have $t_{1,k} = n_1 + 1$. By the same reason, the relations (7.12) and (7.13) force to have $t_{i,k} = n_1 + i$ for $1 \le i \le n_1$. Denote by k_1 the integer such that $t_{1,k+2-k_1} = (k_1 - 1)n_1 + 1$ and $t_{1,k+1-k_1} > k_1 n_1 + 1$. Now, we must have $t_{n_1+1,k+1} = k_1 n_1 + 1$ and, since we still have the same relations (7.12), we must fill in the same way the next $n_1 \times k_1$ box, that is, the $t_{i,j}$ with $n_1 + 1 \le i \le 2n_1$ and $k_1 \le j \le k+2$. We will repeat this n_2 times, where n_2 is the integer such that $t_{1,k+1-k_1} = k_1 n_1 n_2 + 1$. Pushing forward the same argument and strongly using the fact that we have a bijection between the integers $t \in [1,(k+1)(n-k)]$ and the values of the $t_{i,j}$ inside B, we get the existence of integers $n_1, \dots, n_s$ and $k_1, \dots, k_s$ such that

$$k + 1 = k_1 \cdot k_2 \cdots k_s,$$
$$n - k = n_1 \cdot n_2 \cdots n_s,$$

with $n_l \cdot k_l \neq 1$, $1 \leq l \leq s$. Moreover, according to the values that the $t_{i,j}$ have taken, the relations (7.12) also give us $\mathcal{E} = \mathcal{E}^{(k_1,\dots,k_s)}_{(n_1,\dots,n_s)}$. □

Remark 7.2.6. (1) The reader may remember that in Section 2.4, for a fixed Grassmann variety $\mathrm{Gr}(k,n)$, we associated to any initialized irreducible homogenous bundle a normalized $(n-k,k+1)$-step matrix (see Definition 2.4.7 and Remark 2.4.8). Moreover, in Theorem 2.4.9 we characterized which step matrices correspond to aCM initialized irreducible homogenous aCM bundles. Since the largest entry of a step-matrix associated to an initialized irreducible homogeneous bundle is b_1+n, it turns out that irreducible homogeneous Ulrich bundles correspond to step-matrices with pairwise distinct entries.

(2) As we saw in Example 2.4.10 (1), the vector bundle $\mathcal{F} = \Sigma^{(8,8,2,2)}\mathcal{Q} \otimes \Sigma^{(2,2,1,0,0)}\mathcal{S}^\vee$ on $\mathrm{Gr}(3,8)$ is an aCM bundle. It is not Ulrich because its associated normalized step matrix has repeated entries.

(3) We mention in (2.4.1) that irreducible homogeneous bundles $\mathcal{E} = \Sigma^{(b_1,\dots,b_{k+1})}\mathcal{Q} \otimes \Sigma^{(a_1,\dots,a_s)}\mathcal{S}^\vee$ are simple. Therefore, a fortiori, they are indecomposable on $\mathrm{Gr}(k,n)$. So, all Ulrich bundles so far constructed are indecomposable.

Example 7.2.7. (1) Let us consider the Grassmann variety $\mathrm{Gr}(1,25)$. By Theorem 7.2.5 (see also Corollary 7.2.9 (1)), it supports eight initialized irreducible homogeneous Ulrich bundles. Indeed, using the notation $(a^n,b^m) := (\overbrace{a,\dots,a}^{n},\overbrace{b,\dots,b}^{m})$, they are:

$$\begin{aligned}
\mathcal{E}^{(2)}_{(24)} &= \Sigma^{(23)}\mathcal{Q},\\
\mathcal{E}^{(2,1)}_{(12,2)} &= \Sigma^{(23,12)}\mathcal{Q} \otimes \Sigma^{(12^{12},0^{12})}\mathcal{S}^\vee,\\
\mathcal{E}^{(2,1)}_{(8,3)} &= \Sigma^{(23,16)}\mathcal{Q} \otimes \Sigma^{(16^8,8^8,0^8)}\mathcal{S}^\vee,\\
\mathcal{E}^{(2,1)}_{(6,4)} &= \Sigma^{(23,18)}\mathcal{Q} \otimes \Sigma^{(18^6,12^6,6^6,0^6)}\mathcal{S}^\vee,\\
\mathcal{E}^{(2,1)}_{(4,6)} &= \Sigma^{(23,20)}\mathcal{Q} \otimes \Sigma^{(20^4,16^4,12^4,8^4,4^4,0^4)}\mathcal{S}^\vee,\\
\mathcal{E}^{(2,1)}_{(3,8)} &= \Sigma^{(23,21)}\mathcal{Q} \otimes \Sigma^{(21^3,18^3,15^3,12^3,9^3,6^3,3^3,0^3)}\mathcal{S}^\vee,\\
\mathcal{E}^{(2,1)}_{(4,6)} &= \Sigma^{(23,22)}\mathcal{Q} \otimes \Sigma^{(22^2,20^2,18^2,16^2,14^2,12^2,10^2,8^2,6^2,4^2,2^2,0^2)}\mathcal{S}^\vee, \quad \text{and}\\
\mathcal{E}^{(2,1)}_{(1,24)} &= \Sigma^{(23,23)}\mathcal{Q} \otimes \Sigma^{(23,22,21,20,19,18,17,16,15,14,13,12,11,10,9,8,7,6,5,4,3,2,1,0)}\mathcal{S}^\vee.
\end{aligned}$$

(2) On $\mathrm{Gr}(11,29)$, the sequences $(2,2,3)$ and $(2,3,3)$, with $12 = 2\cdot 2\cdot 3$ and $18 = 2\cdot 3\cdot 3$, give the Ulrich bundle

$$\Sigma^{(b_1,\dots,b_{12})}\mathcal{Q} \otimes \Sigma^{(a_1,\dots,a_{18})}\mathcal{S}^\vee,$$

with

$$(b_1,\dots,b_{12}) = (187,186,177,176,119,118,109,108,51,50,41,40)$$

and

$$(a_1, \dots, a_{18}) = (40, 40, 38, 38, 36, 36, 22, 22, 20, 20, 18, 18, 4, 4, 2, 2, 0, 0),$$

meanwhile the sequences (2, 3, 2) and (2, 3, 3), with $12 = 2 \cdot 3 \cdot 2$ and $18 = 2 \cdot 3 \cdot 3$, give the Ulrich bundle

$$\Sigma^{(b_1,\dots,b_{12})}\mathcal{Q} \otimes \Sigma^{(a_1,\dots,a_{18})}\mathcal{S}^\vee,$$

with

$$(b_1, \dots, b_{12}) = (187, 186, 177, 176, 167, 166, 85, 84, 75, 74, 65, 64)$$

and

$$(a_1, \dots, a_{18}) = (64, 64, 62, 62, 60, 60, 34, 34, 32, 32, 30, 30, 4, 4, 2, 2, 0, 0).$$

This example lets us stress that, in fact, the same integers placed in different order inside the sequence give different Ulrich bundles.

In some particular cases, we can give an explicit expression for the number of irreducible homogeneous Ulrich bundles on the Grassmann variety $\mathrm{Gr}(k, n)$. In order to this, let us recall the following classical arithmetic functions.

Definition 7.2.8. Given integers $k, n \in \mathbb{Z}$, the *k-divisor function* $\sigma_k(n)$ is defined as

$$\sigma_k(n) := \sum_{\substack{d \in \mathbb{Z}_{>0} \\ d|n}} d^k.$$

In particular, $\sigma_0(n)$ counts the number of (positive) divisors of n.

Corollary 7.2.9. *Let* $(\mathrm{Gr}(k, n), \mathcal{O}_G(1))$ *be the Grassmann variety.*

(1) *If $k + 1$ is a prime number, then there are exactly $\sigma_0(n - k)$ irreducible homogeneous Ulrich bundles.*
(2) *If $n - k$ is a prime number, then there are exactly $\sigma_0(k + 1)$ irreducible homogeneous Ulrich bundles.*

Proof. (1) As $k + 1$ is a prime number, we only have two ordered sequences of integers $(k_1, \dots, k_s)$ such that $k+1 = k_1 \cdots k_s$, namely $(k_1) = (k+1)$ of length 1 and $(k_1, k_2) = (k+1, 1)$ of length 2. Hence the only sequences we can consider verifying $n - k = n_1 \cdots n_s$ are $(n_1) = (n-k)$ of length 1 and $(n_1, n_2) = (a_i, b_i)$ of length 2 for any divisor a_i, $1 \le a_i < n-k$ of $n - k$. Applying Theorem 7.2.5, we get that there are exactly $\sigma_0(n - k)$ irreducible homogeneous Ulrich bundles.

(2) The proof of (2) is completely analogous, and we leave it to the reader. □

As a consequence of our main result Theorem 7.2.5, we are able determine the smallest possible rank for irreducible homogeneous Ulrich bundles on $\mathrm{Gr}(k,n)$.

Corollary 7.2.10. *Let* $(\mathrm{Gr}(k,n), \mathcal{O}_G(1))$ *be the Grassmann variety. Then, the smallest possible rank of an irreducible homogeneous Ulrich bundle on* $\mathrm{Gr}(k,n)$ *is*

$$\frac{\prod_{1\le i<j\le k+1}(j-i)(n-k)}{k!(k-1)!\cdots 2!}.$$

Proof. By Theorem 7.2.5, we must look for the pair of ordered sequences $(k_1,\dots,k_s)$ and $(n_1,\dots,n_s)$ such that the corresponding Ulrich bundle $\mathcal{E}^{(k_1,\dots,k_s)}_{(n_1,\dots,n_s)}$ has the smallest possible rank. By Lemma 2.4.3, this is the case for the sequences $(k_1)=(k+1)$, $(n_1)=(n-k)$, and the rank of $\mathcal{E}^{(k+1)}_{(n-k)}$ is exactly

$$\frac{\prod_{1\le i<j\le k+1}(j-i)(n-k)}{k!(k-1)!\cdots 2!}.$$ □

We end this section with another interesting consequence of Theorem 7.2.5.

Corollary 7.2.11. *Any Ulrich bundle on* $\mathrm{Gr}(k,n)$ *has slope*

$$\mu = \frac{k(n-k-1)}{2}d,$$

with $d = \deg(\mathrm{Gr}(k,n)) = \frac{((k+1)(n-k))!k!(k-1)!\cdots 2!}{n!(n-1)!\cdots(n-k)!}$.

Proof. By Theorem 7.2.5, $\mathcal{E}^{(k+1)}_{(n-k)} = \Sigma^{(b_1,\dots,b_k)}\mathcal{Q}$, with $b_i = (k-i+1)(n-k-1)$ for $1\le i\le k$, is an Ulrich bundle. Hence, it is enough to compute the slope of $\mathcal{E}^{(k+1)}_{(n-k)}$. By Lemma 7.2.1, the slope of $\mathcal{E}^{(k+1)}_{(n-k)}$ turns out to be

$$\mu(\mathcal{E}^{(k+1)}_{(n-k)}) = \mu(\Sigma^{(b_1,\dots,b_k)}\mathcal{Q}) = \frac{\sum_{i=1}^{k+1}(k-i+1)(n-k-1)}{k+1}d = \frac{k(n-k-1)}{2}d.$$

The result now follows from the well-known fact that all Ulrich bundles on $\mathrm{Gr}(k,n)$ have the same slope. □

Notice that, in spite of the fact that the slope of an Ulrich bundle on $\mathrm{Gr}(k,n)$ is determined, the uniqueness of Ulrich bundles is false, as it has been shown in Theorem 7.2.5.

It is also worthwhile to point out that we have characterized all irreducible homogeneous Ulrich bundles on $\mathrm{Gr}(k,n)$ but the study of nonhomogeneous Ulrich bundles on $\mathrm{Gr}(k,n)$ is completely open.

Problem 7.2.12. *Determine the existence of nonhomogeneous Ulrich bundles on* $\mathrm{Gr}(k,n)$*, especially of low rank.*

7.3 Ulrich bundles on flag varieties[1]

This section is entirely devoted to studying irreducible homogeneous vector bundles on complete and partial flag varieties arising from Schur functors. We will demonstrate that a partial flag variety with three or more steps does not admit an irreducible homogeneous Ulrich bundle. We will also classify irreducible homogeneous Ulrich bundles on two-step flag varieties $F(1, n-1; n)$, $F(2, n-1; n)$, $F(2, n-2; n)$, $F(k, k+1; n)$, and $F(k, k+2; n)$, and we will end this section with a conjectural description of the two-step flag varieties which do not admit irreducible homogeneous Ulrich bundles.

Let us start recalling the basic facts on flag varieties needed in the sequel. Let V be a $\mathbf{k}$-vector space of dimension n and $0 < k_1 < \cdots < k_r < n$ an increasing sequence of integers. For convenience, set $k_0 = 0$ and $k_{r+1} = n$. The r-step partial flag variety $F(k_1, \dots, k_r; n)$ parameterizes partial flags

$$W_1 \subset W_2 \subset \cdots \subset W_r \subset V,$$

where W_i is a vector subspace of V of dimension k_i for $1 \le i \le r$. The partial flag variety $F(k_1, \dots, k_r; n)$ is said to be complete if $r = n-1$ and $k_i = i$ for $1 \le i \le n$. Given any set of indices $1 \le i_1 < i_2 < \cdots < i_s \le r$, there is a natural projection

$$\pi_{i_1,\dots,i_s} : F(k_1, \dots, k_r; n) \longrightarrow F(k_{i_1}, \dots, k_{i_s}; n).$$

In particular, the partial flag variety $F(k_1, \dots, k_r; n)$ can be realized as an iterated Grassmannian bundle

$$F(k_1, \dots, k_r; n) \longrightarrow F(k_2, \dots, k_r; n) \longrightarrow \cdots \longrightarrow \mathrm{Gr}^a(k_r, n)$$

where the Grassmann variety $\mathrm{Gr}^a(k_r, n)$ parameterizes vector subspaces $\mathbf{k}^{k_r}$ of dimension k_r of $\mathbf{k}^n$. From this we immediate deduce that

$$N := \dim F(k_1, \dots, k_r; n) = \sum_{i=1}^{r} k_i(k_{i+1} - k_i).$$

The partial flag variety has r natural projections to Grassmann varieties

$$\pi_i : F(k_1, \dots, k_r; n) \longrightarrow \mathrm{Gr}^a(k_i, n)$$

1 The authors wish to acknowledge the contribution of Professor Izzet Coskun, Professor Jack Huizenga and Dr. Matthew Woolf for their collaboration on the paper *Ulrich Schur bundles on flag varieties*, J. of Algebra, 474 (2017), 49–96 [61]. The material in Section 7.3 is a reprint of this material. In addition, the authors thank Professor Jack Huizenga for creating all of the original illlustrations. We gratefully acknowledge the permission from the publisher Elsevier for including this material.

and the Picard group of $\mathrm{F}(k_1,\dots,k_r;n)$ is generated by $\mathcal{L}_i = \pi_i^*\mathcal{O}_{\mathrm{Gr}^a(k_i,n)}(1)$. A line bundle $\mathcal{L}$ on $\mathrm{F}(k_1,\dots,k_r;n)$ is ample if and only if $\mathcal{L} \cong \mathcal{L}_{a_1,\dots,a_r} = \mathcal{L}_1^{\otimes a_1} \otimes \cdots \otimes \mathcal{L}_r^{\otimes a_r}$ with $a_i > 0$ for every $1 \le i \le r$. In particular, the minimal ample line bundle on the partial flag variety is $\mathcal{O}_{\mathrm{F}}(1) := \mathcal{L}_{1,\dots,1}$ where for short F denotes the corresponding flag variety $\mathrm{F}(k_1,\dots,k_r;n)$. The degree of a partial flag variety was computed in [104], and we have

Proposition 7.3.1. *The degree of the partial flag variety* $\mathrm{F}(k_1,\dots,k_r;n)$ *embedded by the line bundle* $\mathcal{L}_{a_1,\dots,a_r}$ *is*

$$N!\frac{\prod_{1\le i\le j\le r} b_{ij}^{(k_i-k_{i-1})(k_{j+1}-k_j)}}{\prod_{s=1}^{r}\prod_{k_{s-1}<i\le k_s}\frac{(n-i)!}{(k_s-i)!}} \tag{7.14}$$

where $b_{ij} = \sum_{k=i}^{j} a_k$.

We record a few special cases of the degree formula (7.14).

Remark 7.3.2. The degree of $\mathrm{Gr}^a(k,n)$ under the bth power of the Plücker embedding is

$$(k(n-k))!\,b^{k(n-k)}\prod_{1\le i\le k}\frac{(k-i)!}{(n-i)!}.$$

The degree of $\mathrm{F}(k_1,\dots,k_r;n)$ under the line bundle $\mathcal{L}_{1,\dots,1}$ is

$$N!\frac{\prod_{1\le i\le j\le r}(j-i+1)^{(k_i-k_{i-1})(k_{j+1}-k_j)}}{\prod_{s=1}^{r}\prod_{k_{s-1}<i\le k_s}\frac{(n-i)!}{(k_s-i)!}}.$$

In particular, specializing to the case $\mathrm{F}(1,n-1;n)$, we obtain

$$(2n-3)!\frac{2}{(n-1)!(n-2)!},$$

and specializing to the case $\mathrm{F}(1,n-2,n)$, we obtain

$$(3n-7)!\frac{4}{(n-1)!(n-2)!(n-3)!}.$$

The partial flag variety $\mathrm{F}(k_1,\dots,k_r;n)$ is endowed with a collection of tautological subbundles

$$0 = \mathcal{T}_0 \subset \mathcal{T}_1 \subset \mathcal{T}_2 \subset \cdots \subset \mathcal{T}_{r+1} = \underline{\mathcal{V}},$$

where $\mathcal{T}_i$ is a vector bundle of rank k_i and $\underline{\mathcal{V}}$ is the trivial vector bundle of rank n. Set $\mathcal{U}_i = \mathcal{T}_i/\mathcal{T}_{i-1}$.

Definition 7.3.3. Let $\lambda = (\lambda_1|\cdots|\lambda_{r+1})$ be a concatenation of partitions λ_i, where λ_i has length $k_i - k_{i-1}$. Then the *Schur bundle* $\mathcal{E}_\lambda$ on $\mathrm{F}(k_1,\dots,k_r;n)$ is

$$\mathcal{E}_\lambda = \Sigma^{\lambda_1}\mathcal{U}_1^\vee \otimes \Sigma^{\lambda_2}\mathcal{U}_2^\vee \otimes \cdots \otimes \Sigma^{\lambda_{r+1}}\mathcal{U}_{r+1}^\vee,$$

where Σ^{λ_i} denotes the Schur functor of type λ_i.

Remark 7.3.4. The 1-step flag varieties correspond to Grassmann varieties that have been the object of study in Sections 2.4 and 7.2. The reader may have observed that for that case in Remark 2.4.1 we took as generators of the Schur bundles the dual vector bundles $\mathcal{S}^\vee \cong \mathcal{U}_1$ and $\mathcal{Q} \cong \mathcal{U}_2$. This choice is completely harmless since it is well-known from representation theory that $\mathcal{U}_i^\vee \cong \Sigma^{(0,\dots,0,-1)}\mathcal{U}_i$.

Schur bundles can be characterized as the pushforwards of line bundles on complete flag varieties (see [124, Section 2.5]). Since $\mathcal{U}_1,\dots,\mathcal{U}_{r+1}$ give a filtration of the trivial bundle $\underline{V}$, the tensor product of their determinants is trivial by the splitting principle (see [108, Appendix A3]). Hence, adding a constant integer to all the entries of λ corresponds to tensoring the bundle $\mathcal{E}_\lambda$ with a power of the trivial line bundle and does not change its isomorphism class. We will say that two partitions λ and λ' are *equivalent* if they differ by adding a constant to all the entries, and then $\mathcal{E}_\lambda \cong \mathcal{E}_{\lambda'}$. There is also no harm in allowing negative integers in λ, since adding a large constant will make all entries positive.

For studying Ulrich bundles, it is important to know the effect of tensoring $\mathcal{E}_\lambda$ by an ample line bundle $\mathcal{L}_{a_1,\dots,a_r}$. By the Littlewood–Richardson rule (see [98, Appendix A]), adding a constant sequence to λ_i corresponds to tensoring the bundle by the determinant of $\mathcal{U}_i$. Since the determinant of $\mathcal{U}_i$ is $\mathcal{L}_i \otimes \mathcal{L}_{i-1}^{-1}$, we have

$$\mathcal{E}_\lambda \otimes \mathcal{L}_{a_1,\dots,a_r} \cong \mathcal{E}_{\lambda+\psi_{a_1,\dots,a_r}},$$

where $\psi_{a_1,\dots,a_r}$ is the dominant weight corresponding to $\mathcal{L}_{a_1,\dots,a_r}$.

In Chapter 2, Theorem 2.4.4, we applied the Borel–Bott–Weil theorem to compute the cohomology of irreducible homogeneous bundles on $\mathrm{Gr}(k,n)$. Let us now see how the Borel–Bott–Weil theorem will allow us to compute the cohomology of the Schur bundle $\mathcal{E}_\lambda$. To this end, let $\rho = (n, n-1, \dots, 2, 1)$. We say that λ is *singular* if $\lambda+\rho$ has repeated entries. Otherwise, we say that λ is *regular*. Let s be the permutation in $\mathfrak{S}_n$ that lists the entries in $\lambda+\rho$ in decreasing order. Let $q(\lambda)$ denote the minimal length of the permutation s. Keeping these notations, Borel–Bott–Weil theorem (see [204]) can be stated in the following way:

Theorem 7.3.5. *Let $\mathcal{E}_\lambda$ be a Schur bundle on the flag variety $\mathrm{F}(k_1,\dots,k_r;n)$.*
(1) *If λ is singular, then $\mathrm{H}^i(\mathcal{E}_\lambda) = 0$ for every i.*
(2) *If λ is regular, then $\mathrm{H}^i(\mathcal{E}_\lambda) = 0$ for $i \neq q(\lambda)$ and $\mathrm{H}^{q(\lambda)}(\mathcal{E}_\lambda) = \Sigma^{s(\lambda+\rho)-\rho}V^\vee$.*

We will deal with *initialized* Schur bundles $\mathcal{E}_\lambda$. Therefore, $\mathrm{H}^0(\mathcal{E}_\lambda) \neq 0$ and $\mathrm{H}^0(\mathcal{E}_\lambda(-1)) = 0$. In this case, by the Borel–Bott–Weil theorem, the sequence of integers $\lambda = (\lambda_1|\cdots|\lambda_{r+1})$ is weakly decreasing and λ can be thought of as a partition. Moreover, the Young diagram of λ is the concatenation of the Young diagrams of the λ_i.

We recall how to compute the dimension of a representation $\Sigma^\mu V$ of $\mathrm{GL}(n)$ (see [98] for more details). Let $Y(\mu)$ be the Young diagram corresponding to the partition μ. The hook-length $\mathrm{hook}(i,j)$ is the number of boxes in the same row to the right of the box (i,j) and in the same column below the box (i,j) (including the box itself). The dimension of the representation $\Sigma^{\mu}V$ is given by

$$\prod_{(i,j)\in Y(\mu)} \frac{n+j-i}{\mathrm{hook}(i,j)},$$

which is always nonzero. In particular, the rank of the vector bundle $\mathcal{E}_{\lambda}$ is given by

$$\prod_{s=1}^{r+1} \prod_{(i,j)\in Y(\lambda_s)} \frac{k_s-k_{s-1}+j-i}{\mathrm{hook}(i,j)}.$$

Specializing this formula to the case of $\mathrm{Gr}^{a}(k,n)$, we get the expression given in Lemma 2.4.3.

Our next goal it to rephrase combinatorially the condition that a Schur bundle $\mathcal{E}_{\lambda}$ on the partial flag variety $\mathrm{F}(k_1,\ldots,k_r;n)$ is an Ulrich bundle.

Let $\mathcal{E}_{\lambda}$ be a Schur bundle on $\mathrm{F}(k_1,\ldots,k_r;n)$ associated to the concatenated sequence $\lambda = (\lambda_1|\cdots|\lambda_{r+1})$ of partitions of length k_i-k_{i-1}. So λ has total length n. We write $P=\lambda+\rho$, where $\rho=(n,n-1,\ldots,2,1)$. For concreteness, we explicitly write P as

$$P=(a_1^1,\ldots,a_{l_1}^1|a_1^2,\ldots,a_{l_2}^2|\cdots|a_1^{r+1},\ldots,a_{l_{r+1}}^{r+1});$$

here $l_i=k_i-k_{i-1}$. By Theorem 7.3.5, $\mathrm{H}^0(\mathcal{E}_{\lambda})\neq 0$ if and only if the sequence of entries of P is strictly decreasing.

For any integer $t\geq 0$, we define the *time evolution* of the partition P by

$$P(t):=(a_1^1-tr,\ldots,a_{l_1}^1-tr|a_1^2-t(r-1),\ldots,a_{l_2}^2-t(r-1)|\cdots|a_1^{r+1},\ldots,a_{l_{r+1}}^{r+1})$$

by subtracting $t(r+1-i)$ from every entry in the ith part, and analogously we define partitions $\lambda(t)$. We will abuse terminology and call the data $P(t)$ a partition, even though its entries may no longer be weakly decreasing. We view the entries of $P(t)$ as being functions of t and define

$$a_k^i(t)=a_k^i-t(r+1-i),$$

so that

$$P(t)=(a_1^1(t),\ldots,a_{l_1}^1(t)|\cdots|a_1^{r+1}(t),\ldots,a_{l_{r+1}}^{r+1}(t)).$$

Then, by Theorem 7.3.5, $\mathcal{E}_{\lambda}(-t)=\mathcal{E}_{\lambda(t)}$ has $\mathrm{H}^i(\mathcal{E}_{\lambda}(-t))=0$ for all i precisely when there is a repeated entry in the partition $P(t)$. Therefore $\mathcal{E}_{\lambda}$ is an initialized Ulrich bundle if and only if $P(t)$ has a repeated entry for all integers $t\in[N]$, where $N=\dim \mathrm{F}(k_1,\ldots,k_r;n)$.

The above observation suggests the following pure combinatorial definitions.

Definition 7.3.6. Let the partition

$$P = (a_1^1, \dots, a_{l_1}^1 | a_1^2, \dots, a_{l_2}^2 | \cdots | a_1^{r+1}, \dots, a_{l_{r+1}}^{r+1})$$

consist of a decreasing sequence of integers arranged into $r + 1$ blocks, with $l_i > 0$ for all i. The *type* of the partition is $(l_1, \dots, l_{r+1})$. The *dimension* of the partition is $N = \sum_{k<h} l_k l_h$.

The partition P is *Ulrich* if for all $t \in [N] = \{1, \dots, N\}$ (usually we say for any time $t \in [N]$) the partition $P(t)$ has a repeated entry.

Two partitions are *equivalent* if they differ by adding a constant to all the entries. When we count Ulrich partitions, it is always modulo equivalence.

We make explicit the correspondence between the algebro-geometric and combinatorial problems we are interested in.

Proposition 7.3.7. *There is a bijection $\lambda \mapsto \lambda + \rho = P$ between equivalence classes of partitions λ such that the Schur bundle $\mathcal{E}_\lambda$ on $\mathrm{F}(k_1, \dots, k_r; n)$ of dimension N is Ulrich with respect to $\mathcal{O}_{\mathrm{F}}(1)$ and equivalence classes of Ulrich partitions P of dimension N of type $(k_1, k_2 - k_1, \dots, k_r - k_{r-1}, n - k_r)$.*

So our goal is to determine the possible types of Ulrich partitions, and, when there exists an Ulrich partition of a given type, to classify all such partitions up to equivalence.

Remark 7.3.8. (1) The dimension N of a partition P is the number of pairs of entries coming from different blocks. For there to be a repeated entry in $P(t)$ at every time $t \in [N]$, it is therefore necessary and sufficient that every pair of entries meets at some time $t \in [N]$ and that no two pairs of entries meet at the same time.

(2) There is no harm in considering lists P where some of the entries are negative since the notion of Ulrich partition is invariant under equivalence. It will often be convenient to change the definition of $P(t)$ slightly by subtracting a different consecutive range of integers from each part. For example, in the two-step flag variety case where the partition P has 3 parts it will be convenient to subtract $1, 0, -1$ from the three parts, so that the middle block is constant with time. We will make our choice of normalization clear each time we need.

We easily derive simple congruences among the entries of Ulrich partitions.

Lemma 7.3.9. *Let $(a_i^1 | a_i^2 | \cdots | a_i^s)$ be an Ulrich partition. Suppose $1 \le j \le k \le s$. Then all the entries a_i^j in the j-block are congruent to all the entries a_i^k in the k-block modulo $k - j$.*

Proof. We normalize the evolution of the Ulrich partition so that we keep all the entries a_i^j in the j-block fixed and add $k - j$ to all the entries a_i^k in the k-block. This preserves the congruence classes of the a_i^j and a_i^k modulo $k - j$. In order for them to collide at some time, they must all have the same congruence class. □

Let us introduce by means of an example the so called *time evolution* diagram, a tool that we will use to visualize the sequence of partitions $P(t)$ and to check the Ulrich condition.

Example 7.3.10. Consider the partition $P = (5|3, -1, -2, -4|-5) = (a|b_1, b_2, b_3, b_4|c)$ of type $(1, 4, 1)$. We normalize $P(t)$ by subtracting $1, 0, -1$ at each time. The dimension is $N = 9$, and this partition is Ulrich. To visualize the Ulrich condition, we draw a *time evolution diagram* showing the positions of the entries of $P(t)$ in Figure 7.1. Along the first row, we place the entries a, b, b, b, b, c in the corresponding positions; we provide numerical positions above some of the entries for reference. For each time $t \in [N]$, we similarly place the entries of $P(t)$ in the corresponding row. When two entries intersect, we put a box around them to highlight the intersection. The Ulrich condition means that for every time $t \in [N]$ there is exactly one box in the corresponding row. We also display $P(N+1)$ at time $N+1$, which is relevant to the computation of the *dual partition* P^* (see Definition 7.3.11).

	5					0					−5
t	a		b				b	b		b	c
1		a	b				b	b		$\boxed{bc}$	
2			$\boxed{ab}$				b	b	c	b	
3			b	a			b	$\boxed{bc}$		b	
4			b		a		$\boxed{bc}$	b		b	
5			b			$\boxed{ac}$	b	b		b	
6			b		c		$\boxed{ab}$	b		b	
7			b	c			b	$\boxed{ab}$		b	
8			$\boxed{bc}$				b	b	a	b	
9		c	b				b	b		$\boxed{ab}$	
	c		b				b	b		b	a

Figure 7.1: Time evolution diagram of the Ulrich partition $(5|3, -1, -2, -4|-5)$ of type $(1, 4, 1)$. See Example 7.3.10.

Replacing P by the equivalent partition $(11|9, 5, 4, 2|1)$ and putting $\lambda = P - \rho$, we have $\lambda = (6|5, 2, 2, 1|1) = (\lambda_1|\lambda_2|\lambda_3)$. The corresponding initialized Ulrich bundle on $\mathrm{F}(1, 5; 6)$ is

$$\mathcal{E}_\lambda = \Sigma^{\lambda_1}\mathcal{U}_1^\vee \otimes \Sigma^{\lambda_2}\mathcal{U}_2^\vee \otimes \Sigma^{\lambda_3}\mathcal{U}_3^\vee.$$

A straightforward computation with the hook-length formula shows that $\operatorname{rank}(\mathcal{E}_\lambda) = 70$ and $\mathrm{h}^0(\mathcal{E}_\lambda) = 17640 = \operatorname{rank}(\mathcal{E}_\lambda)\deg(\mathrm{F}(1, 5; 6))$.

There are certain symmetries satisfied by Ulrich partitions which motivate the following definition.

Definition 7.3.11. Let $P=(a_1^1,\dots,a_{l_1}^1|\cdots|a_1^{r+1},\dots,a_{l_{r+1}}^{r+1})$ be a partition of type $(l_1,\dots,l_{r+1})$. We define the *symmetric partition*

$$P^s=(-a_{l_{r+1}}^{r+1},\dots,-a_1^{r+1}|\cdots|-a_{l_1}^1,\dots,-a_1^1)$$

of the reversed type $(l_{r+1},\dots,l_1)$ multiplying all the entries of P by -1 and reversing the order in which they are written. It is clear that if P is Ulrich then P^s is also Ulrich. This also can be seen from the fact that $F(k_1,\dots,k_r;n)\cong F(n-k_r,\dots,n-k_1;n)$. The time evolution diagram of P^s is flipped about a vertical axis.

Let $P=(a_1^1,\dots,a_{l_1}^1|\cdots|a_1^{r+1},\dots,a_{l_{r+1}}^{r+1})$ be an Ulrich partition of type $(l_1,\dots,l_{r+1})$. We define the *dual partition*

$$P^*=(a_1^{r+1}(N+1),\dots,a_{l_{r+1}}^{r+1}(N+1)|\cdots|a_1^1(N+1),\dots,a_{l_1}^1(N+1)).$$

The integers which appear in P^* are the integers in $P(N+1)$ reordered them to be decreasing. In the time evolution diagram for P, P^* can be read from the final row below the horizontal line. Then P^* is Ulrich of type $(l_{r+1},\dots,l_1)$. Its time evolution diagram is obtained from the diagram for P by flipping about a horizontal axis $t=(N+1)/2$.

The partition $(P^S)^*$ has the same type as P, and we call this partition the *symmetric dual*. In particular, if there is a unique Ulrich partition of a given type then it is equal to its symmetric dual.

Example 7.3.12. (1) If $P=(5|3,-1,-2,-4|-5)$ then $P^S=(5|4,2,1,-3|-5)$. Both partitions P and P^S are self-dual.

(2) If the flag $\mathrm{F}(k_1,\dots,k_r;n)$ carries an Ulrich Schur bundle $\mathcal{E}_\lambda$, then $\mathcal{E}_{P^s-\rho}$ and $\mathcal{E}_{P^*-\rho}$ are Ulrich bundles on the dual flag $\mathrm{F}(n-k_r,\dots,n-k_1;n)$, while $\mathcal{E}_{(P^s)^*-\rho}$ is an Ulrich bundle on the original flag $\mathrm{F}(k_1,\dots,k_r;n)$. Notice that $\mathcal{E}_{(P^s)^*-\rho}$ is the Ulrich dual of $\mathcal{E}_\lambda$.

Our first result states that there are no equivariant Ulrich bundles on flag varieties with three or more steps.

Proposition 7.3.13. *There are no Ulrich partitions with 4 or more blocks.*

Proof. We will split the proof into three cases.

Case 1. Ulrich partitions with 4 blocks.

Suppose there is an Ulrich partition $(a_1,\dots,a_j|b_1,\dots,b_k|c_1,\dots,c_l|d_1,\dots,d_m)$. By Lemma 7.3.9, the parities of $a_i(t)$ and $c_h(t)$ (respectively, $b_i(t)$ and $d_h(t)$) are equal at all times. On the other hand, entries in adjacent blocks have the same parity at odd times t and opposite parities at even times t. Consequently, when t is even, the only collisions can be ac or bd collisions. Since the last intersection at time $t=N$ has to be between entries from adjacent blocks, we conclude that the dimension N is odd.

By symmetry (see Definition 7.3.11), we can assume that at time $t=2$, a_j and c_1 collide. We normalize the evolution of the partition and we subtract $2,1,0,-1$, respectively, from the A-, B-, C-, and D-blocks. Consequently, the first three collisions are

$$a_jb_1,\ a_jc_1,\ b_1c_1 \quad \text{or} \quad b_1c_1,\ a_jc_1,\ a_jb_1.$$

The corresponding sequences at time $t = 0$ look like

$$\cdots a_j\ b_1 \times\times c_1\ \times\times\cdots \quad \text{or} \quad \cdots a_j\ \times\times b_1\ c_1\ \times\times\cdots.$$

In either case, we conclude that the length k of the B-block is one since by time $t = 3$, a_j must have passed through the entire B-block. From now on we denote the unique entry in the B-block by b.

Similarly, $t = N - 1$ is an even time. Hence, the only possible collisions are ac or bd. If b and d_m collide, then the last three collisions at times $N - 2, N - 1$, and N are

$$c_1 d_m,\ bd_m,\ bc_1 \quad \text{or} \quad bc_1,\ bd_m,\ d_m c_1.$$

In particular, we conclude that the length of the C-block is also one. Denote the lengths of the A- and D-blocks by j and m, respectively. The dimension of the flag variety is

$$N = 2j + jm + 2m + 1.$$

Since the collisions at even times are ac or bd, the dimension can be at most one more than twice the total number $j + m$ of possible ac and bd collisions. We, therefore, have the inequality

$$2j + jm + 2m \leq 2(j + m).$$

This implies that $jm \leq 0$, which is absurd.

We conclude that at time $t = N - 1$, the intersection must be between a_1 and c_l. Hence, the last three intersections at times $N - 2, N - 1, N$ are one of

$$a_1 b,\ a_1 c_l,\ bc_l \quad \text{or} \quad bc_l,\ a_1 c_l,\ a_1 b.$$

Normalize the position of the sequence so that $b = y$ and $d_1 = -y$. Then c_1 equals either $y - 1$ or $y - 3$. Furthermore, there cannot be any c's in positions $-y + 1$, $y + 2$, or $-y + 3$ since d_1 is at these positions at times $1, 2, 3$ when there are other collisions. In either case, d_1 intersects c_1 at a time $t_0 \geq 2y - 3$. On the other hand, b intersects c_l at some time $t_1 < 2y - 3$. Hence, the intersection bc_l happens before the intersection $c_1 d_1$. This is a contradiction, since bc_l is one of the last three intersections and $c_1 d_1$ is not. We conclude that there are no Ulrich partitions for three step flag varieties with respect to $\mathcal{O}_{\mathcal{F}}(1)$.

Case 2. Ulrich partitions with 5 blocks.

Suppose there is an Ulrich partition $(a_1, \ldots, a_j | b_1, \ldots, b_k | c_1, \ldots, c_l | d_1, \ldots, d_m | e_1, \ldots, e_p)$. We normalize the evolution so that at each time we subtract $2, 1, 0, -1, -2$ from the entries in the A-, B-, C-, D-, and E-blocks, respectively. By parity, at time $t = 2$, we must have an ac, bd, or ce collision. Suppose we have a bd collision. As before, the length l of the C-block is 1, so we denote its entry by c. Then the first three collisions are either

$$b_k c,\ b_k d_1,\ cd_1 \quad \text{or} \quad cd_1,\ b_k d_1,\ b_k c,$$

corresponding to the following sequences at time $t = 0$:

$$\cdots b_k\, c \times \times d_1 \cdots \quad \text{or} \quad \cdots b_k \times \times c\, d_1 \cdots.$$

By Lemma 7.3.9, the entries in the B- (respectively, D-) block are the same modulo 6, so the distances between consecutive entries are at least 6. Similarly, the consecutive entries in the A- (respectively, E-) blocks are at least 12 apart. At time $t = 4$, the collision cannot be a bd collision since then the distance between two consecutive entries in either the B- or D-blocks would be only 4 instead of 6. Hence, the collision at time $t = 4$ must be either $a_j c$ or ce_1. However, none of these are possible. In the first case, a_j never collides with d_1 and in the latter case e_1 never collides with b_k. We conclude that the first collision cannot be a bd collision.

The collision at time $t = 2$ must be an ac or ce collision. Without loss of generality, we may assume it is ac. Therefore, there is only one b and b collides with c_1 at time 1 or 3. As in the case of three-step flag varieties, the dimension N of the flag variety must be odd. Hence, at time $N - 1$ we must have an ac, bd or ce collision. By symmetry, we see that the intersection at time $t = N - 1$ cannot be a bd collision, hence must be an ac or ce collision. If it were an ac collision, then b would have to intersect c_l at time N or $N - 2$. Since the time it takes for d_1 to intersect c_1 is longer than the time it takes b to intersect c_l (as in the proof of Case 1), we conclude that the intersection at time $t = N - 1$ must be $c_1 e_p$ and there can be only one entry in the D-block.

We have concluded that the sequence looks like $(a_1, \dots, a_j | b | c_1, \dots, c_l | d | e_1, \dots, e_p)$. The first three intersections are either

$$\text{Case I: } a_j b,\ a_j c_1,\ bc_1 \quad \text{or} \quad \text{Case II: } bc_1,\ a_j c_1,\ a_j b,$$

and the last three intersections are either

$$\text{Case A: } de_p,\ c_1 e_p,\ c_1 d \quad \text{or} \quad \text{Case B: } c_1 d,\ c_1 e_p,\ de_p.$$

For the rest of the argument, we normalize positions so that $b = y$ and $d = -y$. Then at time y we have the bd collision at position 0.

We first show that Case IB is impossible. We compute $c_1 = y - 3$, and the equality

$$c_1 = c_1(N - 2) = d(N - 2) = d + N - 2$$

gives $N = 2y - 1$. Similarly, from $e_p(N) = d(N)$ we find $e_p = -3y + 1$. As $a_j = y + 1$, we conclude the intersection $a_j e_p$ happens at time y. This contradicts that b and d already collide at time y. Symmetrically, Case IIA is also impossible since the intersection $a_1 e_1$ happens at time y. We conclude that we must be in Case IA or Case IIB. These cases are the same under duality (see Definition 7.3.11). Consequently, it suffices to analyze Case IA.

In Case IA, we first prove that at time $t = 0$ the sequence must look like

$$\cdots a_j\, b \times \times c_1 \times \times \times\ c_2 \times \times \times \cdots \times \times \times \times \times\ d \times \times \times \times \times \cdots, \tag{$*$}$$

where $\times$ denotes a position unoccupied by any entries a, b, c, d, e. At time $t = 1, 2, 3$, the collisions are $a_j b, a_j c_1$, and bc_1. Since at these times d is at positions $-y+1, -y+2, -y+3$, there cannot be any entries of the C-block in these positions. Furthermore, there cannot be any entries of the C-block in position $y-5$ because a_j is at position $y-5$ at time $t = 3$ while the collision bc_1 occurs. By Lemma 7.3.9, the distances between the entries in the C-block are even. Hence, there cannot be any entries of the C-block at positions $y - 2\alpha$ for any integer α. At time $t = 4$, a collision of the form ac, bd, or ce must occur. Since e_1 has not passed through d and $b(4) > a_j(4) > d(4)$, the only possible collision is $a_j c_2$. Hence, $c_2 = y - 7$.

As in Case IB, we now compute that the dimension of the flag variety is $N = 2y - 3$. At times $N-2$, $N-1$, and N, the collisions are specified and do not involve b. At these times b is at positions $-y+5, -y+4, -y+3$. Hence, we conclude that there cannot be any entries of the C-block at positions $-y+1, \dots, -y+5$. Since $e_p(N-2) = d(N-2)$, we have $e_p = -3y+5$. The $c_2 d$ collision happens at time $2y-7$ and $e_p(2y-7) = y-9$, so there cannot be a c at position $y-9$ either. Combining all these observations, we recover the claimed shape of the sequence at time $t = 0$.

Next consider the position of e_1. Since there are intersections for times $1 \le t \le 4$, we find $e_1 \le -y-5$. To complete the proof of the claim that the entries of the partition are as in $(*)$, we only need to show that, in fact, $e_1 < -y-5$. So suppose $e_1 = -y-5$. Since $c_2 = y-7$, the intersection $c_2 e_1$ occurs at time $y-1$. However, since $a_j = y+1$ and $e_p = -3y+5$, the intersection $a_j e_p$ also occurs at time $y-1$. This contradiction shows that $e_1 < -y-5$, and establishes the pattern $(*)$ for the entries.

We finally obtain a contradiction by showing that there is no possible collision at time $t = 5$. The collision cannot be a de collision since $e_1 < -y-5$. Since nothing has passed through the d before time 5, the time $t = 5$ intersection cannot involve an e.

Since there are no c entries at position $-y+5$, the collision cannot be of the form cd. Since there are no c entries at position $y-5$, the collision cannot be of the form bc. Since b has not passed though c_2, the collision cannot be of the form bd. Since there are no c entries in position $y-9$, the collision cannot be $a_j c$. Since the entries in the A-block are at least 12 apart, the collision cannot be $a_{j-1} c_1$. We conclude that the collision is not of the form ac. For the same reason, the collision cannot be of the form ab.

This only leaves the possibility of the collision being between a_1 and d. This implies $y = 7$, so $c_2 = 0$ and there is a triple intersection by b, c_2, and d at time 7. This concludes the proof that there cannot be an Ulrich partition for four-step flag varieties.

Case 3. Ulrich partitions with 6 or more blocks.

Suppose there is an Ulrich partition $(a_1^1, \dots, a_{l_1}^1 | \cdots | a_1^s, \dots, a_{l_s}^s)$ with $s \ge 6$. Since $s \ge 6$, we have that $\max\{k-1, s-k\} \ge 3$ for every $1 \le k \le s$. By Lemma 7.3.9, we conclude that the entries a_i^k in the k-block all have the same congruence class modulo 6. In particular, they all have the same parity.

At time $t = 1$, two adjacent blocks must meet, say $a^k_{l_k}(1) = a^{k+1}_1(1)$. This forces $a^k_{l_k}$ and a^{k+1}_1 to have opposite parities. If $k - j$ is even, then the entries $a^k_i(t)$ and $a^j_h(t)$ in the k- and j-blocks have the same parities at all times t. On the other hand, if $k - j$ is odd, then the entries in the k- and j-blocks have the same parity at odd times and opposite parities at even times t. Consequently, at even times the intersections must occur between entries in nonadjacent blocks.

Suppose that at time $t = 2$, $a^j_{l_j}$ and a^{j+2}_1 collide. Normalize the evolution of the sequence so that $1, 0, -1$ is subtracted from the j-, $(j+1)$-, and $(j+2)$-blocks, respectively. Then, up to switching their order, at times $t = 1, 3$, the collisions must be $a^j_{l_j} a^{j+1}_1$ and $a^{j+1}_1 a^{j+2}_1$ depending on the sequence at time $t = 0$. We observe that the possible positions of the entries at time $t = 0$ are

$$\cdots \times \times a^j_{l_j}\, a^{j+1}_1 \times \times a^{j+2}_1 \cdots \quad \text{or} \quad \cdots \times \times a^j_{l_j} \times \times a^{j+1}_1\, a^{j+2}_1 \cdots,$$

where we denote the positions unoccupied by an entry by $\times$. In particular, after normalizing the positions, the triple $(a^j_{l_j}, a^{j+1}_1, a^{j+2}_1)$ either equals $(1, 0, -3)$ or $(3, 0, -1)$. Furthermore, there is a unique entry in the $(j+1)$-part, $l_{j+1} = 1$.

At time $t = 4$, the collision is between entries from nonadjacent blocks. For $k \neq j$, if $a^k_{l_k}(4) = a^{k+2}_1(4)$, then $a^k_{l_k}$ passes through the entries in the $(k+1)$-block without intersecting them, contradicting that the partition is Ulrich. Hence, the only possible collisions at time $t = 4$ are either $a^j_{l_j} a^{j+2}_2$ or $a^j_{l_j - 1} a^{j+2}_1$. However, these are both impossible since they would give a separation of only four between either $a^j_{l_j}$ and $a^j_{l_j-1}$ or a^{j+2}_1 and a^{j+2}_2. This contradicts the fact that entries in each block are congruent modulo 6, and we conclude that there cannot be any Ulrich partitions with $s \geq 6$.

Therefore, as a consequence of the three cases, we have seen that there are no Ulrich partitions with 4 or more blocks. □

Theorem 7.3.14. *A flag variety* $(\mathrm{F}(k_1, \ldots, k_s), \mathcal{O}_{\mathrm{F}}(1))$*, with* $s \geq 3$*, does not have a Schur Ulrich bundle* $\mathcal{E}_\lambda$.

Proof. It immediately follows from Propositions 7.3.13 and 7.3.7. □

As a special case of Theorem 7.3.14, we obtain the following corollary.

Corollary 7.3.15. *The complete flag variety* $(\mathrm{F}(1, 2, 3, \ldots, n-1; n), \mathcal{O}_{\mathrm{F}}(1))$ *does not admit a Schur Ulrich bundle for* $n \geq 4$.

We will focus on the classification of Ulrich partitions with three parts, which corresponds to Ulrich bundles $\mathcal{E}_\lambda$ on two-step flag varieties with respect to $\mathcal{O}_{\mathrm{F}}(1)$. So, from now on, $\mathrm{F}(p, q; v)$ denotes a two-step flag variety. Let $(\alpha, \beta, \gamma) = (p, q-p, v-q)$ be the type of a partition. We will classify:

(1) Ulrich partitions where $\beta \leq 2$ but α and γ are arbitrary; and
(2) Ulrich partitions where $\alpha \leq 2$ and $\gamma \leq 2$, but β is arbitrary.

We begin by analyzing the simplest case of $(1, n, 1)$ partitions, which correspond to Schur bundles on $\mathrm{F}(1, n+1; n+2)$.

Theorem 7.3.16. *There are 2^n Ulrich partitions of type $(1, n, 1)$ for any $n \geq 1$ (up to equivalence).*

Proof. Indeed, suppose $(a|b_1, \dots, b_n|c)$ is an Ulrich partition. Since a and c have the same parity, we may normalize $a = y + 1$, $c = -y - 1$ for some integer $y \geq 0$. At time $y + 1$, we have the ac intersection. Before time $y + 1$, every intersection is of type ab or bc. Thus, for every position $p \in [y]$, exactly one of p or $-p$ is in $B = \{b_1, \dots, b_n\}$. We conclude $y = n$, and if $B' \subset [n]$ is any subset then there is a unique Ulrich partition $(n+1|B|{-n}-1)$ with $B \cap [n] = B'$. Thus equivalence classes of Ulrich partitions of type $(1, n, 1)$ are in bijective correspondence with subsets of $[n]$. □

Example 7.3.17. The Ulrich partition $(5|3, -1, -2, -4|-5)$ corresponds to $B' = \{3\} \subset [4]$ in the previous construction and it was already studied in Example 7.3.10. There are exactly 16 Ulrich partitions of type $(1, 4, 1)$. They are:

$$
\begin{array}{lll}
(5|1,-2,-3,-4|-5), & (5|2,-1,-3,-4|-5), & (5|3,-1,-2,-4|-5),\\
(5|4,-1,-2,-3|-5), & (5|2,1,-3,-4|-5), & (5|3,1,-2,-4|-5),\\
(5|4,1,-2,-3|-5), & (5|3,2,-1,-4|-5), & (5|4,2,-1,-3|-5),\\
(5|4,3,-1,-2|-5), & (5|3,2,1,-4|-5), & (5|4,2,1,-3|-5),\\
(5|4,3,1,-2|-5), & (5|4,3,2,-1|-5), & (5|4,3,2,1|-5),\\
(5|-1,-2,-3,-4|-5). & &
\end{array}
$$

We now assume $\beta \leq 2$ and classify all Ulrich partitions $(\alpha, 1, \gamma)$ and $(\alpha, 2, \gamma)$. Let us first assume $\beta = 1$.

Theorem 7.3.18. *For every $m \geq 0$, there is a unique Ulrich partition of type $(2, 1, k)$, where $k = (4^{m+1} - 1)/3$. Up to symmetry, any Ulrich partition of type $(\alpha, 1, \gamma)$ is of this form.*

Proof. We first rephrase our combinatorial problem. Given a partition

$$(a_1, \dots, a_\alpha|b|c_1, \dots, c_\gamma),$$

we first normalize $b = 0$ (so all c_j are negative). The entry a_i meets $b = 0$ at time $t = a_i$ and meets c_j at time $\frac{1}{2}(a_i - c_j)$, while c_j meets b at time $t = -c_j$. In order for the a's to meet the c's at integral times, they must all have the same parity. Furthermore, they must all be odd, for otherwise at time $t = 1$ no two entries would meet.

Let $A = \{a_1, \dots, a_\alpha\}$ and $C = \{-c_1, \dots, -c_\gamma\}$. Then the Ulrich condition says precisely that $A + C := \{a + c : a \in A, c \in C\}$ has size $\alpha\gamma$ and the set $[N]$ of times is a disjoint union

$$[N] = A \amalg C \amalg \tfrac{1}{2}(A + C).$$

Equivalently, by subtracting 1 from every element of A and C and putting $N' = N - 1$, we see that the existence of an Ulrich partition of type $(\alpha, 1, \gamma)$ is equivalent to the existence of subsets of even numbers $A', C' \subset [0, N']$ of sizes α and γ such that $[0, N']$ is a disjoint union

$$[0, N'] = A' \amalg C' \amalg \tfrac{1}{2}(A' + C').$$

The numerics force $A' + C'$ to have size $\alpha\gamma$. This decomposition is the problem we will study for the rest of the proof.

We now claim that if $(\alpha, 1, \gamma)$ is the type of an Ulrich partition then at least one of α, γ is 2. Suppose $A', C' \subset [0, N']$ correspond to such an Ulrich partition, and put $S' = \frac{1}{2}(A' + C')$. Since A' and C' are disjoint, we have $0 \notin S'$, and therefore either $0 \in A'$ or $0 \in C'$. By symmetry we may assume $0 \in A'$. We must have $1 \in S'$ since A', C' consist of even numbers, and this forces $2 \in C'$.

The number N' must be even since A' and C' consist of even numbers no larger than N'. Then N' must actually be in A' or C'. If $N' \in C'$ then, since $N' - 1$ must be in S', we have to have $N' - 2 \in A'$. This is a contradiction since $0 + N' = 2 + (N' - 2)$ gives a repetition in $A' + C'$. We conclude $N' \in A'$ and similarly $N' - 2 \in C'$.

We next show by induction that if k is an integer with $k \geq 0$ and $4k + 2 < N'$ then $4k + 2 \in C'$. We already know $2 \in C'$. Suppose $4k + 2 \in C'$ for all $k \in [0, k_0)$. Then $4k+2 \notin A'$ for all $k \in [0, k_0)$. Furthermore, $4k+4 \notin A'$ for all $k \in [0, k_0)$, for if $4k+4 \in A'$ then $(4k + 4) + (N' - 2) = N' + (4k + 2)$ is a repetition. Thus A' contains no numbers in the interval $(0, 4k_0]$. The odd number $2k_0 + 1$ must be in S', but, since every element of C' is ≥ 2 while the only nonzero elements of A' are at least $4k_0 + 2$, the only possibility is that $4k_0 + 2 \in C'$. By induction, C' must contain all numbers in $(0, N')$ which are $2 \pmod 4$. This argument also shows that $A' = \{0, N'\}$, i. e., $\alpha = 2$.

Suppose N' is not divisible by 4. Reversing the argument in the previous paragraph, we see that C' must consist precisely of all the even numbers in $(0, N')$. In particular, since $N' > 4$ we have $4 \in C'$, so $2 \in S'$; this contradicts that $2 \in C'$. Therefore 4 divides N'.

At this point we conclude that the sets $A', C' \subset [0, N']$ must have a recursive structure. Let $N'' = N'/4$,

$$A'' = \{0, N''\}, \quad C'' = \{c : 4c \in C'\}, \quad C''' = \{c \in [0, N'] : c \equiv 2 \pmod 4\},$$

so that $C' = 4C'' \cup C'''$. Then

$$C''' \cup \tfrac{1}{2}(A' + C''') = \{d \in [0, N'] : d \not\equiv 0 \pmod 4\};$$

furthermore, this union is disjoint and there is no repetition in $A' + C'''$. It then follows that the sets A', C' correspond to an Ulrich partition if and only if

$$[0, N''] = A'' \amalg C'' \amalg \tfrac{1}{2}(A'' + C'').$$

That is, either A'', C'' correspond to an Ulrich partition or C'' is empty, $N'' = 1$, and $A'' = \{0, 1\}$, in which case we originally had $A' = \{0, 4\}$ and $C' = \{2\}$.

Since C'' always has fewer elements than C', we conclude by induction. We may assume C'' has $(4^m - 1)/3$ elements for some $m \geq 1$. Then

$$N'' + 1 = |A''| + |C''| + |A''||C''| = 2 + \frac{(4^m - 1)}{3} + \frac{2(4^m - 1)}{3} = 4^m + 1,$$

so $N'' = 4^m$ and $N' = 4^{m+1}$. From this we similarly conclude $|C'| = (4^{m+1} - 1)/3$, completing the proof. □

Example 7.3.19. In this example we show how we recover the Ulrich partitions from this kind of construction. Let us find the Ulrich partition of type $(2, 1, 5)$ corresponding to $m = 1$. Then $N = 17$, so $N' = 16$, $A' = \{0, 16\}$, and $C' = \{2, 6, 8, 10, 14\}$. Correspondingly, the Ulrich partition is $(17, 1|0|-3, -7, -9, -11, -15)$. We give the time evolution diagram of this partition in Figure 7.2.

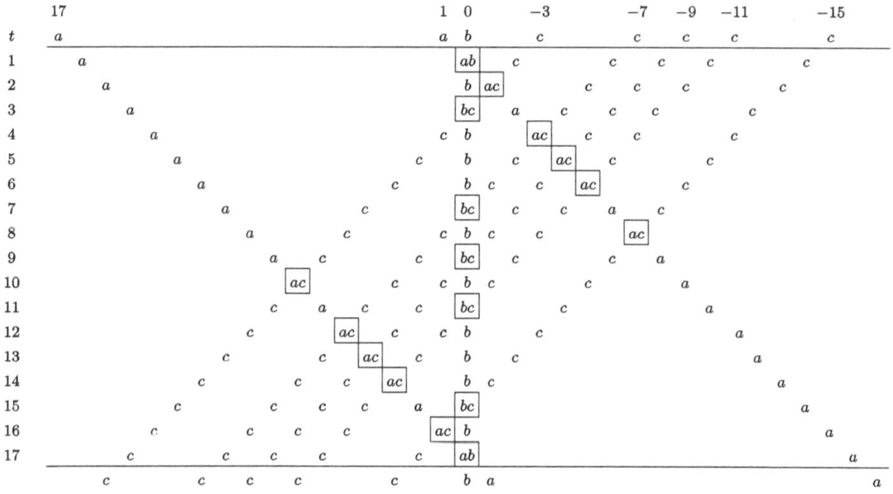

Figure 7.2: Time evolution diagram of the Ulrich partition of type $(2, 1, 5)$.

Next, we classify Ulrich partitions of type $(\alpha, 2, \gamma)$. We will see that there are many examples of such partitions, including a two-parameter infinite family and several sporadic examples. This will follow after a series of preliminary results and constructions as the greedy algorithm, the trapezoid rule and the rectangle rule. Let us start with an algorithm to construct 3-part Ulrich partitions.

The greedy algorithm

We normalize the evolution of partitions by subtracting $1, 0, -1$, respectively, from the three blocks. Fix a finite set $B \subset \mathbb{Z}$, which we view as being the list of entries in the

B-block of a hypothetical Ulrich partition. Let $A \subset \mathbb{Z}$ and $C \subset \mathbb{Z}$ be two finite sets such that $a > b > c$ holds for any $a \in A$, $b \in B$, $c \in C$. We write

$$A(t) = A - t, \qquad B(t) = B, \qquad C(t) = C + t,$$

where the right-hand side is interpreted as adding t or $-t$ at each entry. Therefore, $A(t)$ gives the set of positions of the A-entries of $(A|B|C)$ at time t.

Definition 7.3.20. The partition $(A|B|C)$ is *pre-Ulrich* if
(1) all elements of A and C have the same parity, and
(2) the set $T \subset \mathbb{N}_{>0}$ of times t where two of $A(t)$, $B(t)$, $C(t)$ have a common entry has size N, the dimension of $(A|B|C)$.

Remark 7.3.21. Write $(A'|B'|C') \subset (A|B|C)$ if $A' \subset A$, $B' \subset B$, and $C' \subset C$. Then if $(A|B|C)$ is pre-Ulrich, $(A'|B'|C')$ is pre-Ulrich. Clearly, Ulrich partitions are pre-Ulrich; the Ulrich condition requires that in addition $T = [N]$ is a consecutive range of integers.

Given any pre-Ulrich triple $(A|B|C)$, we introduce two operations for enlarging the A or C part of the triple.

Definition 7.3.22. Let $(A|B|C)$ be a pre-Ulrich partition. Let $t_0 \in \mathbb{N}_{>0}$ be the smallest time not in T.
(1) Put $a_0(t_0) := \max\{B(t_0) \cup C(t_0)\}$ and $a_0 := a_0(t_0) + t_0$. The triple $(A \cup \{a_0\}|B|C)$ is *obtained by adding a new a.*
(2) Put $c_0(t_0) := \min\{A(t_0) \cup B(t_0)\}$ and $c_0 := c_0(t_0) - t_0$. The triple $(A|B|C \cup \{c_0\})$ is *obtained by adding a new c.*

Remark 7.3.23. The triple obtained from a pre-Ulrich triple by adding a new a or new c is not necessarily pre-Ulrich, since the new entry may meet old entries at times after t_0 that already have intersections. See the proof of Lemma 7.3.27 for an example of this.

The next result shows that an Ulrich partition $(A|B|C)$ can always be obtained from $(\emptyset|B|\emptyset)$ by greedily adding a's and c's.

Proposition 7.3.24. *If $(A|B|C)$ is an Ulrich triple, then it can be obtained by repeatedly adding new a's and c's to the triple $(\emptyset|B|\emptyset)$. Furthermore, the sequence of a's and c's which are added is uniquely determined.*

Proof. The sequence of a's and c's to add to $(\emptyset|B|\emptyset)$ is readily determined by the pattern of intersections of the Ulrich partition. Let $(A'|B|C') \subset (A|B|C)$ be any triple. Since $(A|B|C)$ is Ulrich, $(A'|B|C')$ is pre-Ulrich. Let $t_0 \in \mathbb{N}_{>0}$ be the smallest time where $(A'|B|C')$ does not have an intersection, and consider the intersection occurring in $(A|B|C)$ at time t_0.

First suppose the intersection at time t_0 occurs between the A- and B-blocks, between an entry $a_0 \in A$ and $b_0 \in B$, so that $a_0(t_0) = b_0(t_0)$. Then $a_0 \notin A'$, since otherwise

$(A'|B|C')$ has an intersection at time t_0. No intersections between a_0 and an entry of B or C' have occurred before time t_0 since all these times already have intersections from $(A'|B|C')$. In order for the intersections between a_0 and the entries of B, C' to occur at time t_0 or later, we must have $a_0(t_0) \geq b(t_0)$ and $a_0(t_0) \geq c(t_0)$ for all $b \in B$ and $c \in C$, and we conclude $a_0(t_0) = \max(B(t_0) \cup C(t_0))$. The triple obtained from $(A'|B|C')$ by adding a new a is precisely $(A' \cup \{a_0\}|B|C') \subset (A|B|C)$.

If instead the intersection at time t_0 occurs between the B- and C-blocks, a symmetric argument shows that the triple $(A', B, C' \cup \{c_0\})$ obtained by adding a new c is contained in (A, B, C).

Finally, suppose the intersection at time t_0 occurs between the A- and C-blocks, and let $a_0 \in A$ and $c_0 \in C$ be such that $a_0(t_0) = c_0(t_0)$. By the choice of t_0, it can't be the case that both $a_0 \in A'$ and $c_0 \in C'$; we claim that in fact exactly one of $a_0 \in A'$ or $c_0 \in C'$ holds. Suppose $a_0 \notin A'$. As before, this implies that $a_0(t_0) \geq b(t_0)$ and $a_0(t_0) \geq c(t_0)$ for all $b \in B$ and $c \in C'$, so $a_0(t_0) = \max(B(t_0) \cup C(t_0))$. In fact, $a_0(t_0) > b(t_0)$ for all $b \in B$ since a_0 has not yet met the B-block at time t_0; fix some $b_0 \in B$. Since $c_0(t_0) = a_0(t_0) > b_0(t_0)$, the intersection between c_0 and b_0 must have occurred at a time before t_0. This implies $c_0 \in C'$. Thus if $a_0 \notin A'$, we conclude that the triple obtained from $(A'|B|C')$ by adding a new a is $(A' \cup \{a_0\}|B|C') \subset (A|B|C)$; similarly, if $c_0 \notin C'$, then the triple obtained from $(A'|B|C')$ by adding a new c is $(A'|B|C' \cup \{c_0\}) \subset (A|B|C)$.

Starting from $(\emptyset|B|\emptyset)$, we can now construct a chain

$$(\emptyset|B|\emptyset) \subset (A_1|B|C_1) \subset \cdots \subset (A_n|B|C_n) = (A|B|C)$$

of pre-Ulrich partitions where each triple differs from the previous one by adding an a or a c. Uniqueness is clear. □

Duality and the trapezoid rule

The greedy algorithm allows us to determine the structure of an Ulrich partition $(A|B|C)$ for early times t. Recall that the dual

$$(A|B|C)^* := (A^*|B^*|C^*) := \big(C(N+1), B(N+1), A(N+1)\big)$$

is also an Ulrich partition. This fact will allow us to determine the structure of an Ulrich partition at late (that is close to N) times. *The trapezoid rule* allows us to combine information about a partition and its dual to say something at times close to $N/2$. The next result explains the trapezoid rule.

Lemma 7.3.25. *Let $(A|B|C)$ be an Ulrich partition and $(A^*|B^*|C^*)$ its dual. If there exist $a \in A$, $a^* \in A^*$, $c \in C$, and $c^* \in C^*$ such that $a^* - a = c - c^*$, then $c(N+1) = a^*$ and $a(N+1) = c^*$. In particular, $N+1 = a^* - c = a - c^*$.*

Proof. The assumptions give that $c' := a^* - (N+1) \in C$ and $a' := c^* + (N+1) \in A$. Then a meets c' at time $\frac{1}{2}(a - c') = \frac{1}{2}(a - a^* + (N+1))$ and a' meets c at time $\frac{1}{2}(a' - c) =$

$\frac{1}{2}(c^* - c + (N+1))$. The hypothesis is that these times are equal; thus, since $(A|B|C)$ is Ulrich, we must have $a = a'$ and $c = c'$. □

The name of the trapezoid rule comes from the following geometric interpretation using time evolution diagrams. Suppose we can find a, a^*, c, c^* as in the trapezoid rule. View a and c as being entries of $(A|B|C)$ at time 0, and view a^* and c^* as being entries of $(C|B|A)$ at time $N+1$. The plane trapezoid with vertices at $(a, 0)$, $(c, 0)$, $(c^*, N+1)$, and $(a^*, N+1)$ is horizontally symmetric by assumption. The conclusion of the trapezoid rule is that this trapezoid has diagonals which meet orthogonally. See Figure 7.3.

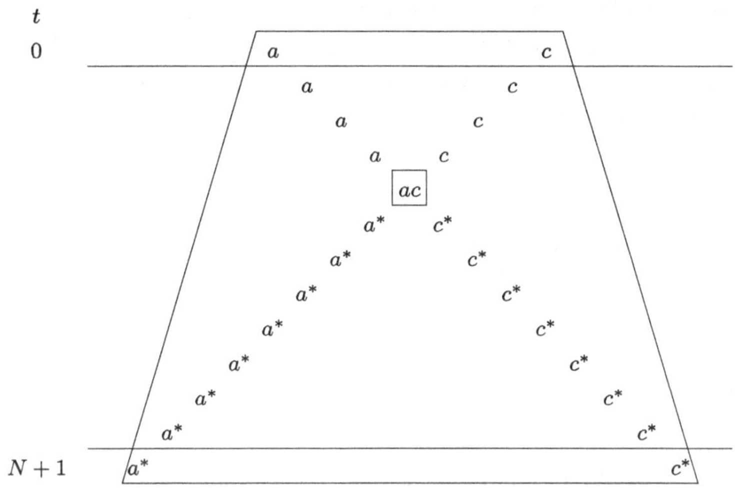

Figure 7.3: Graphical depiction of the trapezoid rule. If $(A|B|C)$ is Ulrich and $a \in A$, $a^* \in A^*$, $c \in C$, and $c^* \in C^*$ can be chosen such that the trapezoid displayed above is horizontally symmetric, then the diagonals meet orthogonally. Equivalently, $N+1 = a^* - c = a - c^*$.

Rectangle rule

This rule allows us to exclude many combinations of configurations of intersections for the early and late times and the next result describes how it works.

Lemma 7.3.26. *Let $(A|B|C)$ be Ulrich, and let $(A^*|B^*|C^*)$ be its dual. It is not possible that A and A^* share (at least) two entries. The same holds for C.*

Proof. The hypotheses imply that there are two entries in A with the same gap between them as two entries in C; there will necessarily be a multiple intersection at some time. □

In order to classify all Ulrich partitions $(A|b_1, b_2|C)$ of type $(\alpha, 2, \gamma)$, first of all we will restricts the value of the gap $b_1 - b_2$ in the B-block.

Lemma 7.3.27. *If* $(A|b_1, b_2|C)$ *is an Ulrich partition then* $b_1 - b_2 \in \{1, 3, 5\}$.

Proof. First, let us show that $b_1 - b_2$ is odd. By way of contradiction, suppose $b_1 = k$ and $b_2 = -k$. We consider what happens when we add new a's and c's to the pre-Ulrich triple $(\emptyset|k, -k|\emptyset)$. Without loss of generality, we first add a new a, giving the triple $(k+1|k, -k|\emptyset)$. This triple has intersections at times 1 and $2k+1$; in particular, there is no intersection at time 2. If we add a new a to this triple, we get $(k+2, k+1|k, -k|\emptyset)$, while if we add a new c, we get $(k+1|k, -k|-k-2)$. Neither triple is pre-Ulrich, since they both violate the parity requirement.

Next suppose that $b_1 - b_2$ is odd and ≥ 7. We consider adding a's and c's to $(\emptyset|k, -k-1|\emptyset)$. Without loss of generality, we first add an a at time 1 to get $(k+1|k, -k-1|\emptyset)$. The triple is no longer pre-Ulrich if we add an a at time 2, so we add a c at time 2 and get $(k+1|k, -k-1|-k-3)$. At times 3 and 4 (which do not have intersections yet since $k \geq 3$) the parity condition implies that we must add an a and then another c, yielding

$$(k+3, k+1|k, -k-1|-k-3, -k-5).$$

This partition is no longer pre-Ulrich since both the A- and C-blocks have entries which are 2 apart; there is a multiple intersection at time $k+3$. □

The classification of Ulrich partitions $(A|b_1, b_2|C)$ of type $(\alpha, 2, \gamma)$ changes considerably according to the value of $b_1 - b_2$. We will analyze each case separately.

Case $b_1 - b_2 = 5$

We fix $B = \{5, 0\}$. Let $(A|B|C)$ be an Ulrich partition, and assume that the intersection at time $t = 1$ occurs between the A- and B-blocks. In particular, $6 \in A$. After these choices, we prove that there is a unique such partition.

Example 7.3.28. The partition $(8, 6|5, 0|-2)$ is Ulrich. It is obtained from $(\emptyset|5, 0|\emptyset)$ by adding an a, then a c, then an a. See Figure 7.4.

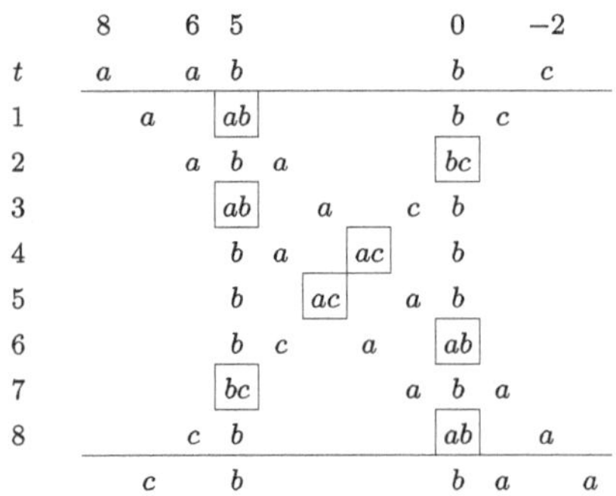

Figure 7.4: Time evolution diagram of the Ulrich partition $(8, 6|5, 0|-2)$ of type $(2, 2, 1)$.

Lemma 7.3.29. *The partition* $(A|B|C)$ *contains* $(8,6|5,0|-2)$.

Proof. Consider the sequence of a's and c's which must be added to $(\emptyset|5,0|\emptyset)$ to produce $(A|B|C)$. By assumption, at time $t = 1$ an a must be added, giving $(6|5,0|\emptyset)$. By parity, at time $t = 2$ a c is added, giving $(6|5,0|-2)$. Again by parity, at time $t = 3$ an a is added, yielding $(8,6|5,0|-2)$ as required. □

As a first application of duality and the trapezoid rule we get:

Proposition 7.3.30. *The partition* $(A|B|C)$ *equals* $(8,6|5,0|-2)$.

Proof. By Lemma 7.3.29, we have $(8,6|5,0|-2) \subset (A|5,0|C)$. The dual partition $(A^*|B^*|C^*)$ is Ulrich. Thus, by Lemma 7.3.29, it contains either $(8,6|5,0|-2)$ or the symmetric partition $(7|5,0|-1,-3)$, according to whether the time $t = 1$ intersection occurs between the first two or last two blocks. The first possibility is ruled out by the rectangle rule (Lemma 7.3.26). Likewise, if it contains $(7|5,0|-1,-3)$, then since $7-6 = -2-(-3)$ the trapezoid rule (Lemma 7.3.25) shows that $N = 8$. Since $(8,6|5,0|-2)$ has dimension 8, this implies $(A|B|C) = (8,6|5,0|-2)$. □

Case $b_1 - b_2 = 3$

The classification here is substantially more complicated than when $b_1 - b_2 = 5$. We normalize $B = \{3,0\}$, and assume the first intersection occurs between the A- and B-blocks, so $4 \in A$. There are three such examples of Ulrich partitions.

Example 7.3.31. The partition $(4|3,0|-2)$ is Ulrich. It is obtained from $(\emptyset|B|\emptyset)$ by adding an a and then a c. This is an example of a partition obtained from Theorem 7.3.16.

Example 7.3.32. The partition $(12,4|3,0|-2,-8)$ is Ulrich. It is obtained from $(\emptyset|3,0|\emptyset)$ by adding a, c, c, a. See Figure 7.5.

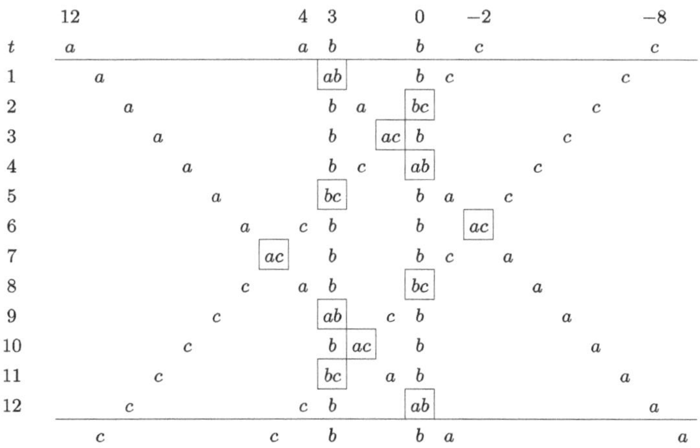

Figure 7.5: Time evolution diagram of the Ulrich partition $(12,4|3,0|-2,-8)$ of type $(2,2,2)$.

Example 7.3.33. The partition $(16, 10, 4|3, 0|-2, -12)$ is Ulrich. It is obtained from $(\emptyset|3, 0|\emptyset)$ by adding in order a, c, a, c, a. See Figure 7.6.

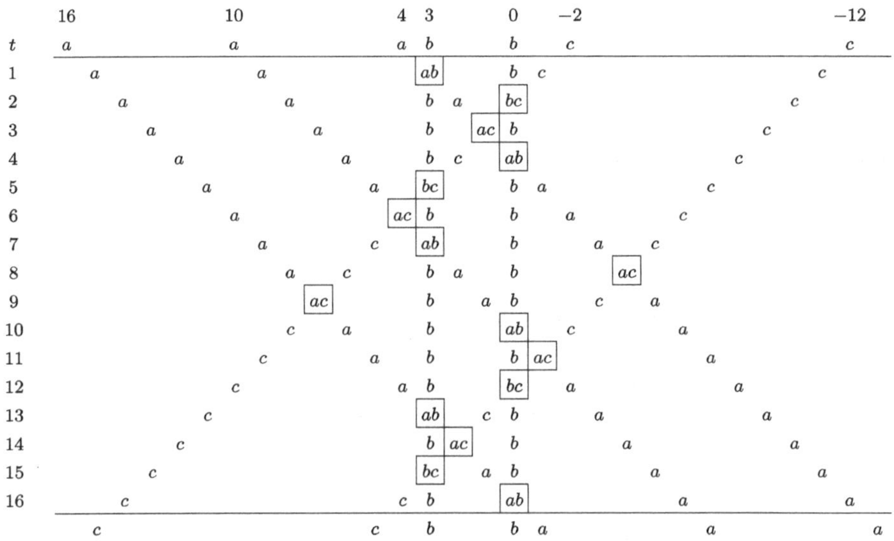

Figure 7.6: Time evolution diagram of the Ulrich partition $(16, 10, 4|3, 0|-2, -12)$ of type $(3, 2, 2)$.

By comparison with the $b_1 - b_2 = 5$ case, there is less forced structure to the early intersections for an Ulrich triple in this case. We let $\sigma \in \{a, c\}^*$ be the string of a's and c's which must be added to $(\emptyset|B|\emptyset)$ to yield the Ulrich triple $(A|B|C)$. We write $|\sigma|$ for the length of σ.

Lemma 7.3.34. *Assume that $(A|B|C)$ is not the type $(1, 2, 1)$ partition of Example 7.3.31 which means that $\sigma \neq ac$. Then σ begins with one of the strings*
1. *acaaa, so that* $(16, 14, 10, 4|3, 0|-2) \subset (A|B|C)$,
2. *acaca, so that* $(16, 10, 4|3, 0|-2, -12) \subset (A|B|C)$,
3. *acca, so that* $(12, 4|3, 0|-2, -8) \subset (A|B|C)$, *or*
4. *acccc, so that* $(4|3, 0|-2, -8, -10, -14) \subset (A|B|C)$.

Proof. By assumption the first letter of σ is a, and by parity the second letter must be c. Observe that the pre-Ulrich partitions $(10, 4|3, 0|-2)$ and $(4|3, 0|-2, -8)$ corresponding to the strings *aca* and *acc* are both not Ulrich, so $|\sigma| \geq 4$. To prove the lemma, we must show that unless σ begins with *acca* then $|\sigma| \geq 5$ and the fifth letter in σ is determined by the first four.

Suppose σ begins with *acaa*. This gives $(14, 10, 4|3, 0|-2) \subset (A|B|C)$. The first time where $(14, 10, 4|3, 0|-2)$ has no intersection is time $t = 9$, and this triple is not Ulrich so

$|\sigma| \geq 5$. Adding a c would yield $(14, 10, 4|3, 0|-2, -14)$, but this is not pre-Ulrich since there is a multiple intersection at time $t = 14$. Thus σ must begin with *acaaa*.

If instead σ begins with *acac*, then $(10, 4|3, 0|-2, -12) \subset (A|B|C)$, and this triple is not Ulrich so $|\sigma| \geq 5$. Adding a c would yield $(10, 4|3, 0|-2, -12, -14)$, which is not pre-Ulrich since there is a multiple intersection at time $t = 12$. So, σ begins with *acaca*.

Finally, suppose σ begins with *accc*, so $(4|3, 0|-2, -8, -10) \subset (A|B|C)$. This is not Ulrich, and adding an a gives $(16, 4|3, 0|-2, -8, -10)$. Considering time 13 shows this is not pre-Ulrich, so σ begins with *acccc*. □

We now treat each case of Lemma 7.3.34 in further detail. First of all, we will see that the Case (4) never arises.

Proposition 7.3.35. *If $(A|B|C)$ is an Ulrich partition normalized so that $B = \{3, 0\}$, then the word σ cannot begin with acccc.*

Proof. If σ begins with *acccc* then $(4|3, 0|-2, -8, -10, -14) \subset (A|B|C)$, so $N \geq 17$ since there is an intersection at time 17 in this subpartition. We first study the dual Ulrich partition $(A^*|B^*|C^*)$. By Lemma 7.3.34, this partition contains one of the following 8 partitions, of which the first 7 are easily ruled out:

(1) $(16, 14, 10, 4|3, 0|-2)$. In this case the equality $4 - 4 = -2 - (-2)$ and the trapezoid rule implies $N = 5$, which is absurd.
(2) $(5|3, 0|-1, -7, -11, -13)$. Here the equality $5 - 4 = -10 - (-11)$ and the trapezoid rule gives $N = 14$, a contradiction.
(3) $(16, 10, 4|3, 0|-2, -12)$. The same logic as in (1) applies.
(4) $(15, 5|3, 0|-1, -7, -13)$. This time the equality $15 - 4 = -2 - (-13)$ and the trapezoid rule gives $N = 16$.
(5) $(12, 4|3, 0|-2, -8)$. Same as (1).
(6) $(11, 5|3, 0|-1, -9)$. The equality $5 - 4 = -8 - (-9)$ and the trapezoid rule gives $N = 12$.
(7) $(4|3, 0|-2, -8, -10, -14)$. Same as (1).
(8) $(17, 13, 11, 5|3, 0|-1)$.

We conclude that $(A^*|B^*|C^*)$ must contain $(17, 13, 11, 5|3, 0|-1)$. Note that the sequence of a's and c's which must be added to $(\emptyset, B, \emptyset)$ to arrive at this subpartition is the sequence *caaaa*.

As a first observation, we find $|A| \geq 2$, for if $|A| = 1$ then $N = 4$. Thus there is some $k \geq 4$ such that σ begins with $ac^k a$. By way of contradiction, let $k \geq 4$ be minimal such that there is an Ulrich partition $(A|B|C)$ such that the corresponding word σ begins with $ac^k a$. It follows from minimality and symmetry that the word σ^* corresponding to the dual partition $(A^*|B^*|C^*)$ at least begins with ca^k.

We now compute the partition $(4|3, 0|C_k)$ $(k \geq 1)$ obtained from $(\emptyset|B|\emptyset)$ by adding the letters of the word ac^k. We label the elements of C_k in decreasing order as $C_k = \{c_1, \ldots, c_k\}$, so $C_{k+1} = C_k \cup \{c_{k+1}\}$. We have $c_1 = -2$. Let $T_k \subset \mathbb{N}_{>0}$ be the set of times which are *not* intersection times for the partition $(4|3, 0|C_k)$, so $T_1 = [6, \infty]$. We let

$t_{k+1} = \min T_k$. For $k \geq 1$, when a c is added to $(4|3,0|C_k)$ it is added at time t_{k+1} and meets the A-entry at that time, so $c_{k+1}(t_{k+1}) = 4 - t_{k+1}$ and $c_{k+1} = 4 - 2t_{k+1}$. Computing the sequence of sets C_k is therefore equivalent to computing the sequence of times $\{t_k\}_k$.

The computation of the sequence of times $\{t_k\}_k$ when new c's are added is best explained in terms of a sieve. Adding a new c at time t_{k+1} means we include $c_{k+1} = 4 - 2t_{k+1}$ in C_{k+1}. Then $c_{k+1}(2t_{k+1} - 4) = 0$ and $c_{k+1}(2t_{k+1} - 1) = 3$, so c_{k+1} meets the B-block at times $2t_{k+1} - 4$ and $2t_{k+1} - 1$. Therefore $T_{k+1} = T_k \setminus \{t_{k+1}, 2t_{k+1} - 4, 2t_{k+1} - 1\}$.

We now make this computation explicit. To get started, we have $t_2 = 6$, which sieves out times 8 and 11. We must then include $t_3 = 7$, which sieves times 10 and 13, etc. We include the result of this sieve computation for small times below; ×'s denote times which are sieved out.

6	7	8	9	10	11	12	13	14	15	16	17	18	19	20	21	22
t_2	t_3	×	t_4	×	×	t_5	×	×	t_6	t_7	×	t_8	t_9	×	t_{10}	t_{11}
23	24	25	26	27	28	29	30	31	32	33	34	35	36	37	38	39
×	t_{12}	t_{13}	×	t_{14}	×	×	t_{15}	×	×	t_{16}	×	×	t_{17}	×	×	t_{18}
40	41	42	43	44	45	46	47	48	49	50	51	52	53	54	55	56
×	×	t_{19}	×	×	t_{20}	×	×	t_{21}	×	×	t_{22}	t_{23}	×	t_{24}	t_{25}	×
57	58	59	⋯	95	96	97	98	99	100							
t_{26}	t_{27}	×	⋯	×	t_{52}	t_{53}	×	t_{54}	×							

The result of the computation is easy to describe. First, every time $t \geq 6$ with $t \equiv 0 \pmod 3$ appears in $\{t_k\}$. No times with $t \equiv 2 \pmod 3$ appear. The times congruent to $1 \pmod 3$ may or may not appear, but the pattern with which they appear is simple. The first time (7) congruent to 1 appears, the next 2 times $(10, 13)$ do not, the next 4 times $(16, 19, 22, 25)$ do appear, the next 8 do not, the next 16 do, etc. This description is straightforward to prove, and completely specifies the sequence $\{t_k\}$.

Now suppose that an a, call it a_2, is added to $(4|3,0|C_k)$. Unless k is very special, it turns out that the resulting partition is not pre-Ulrich. We have $a_2(t_{k+1}) = c_1(t_{k+1})$, so $a_2 = 2t_{k+1} - 2$. Since $\{c_1, c_2, c_3\} = \{-2, -8, -10\}$, we find that a_2 will also meet the C-block at times $t_{k+1} + 3$ and $t_{k+1} + 4$. For the resulting partition to be pre-Ulrich, it is therefore necessary that these times are not sieved out. Additionally, if $t_{k+2} = t_{k+1} + 1$ then no intersection with a_2 will occur at time t_{k+2}, so it is necessary to add another a or c at this time; it must be a c that is added, for otherwise we will have two a's which are 2 apart from one another, a contradiction since c_2 and c_3 are already 2 apart from one another. But then the c which is added at time t_{k+2}, call it c_{k+1}, will satisfy $c_{k+1} = 4 - 2t_{k+2} = 2 - 2t_{k+1}$. Then a_2, $0 \in B$, and c_{k+1} all coincide at time $2t_{k+1} - 2$. We finally conclude that the times $t_{k+1} + 1$ must be sieved out and $t_{k+1} + 3$ and $t_{k+1} + 4$

cannot be sieved out in order for the partition to have a chance of being pre-Ulrich. Inspecting the sieve, the only way for this to occur is for t_{k+1} to be one of the times immediately before the t_k's start occurring in pairs, e. g., $t_5 = 12$, $t_{21} = 48$, or $t_{85} = 192$. A straightforward computation shows that these times are precisely the numbers t_{k+1} of the form $3 \cdot 4^m$ with $m \geq 1$, and $k + 1 = \sum_{i=0}^{m} 4^i$.

Finally, suppose $t_{k+1} = 3 \cdot 4^m$ and $a_2 = 2t_{k+1} - 2$. We use the trapezoid rule to show that $(a_2, 4|3, 0|C_k)$ is not contained in any Ulrich partition $(A|B|C)$. Observe that a c was added at time $\frac{1}{2}t_{k+1} + 1$; this time is the last time that was added in the block of pairs of times preceding t_{k+1}. Thus $4 - 2(\frac{1}{2}t_{k+1} + 1) = 2 - t_{k+1} \in C$. Since the dual partition $(A^*|B^*|C^*)$ has corresponding word σ^* beginning with ca^k we symmetrically have $-(2 - t_{k+1}) + 3 = t_{k+1} + 1 \in A^*$ and $-1 \in C^*$. The equality

$$(t_{k+1} + 1) - (2t_{k+1} - 2) = (2 - t_{k+1}) - (-1)$$

and the trapezoid rule prove $a_2(N + 1) = -1$, from which it follows that $|A| = 2$. We also have $N = a_2 = 2t_{k+1} - 2$. At time N we have $a_2(N) = 0 \in B$, so at time $N - 1$ the smallest entry $c_y \in C$ has $c_y(N - 1) = 3$. Then $c_y = 3 - (N - 1) = 4 - 2(t_{k+1} - 1)$. This means that c_y had to be added at time $t_{k+1} - 1$, contradicting that this time was sieved out since it is congruent to 2 (mod 3). □

In the next result, we will see that Case (1) of Lemma 7.3.34 never arises.

Proposition 7.3.36. *If $(A|B|C)$ is an Ulrich partition normalized so that $B = \{3, 0\}$, then the word σ cannot begin with acaaa.*

Proof. It is analogous to the proof of Proposition 7.3.35 and we omit it. □

Proposition 7.3.37. *If σ begins with acaca then σ = acaca and $(A|B|C)$ is the $(3, 2, 2)$ partition of Example 7.3.33.*

Proof. Since σ begins with *acaca*, $(A|B|C)$ contains $(16, 10, 4|3, 0|-2, -12)$, and therefore $N \geq 16$ with equality if and only if σ = *acaca*. By duality, the partition $(A^*|B^*|C^*)$ is Ulrich. By Lemma 7.3.34 and Propositions 7.3.35 and 7.3.36, it contains one of the following partitions:
(1) $(16, 10, 4|3, 0|-2, -12)$; this is impossible by the rectangle rule.
(2) $(15, 5|3, 0|-1, -7, -13)$; in this case, the equality $15 - 10 = -2 - (-7)$ and the trapezoid rule give $N = 16$, so, in fact, σ = *acaca*.
(3) $(12, 4|3, 0|-2, -8)$; the equality $4 - 4 = -2 - (-2)$ and the trapezoid rule give $N = 5$, a contradiction.
(4) $(11, 5|3, 0|-1, -9)$; the equality $11 - 4 = -2 - (-9)$ and the trapezoid rule again gives $N = 12$.

Thus the only possibility is that σ = *acaca*. □

The following result completes the classification when $b_1 - b_2 = 3$.

Proposition 7.3.38. *If σ begins with acca then $\sigma = acca$ and $(A|B|C)$ is the $(2,2,2)$ partition of Example 7.3.32.*

Proof. The Ulrich partition $(A|B|C)$ contains $(12,4|3,0|-2,-8)$, so $N \geq 12$ with equality if and only if $\sigma = acca$. As in the proof of Proposition 7.3.37, the dual $(A^*|B^*|C^*)$ contains one of the following partitions:

(1) $(16,10,4|3,0|-2,-12)$; this is impossible by the trapezoid rule applied to the equality $4-4 = -2-(-2)$.

(2) $(15,5|3,0|-1,-7,-13)$; the equality $5-12 = -8-(-1)$ and the trapezoid rule gives $N = 12$ so $\sigma = acca$ (in fact, this is also a contradiction, since the dual partition has the wrong structure.)

(3) $(12,4|3,0|-2,-8)$; this contradicts the rectangle rule.

(4) $(11,5|3,0|-1,-9)$; the equality $11-4 = -2-(-9)$ and the trapezoid rule give $N = 12$, so $\sigma = acca$.

We conclude that $\sigma = acca$. □

Case $b_1 - b_2 = 1$

In this final case we normalize $B = \{1,0\}$. We first classify an infinite family of examples which serve as primitive building blocks for a further family of examples.

Proposition 7.3.39. *Let $(2|1,0|C_k)$ be the partition obtained from $(2|1,0|\emptyset)$ by adding k c's. This partition is Ulrich if and only if k is a number of the form $k = \sum_{i=0}^{m} 4^i = \frac{1}{3}(4^{m+1}-1)$ for some $m \geq 0$.*

Proof. Write $C_k = \{c_1,\ldots,c_k\}$ with increasing entries. We prove the result by induction on k; clearly, $(2|1,0|C_1) = (2|1,0|-4)$ is Ulrich. Fix some $m_0 \geq 0$ and let $k_0 = \frac{1}{3}(4^{m_0+1}-1)$, and assume $(2|1,0|C_{k_0})$ is Ulrich. The dimension of the flag variety corresponding to the type $(1,2,k_0)$ is $4^{m_0+1}+1$, so the set of times where $(2|1,0|C_{k_0})$ has an intersection is $[1,4^{m_0+1}+1]$.

At time $4^{m_0+1}+2$ the a entry is at position -4^{m_0+1}, so adding a new c to $(2|1,0|C_{k_0})$ gives $c_{k_0+1}(4^{m_0+1}+2) = -4^{m_0+1}$, i.e., $c_{k_0+1} = -2\cdot 4^{m_0+1}-2$. This entry intersects the B-block at times $2\cdot 4^{m_0+1}+2$ and $2\cdot 4^{m_0+1}+3$. Continuing inductively, if ℓ c's are added to $(2|1,0|C_{k_0})$ and $\ell \leq 4^{m_0+1}$, then $c_{k_0+\ell}(4^{m_0+1}+1+\ell) = -4^{m_0+1}+1-\ell$ so $c_{k_0+\ell} = -2\cdot 4^{m_0+1}-2\ell$, which meets the B-block at times $2\cdot 4^{m_0+1}+2\ell$ and $2\cdot 4^{m_0+1}+1+2\ell$. Then if $\ell < 4^{m_0+1}$, the partition $(2|1,0|C_{k_0+\ell})$ is not Ulrich since there is no intersection at time $2\cdot 4^{m_0+1}+1$. When $\ell = 4^{m_0+1}$, there are intersections between some $c_{k_0+\ell}$ and the a for all $t \in [4^{m_0+1}+2, 2\cdot 4^{m_0+1}+1]$ and between some $c_{k_0+\ell}$ and the B-block for all $t \in [2\cdot 4^{m_0+1}+2, 4^{m_0+2}+1]$. Therefore $(2|1,0|C_{k_0})$ is Ulrich. □

Remark 7.3.40. The sets C_k (equivalently, the numbers c_k) which appear in the proof of Proposition 7.3.39 are computable. Here are the first several terms:

$$-4, \quad -10,-12,-14,-16, \quad -34,-36,\ldots,-64, \quad -130,-132,\ldots,-256, \quad \ldots$$

A decreasing block of 4^k even integers ending in 4^{k+1} is followed by a gap of size $4^{k+1}+2$. In Figure 7.7 we give the time evolution diagram for the partition $(2|1,0|C_5)$ of type $(1,2,5)$.

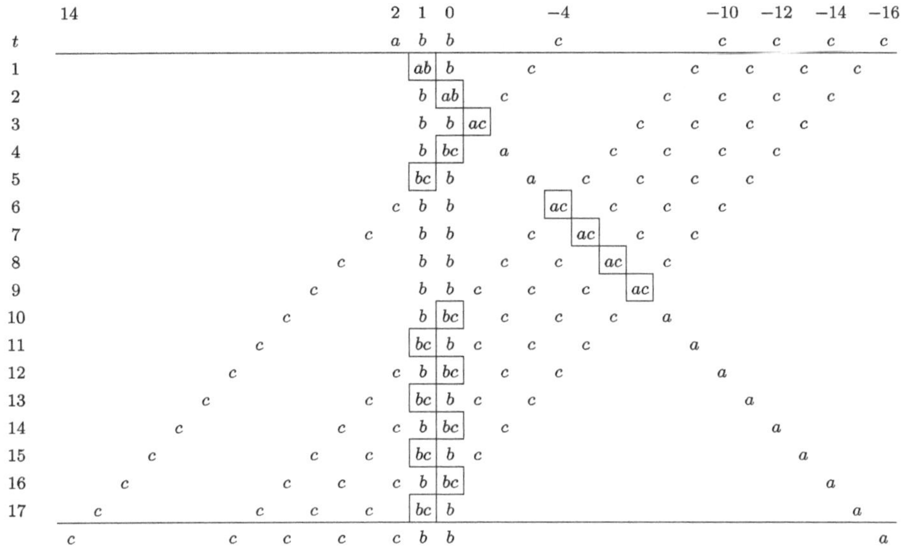

Figure 7.7: Time evolution diagram of the Ulrich partition $(2|1,0|C_5)$ of type $(1,2,5)$.

A careful analysis proves the following result.

Lemma 7.3.41. *Suppose $(A|1,0|C)$ is an Ulrich partition and that the word $\sigma \in \{a,c\}^*$ generating it from $(\emptyset|B|\emptyset)$ begins with acc. Then $|A| = 1$ and $2 \in A$, so $(A|B|C)$ is one of the examples of Proposition 7.3.39.*

Proof. We only sketch the proof since it is similar to other arguments already used. Consider the partition obtained from $(2|1,0|C_k)$ by adding an a, with $k \geq 2$. It is easy to show that, unless k is of the form $k = (4^{m+1} - 1)/3$ (so $m \geq 1$ and the original partition is Ulrich), the resulting partition is not even pre-Ulrich. On the other hand, if k is of this form, then adding a new a_2 at time $t_0 = 4^{m+1} + 2$ will give $a_2 = 2 \cdot 4^{m+1}$. One shows that we will be forced to add new c's at positions $-3 \cdot 4^{m+1}$ and -4^{m+2}. Then the c at position $-3 \cdot 4^{m+1}$ meets $0 \in B$ at the same time as a_2 meets the c at position -4^{m+2}, a contradiction. □

Proposition 7.3.39 and Lemma 7.3.41 allow us to focus on classifying Ulrich partitions $(A|1,0|C)$ with $2 \in A$ and $|A| \geq 2$, since they completely classify Ulrich partitions where $A = \{2\}$.

Lemma 7.3.42. *Let $(A|1,0|C)$ be an Ulrich partition and assume that $2 \in A$. Let $c_1 = \max C$, so that c_1 meets the B-block at time $t_0 := -c_1$. Let $A' \subset A$ be those a's which have met the B-block before time t_0. Then the partition $(A'|1,0|c_1) \subset (A|1,0|C)$ is Ulrich of dimension t_0+1, so the dual $(A'|1,0|c_1)^*$ is one of the Ulrich partitions of Proposition 7.3.39.*

Proof. Write $A = \{a_\alpha, \dots, a_1\}$ and $C = \{c_1, \dots, c_\gamma\}$ in decreasing order, and consider how $(A|1,0|C)$ is built from $(\emptyset|1,0|\emptyset)$ by adding a's and c's; let $\sigma \in \{a,c\}^*$ be the corresponding word. Since the time $t = 1$ intersection is between A and B, σ begins with an a. We can then write $\sigma = a^k c\sigma'$ for some $k \geq 1$ and some word σ'. Then A contains the first k even integers $\{2, \dots, 2k\}$ and the intersections at times $t \in [1, 2k-1]$ occur between the A- and B-blocks. At time $2k$ the entry a_1 is at position $2 - 2k$, so $c_1(2k) = -2k + 2$ and $c_1 = -4k + 2$.

Suppose c_2 meets a_1 at time t_1. We claim $t_1 > t_0$. Indeed, first notice that c_1 meets a_i $(1 \leq i \leq k)$ at time $2k-1+i$, so $t_0 \geq 3k$. Since c_2 meets a_1 at time t_1, it meets a_i $(1 \leq i \leq k)$ at time $t_1 - 1 + i$. But then assuming $3k \leq t_1 \leq t_0 = 4k - 2$, we find that c_2 meets some a_i at the same time as c_1 meets $0 \in B$. We conclude $t_1 > t_0$.

Therefore, if σ contains at least 2 c's, then the second c is added after time t_0. It follows that if $\sigma' = a^k c a^\ell$, where ℓ is the additional number of ℓ's which are added before time t_0, then σ' is an initial segment of σ, the corresponding subpartition is $(A'|1,0|c_1)$, and this partition is Ulrich. The last intersection in this partition occurs between c_1 and $1 \in B$, so its dual is $(2|1,0|A'^*)$ and Proposition 7.3.39 applies. □

In the next example, we will introduce a two-parameter family of Ulrich partitions.

Example 7.3.43. Let $m_1, m_2 \geq 0$, and let $k_i = \frac{1}{3}(4^{m_i+1} - 1)$. Let $(2|1,0|C_{k_1})$ be the Ulrich partition of Proposition 7.3.39, and let $(A_{k_2}|1,0|-1)$ be the partition symmetric to $(2|1,0|C_{k_2})$. There is an Ulrich partition $(A|B|C)$ uniquely specified by the requirements that the type is $(k_1 + k_2, 2, 1)$ and

$$(2|1,0|C_{k_1})^* \subset (A|B|C),$$
$$(A_{k_2}|1,0|-1)^* \subset (A|B|C)^*.$$

The dimension of the type $(k_1 + k_2, 2, 1)$ is $N = 4^{m_1+1} + 4^{m_2+1}$. Let $t_0 = 4^{m_1+1}$ be as in Lemma 7.3.42. For times $t \in [1, t_0]$, the pattern of intersections in $(A|B|C)$ is the same as that of $(2|1,0|C_{k_1})^*$. Applying Lemma 7.3.42 to the dual $(A|B|C)^*$, the corresponding time is $t_0^* = 4^{m_2+1}$, and the pattern of intersections in $(A|B|C)$ for times $t \in [t_0 + 1, N]$ is the same as that of $(A_{k_2}|1,0|-1)$ for times $t \in [1, t_0^*]$. Thus every time $t \in [1, N]$ has an intersection.

Observe that if $m_1 \neq m_2$ then the partitions corresponding to (m_1, m_2) and (m_2, m_1) are distinct but related by the symmetric dual. The partition corresponding to (m_1, m_1) is its own symmetric dual.

For example, consider the case $m_1 = 0$ and $m_2 = 1$. Then we compute

$$\begin{aligned} N &= 20, \\ C_{k_1} &= \{-4\}, \\ A_{k_2} &= \{17, 15, 13, 11, 5\}, \\ (2|1,0|C_{k_1})^* &= (2|1,0|-4), \\ (A_{k_2}|1,0|-1)^* &= (17|1,0|-1,-3,-5,-7,-13), \\ (A|B|C) &= (20,18,16,14,8,2|1,0|-4). \end{aligned}$$

See Figure 7.8 for the time evolution diagram. Note that the examples of Proposition 7.3.39 can be regarded as (duals of) the degenerate case where $m_2 = -1$.

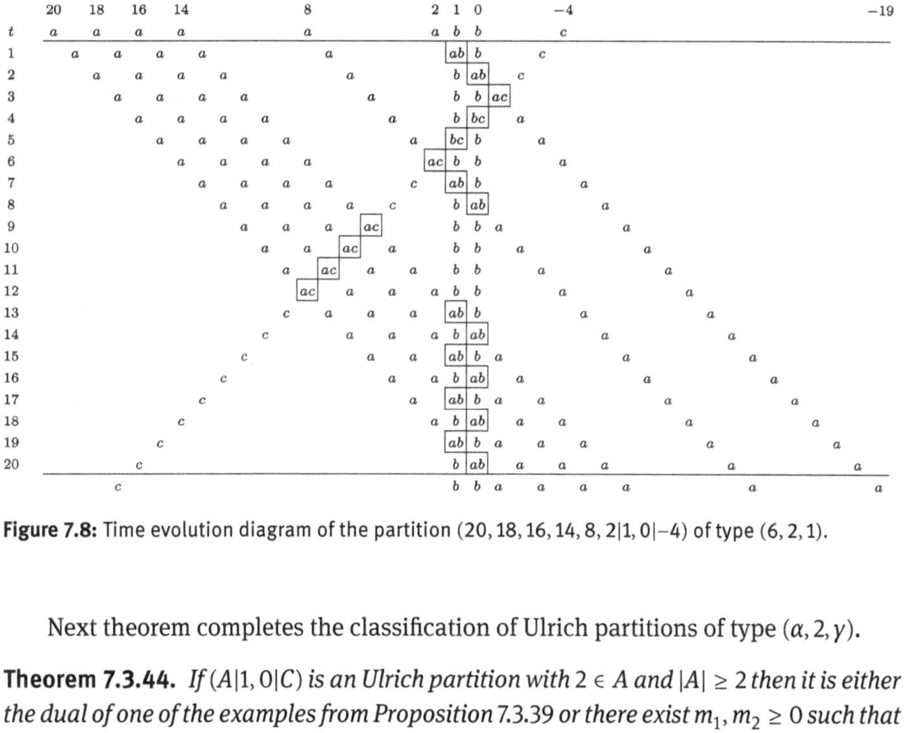

Figure 7.8: Time evolution diagram of the partition $(20, 18, 16, 14, 8, 2|1, 0|-4)$ of type $(6, 2, 1)$.

Next theorem completes the classification of Ulrich partitions of type $(\alpha, 2, \gamma)$.

Theorem 7.3.44. *If $(A|1,0|C)$ is an Ulrich partition with $2 \in A$ and $|A| \geq 2$ then it is either the dual of one of the examples from Proposition 7.3.39 or there exist $m_1, m_2 \geq 0$ such that $(A|1,0|C)$ is the partition corresponding to (m_1, m_2) in Example 7.3.43.*

Proof. Applying Lemma 7.3.42 to $(A|B|C)$ and its dual, we find that there are $m_1, m_2 \geq 0$ and $k_i = \frac{1}{3}(4^{m_i+1} - 1)$ such that $(2|1,0|C_{k_1})^* \subset (A|B|C)$ and either $(2|1,0|C_{k_2})^* \subset (A|B|C)^*$ or $(A_{k_2}|1,0|-1)^* \subset (A|B|C)^*$, according to whether $2 \in A^*$ or $-1 \in C^*$. Without loss of generality, assume $k_1 \leq k_2$.

If $2 \in A^*$ then $(2|1,0|C_{k_2})^* \subset (A|B|C)^*$. Computing the dual of the Ulrich partition $(2|1,0|C_{k_2})$ of dimension $4^{m_2+1} + 1$, we find $-4^{m_2+1} \in C^*$ and $4^{m_2+1} - 2 \in A^*$ (since

$-4 \in C_{k_2}$). Using the containment $(2|1,0|C_{k_1})^* \subset (A|B|C)$, we also have that $-4^{m_1+1} \in C$ and $4^{m_1+1} - 2 \in A$. Then the equality

$$(4^{m_2+1} - 2) - (4^{m_1+1} - 2) = (-4^{m_1+1}) - (-4^{m_2+1})$$

and the trapezoid rule give that $N = 4^{m_1+1} + 4^{m_2+1} - 1$ and $|C| = 1$. This is impossible: the dimension of the type $(k_1 + k_2, 2, 1)$ is $4^{m_1+1} + 4^{m_2+1}$, so no type $(k_3, 2, 1)$ can have dimension $4^{m_1+1} + 4^{m_2+1} - 1$ by considering congruences mod 3. Therefore it must be the case that $-1 \in C^*$.

We now know that $(2|1,0|C_{k_1})^* \subset (A|B|C)$ and $(A_{k_2}|1,0|-1)^* \subset (A|B|C)^*$. Assume both of these containments are proper, since otherwise we are in the case of Proposition 7.3.39. Think of building $(A|B|C)$ from $(\emptyset|B|\emptyset)$ by adding a's and c's. After the word corresponding to $(2|1,0|C_{k_1})^*$ has been added, we must either add an a or a c.

Case 1. Suppose the next letter which is added is an a. We claim that the trapezoid rule implies that $(A|B|C)$ is the partition of Example 7.3.43 corresponding to the integers m_1, m_2. The Ulrich subpartition $(2|1,0|C_{k_1})^*$ has dimension $4^{m_1+1} + 1$, so the new a is added at time $4^{m_1+1} + 2$. The element of C arising from the inclusion $(2|1,0|C_{k_1})^* \subset (A|B|C)$ is $-4^{m_1+1} \in C$, so the new a is at position 2 at time $4^{m_1+1}+2$, and thus $4^{m_1+1}+4 \in A$. On the other hand, the inclusion $(A_{k_2}|1,0|-1)^* \subset (A|B|C)^*$ gives $4^{m_2+1} + 1 \in A^*$ and $5 - (4^{m_2+1} + 2) = -4^{m_2+1} + 3 \in C^*$. We have

$$(4^{m_2+1} + 1) - (4^{m_1+1} + 4) = (-4^{m_1+1}) - (-4^{m_2+1} + 3),$$

so by the trapezoid rule $N = 4^{m_1+1} + 4^{m_2+1}$ and $|C| = 1$. Therefore $(A|B|C)$ is the Example 7.3.43.

Case 2. Suppose the next letter which is added is a c. We claim that this is impossible: there are no such Ulrich partitions. We focus solely on the Ulrich partition $(A|B|C) := (2|1,0|C_{k_1})^*$, as the contradiction arises from this initial segment and not from "global" considerations given by the trapezoid rule. If $m_1 = 0$ then Lemma 7.3.41 gives $|A| = 1$, a contradiction, so we assume $m_1 \geq 1$. Let $(A|B|C \cup \{c_2\})$ be the partition obtained by adding c_2 at time $t_0 := 4^{m_1+1} + 2$. Writing $A = \{a_1, \ldots, a_{k_1}\}$ in increasing order, we have $c_2(t_0) = a_1(t_0) = -4^{m_1+1}$ since $a_1 = 2$. For $1 \leq i \leq 4^{m_1}$ we have $a_i = 2i$, so c_2 meets the A-block for all times $t \in [t_0, t_0 + 4^{m_1} - 1]$.

At time $t_0 + 4^{m_1}$ there is no intersection yet. It is not possible to add a new c at this time. If we were to add some c_3 at time $t_0 + 4^{m_1}$ then this would provide intersections between c_3 and the A-block for the next 4^{m_1} times. However, $a_{4^{m_1}+1} = a_{4^{m_1}} + 4^{m_1} + 2$ meets c_2 at time $t_0 + 4^{m_1} + 2 \cdot 4^{m_1-1}$, which is a time excluded by the intersection of c_3 with the A-block. Thus, at time $t_0 + 4^{m_1}$ we must add a new a, call it a_{k_1+1}. It has

$$a_{k_1+1}(t_0 + 4^{m_1}) = c_1(t_0 + 4^{m_1})$$

so

$$a_{k_1+1} = -4^{m_1+1} + 2(t_0 + 4^{m_1}) = t_0 + 2 \cdot 4^{m_1} + 2.$$

By the same argument as in the last paragraph, no new c's can be added before the time t_1 when c_2 and a_{k_1} meet. At each time in $[t_0, t_1]$ where c_2 does not meet the A-block, a new a *must* be added. This statement amounts to the claim that a_{k_1+1} does not meet the B-block before time t_1. Since $a_{k_1}(t_0) = -4$ and $c_2(t_0) = -4^{m_1+1}$, we have $t_1 = t_0 + 2 \cdot 4^{m_1} - 2$. Thus a_{k_1+1} meets the B-block at times $t_1 + 3$ and $t_1 + 4$.

Finally, we obtain our contradiction at time $t_1 + 1$. No intersection has been scheduled yet. We cannot add some c_3 at this time, since it would still meet the A-block at time $t_1 + 3$ when a_{k_1+1} meets $1 \in B$. Thus we must add an a, call it a', at time $t_1 + 1$. We have $a'(t_1 + 1) = c_1(t_1 + 1)$, so

$$a'(t_0) = a'(t_1 + 1) + (t_1 + 1 - t_0) = -4^{m_1+1} + 2(t_1 + 1) - t_0 = 4^{m_1+1}$$

while $c_2(t_0) = -4^{m_1+1}$. Thus, at time $t_0 + 4^{m_1+1}$, all three of a', $0 \in B$, and c_2 coincide. Therefore, the partition $(A|B|C \cup \{c_2\})$ cannot be extended to an Ulrich partition by adding a's and c's. □

Summarizing, we have proved the following result.

Theorem 7.3.45. *Given integers $m_1, m_2 \geq 0$ denote by $k_i = \frac{1}{3}(4^{m_i+1} - 1)$. There are Ulrich partitions $(A|b_1, b_2|C)$ of the following types:*
(1) $(2, 2, 2)$,
(2) $(2, 2, 3)$,
(3) $(1, 2, k_1)$ *for any* $m_1 \geq 0$, *and*
(4) $(1, 2, k_1 + k_2)$ *for any* $m_1, m_2 \geq 0$.

Unless the partition is of type $(1, 2, 2)$ *or* $(2, 2, 2)$, *it is unique up to taking the symmetric dual. Up to symmetry, any Ulrich partition of type* $(\alpha, 2, \gamma)$ *is one of the above examples.*

Proof. It follows from Lemma 7.3.27 and the case by case analysis done afterwards according to the value of the gap $b_1 - b_2$. □

We next turn our attention to cases where α and γ are small and β is arbitrary. We first classify all Ulrich partitions of type $(1, n, 2)$.

Example 7.3.46. There is a fundamental example of Ulrich partitions of type $(1, n, 2)$ given by the partition

$$(a|b_1, \ldots, b_n|c_1, c_2) = (2n + 1|n - 1, n - 2, \ldots, 1, 0|-1, -2n - 3).$$

We have:
(1) for times $t \in [1, n]$, the c_1 entry meets the B-block,
(2) at time $t = n + 1$, a meets c_1,
(3) for times $t \in [n + 2, 2n + 1]$, a meets the B-block,
(4) when $t = 2n + 2$, a meets c_2, and
(5) for $t \in [2n + 3, 3n + 2]$, c_2 meets the B-block.

Therefore the partition is Ulrich. In Figure 7.9 we present the time evolution diagram of the Ulrich partition $(7|2,1,0|-1,-9)$ of type $(1,3,2)$.

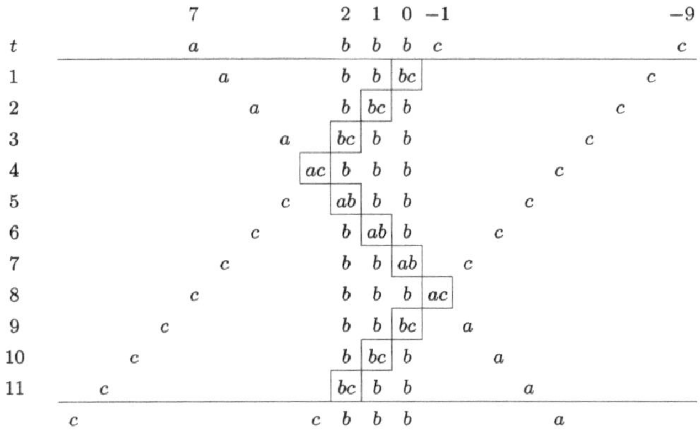

Figure 7.9: Time evolution diagram of the Ulrich partition $(7|2,1,0|-1,-9)$ of type $(1,3,2)$.

To classify Ulrich partition of type $(1,n,2)$ or, equivalently, of type $(2,n,1)$ we have to fix some notation. We consider Ulrich partitions of the form $P=(a_1,a_2|b_1,\dots,b_n|c)$, and normalize the evolution of the partition to subtract $1,0,-1$ from the blocks. We further normalize the positions

$$a_2=y,\quad a_1=y+2m,\quad \text{and}\quad c=-y,$$

so that the intersection a_2c happens at time y and the gap between the a-entries is $2m$.

Definition 7.3.47. For any $m\geq 2$, the *fundamental pattern* F_m *of type* m is the partition of type $(2,m-1,1)$ given by

$$F_m=(3m,m|m-1,m-2,\dots,2,1|-m).$$

The fundamental pattern is Ulrich by Example 7.3.46.

Definition 7.3.48. The *elongation* of a partition

$$P=(a_1,a_2|b_1,\dots,b_n|c)=(y+2m,y|b_1,\dots,b_n|-y)$$

of type $(2,n,1)$ is the partition $E(P)$ of type $(2,n+2m,1)$ obtained by adding two contiguous blocks of length m at the beginning and end of the b sequence and shifting the a and c entries as follows:

$$(y+5m,y+3m|y+3m-1,\dots,y+2m,b_1,\dots,b_n,-y-m,\dots,-y-2m+1|-y-3m).$$

The kth elongation of P is defined inductively by

$$E^k(P) := E(E^{k-1}(P))$$

and $E^0(P) = P$. Notice that it has type $(2, n + 2mk, 1)$.

Example 7.3.49. The fundamental pattern of type 2 is the partition $F_2 = (6, 2|1|-2)$. Its first and second elongations are

$$E(F_2) = (12, 8|7, 6, 1, -4, -5|-8) \quad \text{and}$$
$$E^2(F_2) = (18, 14|13, 12, 7, 6, 1, -4, -5, -10, -11|-14).$$

The fundamental pattern of type 3 is the partition $F_3 = (9, 3|2, 1|-3)$. Its first elongation is

$$E(F_3) = (18, 12|11, 10, 9, 2, 1, -6, -7, -8|-12).$$

See Figure 7.10 for the time evolution diagram of $E(F_2)$.

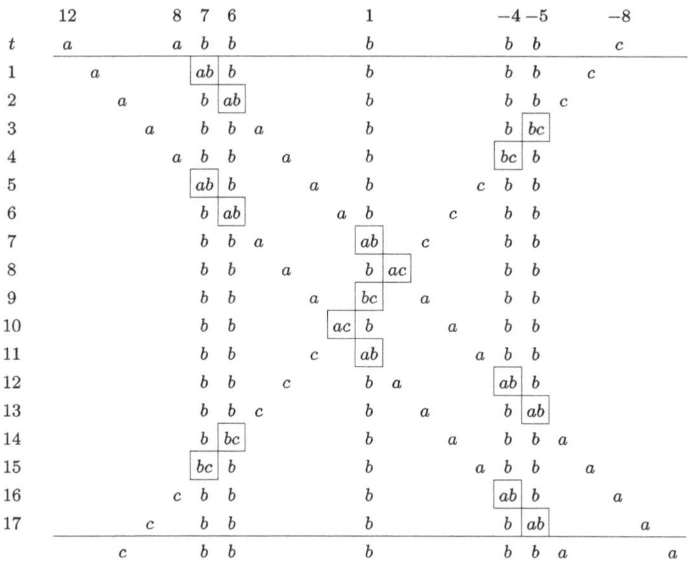

Figure 7.10: Time evolution diagram of the partition $E(F_2)$ of type $(2, 5, 1)$.

Remark 7.3.50. We will also need a degenerate case of the previous definitions. We define the fundamental pattern F_1 to be the partition $(3, 1|\emptyset|-1)$. Its elongation $E(F_1) = (6, 4|3, -2|-4)$ still makes sense. Observe that $E(F_1)$ is the Ulrich partition of Example 7.3.28. To avoid discussing trivialities in the arguments that follow, we generally focus on the $m \geq 2$ case and assure the reader that appropriate arguments work in the $m = 1$ case.

The main theorem concerning Ulrich partitions P of type $(2, n, 1)$ is the following.

Theorem 7.3.51. *A partition P of type $(2, n, 1)$ is Ulrich if and only if there exists $k \geq 0$ and $m > 0$ such that $n = 2mk + m - 1$ and $P = E^k(F_m)$.*

We will prove it as a consequence of a series of results. First, we observe that the partitions $E^k(F_m)$ are indeed Ulrich.

Lemma 7.3.52. *If P is an Ulrich partition of type $(2, n, 1)$ and P has dimension $2y + m - 1$ then $E(P)$ is an Ulrich partition of dimension $2y' + m - 1$, where $y' = y + 3m$. In particular, the partition $E^k(F_m)$ is Ulrich of type $(2, 2mk + m - 1, 1)$.*

Proof. The beginning and ending intersections in $E(P)$ all occur between a's or c's and the new contiguous blocks of b's as follows:
(1) For $t \in [1, m]$, a_2 meets the left new B-block.
(2) For $t \in (m, 2m]$, c meets the right new B-block.
(3) For $t \in (2m, 3m]$, a_1 meets the left new B-block.
(4) For $t \in [2y + 4m, 2y + 5m)$, a_2 meets the right new B-block.
(5) For $t \in [2y + 5m, 2y + 6m)$, c meets the left new B-block.
(6) For $t \in [2y + 6m, 2y + 7m)$, a_1 meets the right new B-block.

Note that P can be obtained from $E(P)$ by shifting to the time $3m$ position $(E(P))(3m)$ and throwing out the two new B blocks. Since P is Ulrich of dimension $2y + m - 1$, we conclude that there are unique intersections in $E(P)$ at all times $t \in (3m, 2y + 4m)$. Clearly, $\dim E(P) = \dim P + 6m = 2y + 7m - 1$, so $E(P)$ is Ulrich.

For the second statement, it suffices to observe that the fundamental partition F_m satisfies the equality $\dim F_m = 2y + m - 1 = 3m - 1$, which is clear. □

Remark 7.3.53. By Lemma 7.3.52, if $P = (y + 2m, y|b_1, \dots, b_n|-y)$ is of the form $E^k(F_m)$ then it satisfies the formula

$$\dim P = 2y + m - 1.$$

For any P, we say that it satisfies the *dimension formula* if the above equality holds. Theorem 7.3.51 in particular claims that the dimension formula holds for any Ulrich partition of type $(2, n, 1)$.

We also observe that Theorem 7.3.51 implies that for any Ulrich P the sequence $b_1, \dots, b_n$ begins with a contiguous block $y - 1, \dots, y - l$ of length exactly l, where l is either m or $m - 1$ depending on whether $k > 0$ or $k = 0$ in the equality $P = E^k(F_m)$.

The next lemma will form the base of an induction to prove Theorem 7.3.51.

Lemma 7.3.54. *Let $P = (a_1, a_2|b_1, \dots, b_n|c) = (y + 2m, y|b_1, \dots, b_n|-y)$ be an Ulrich partition. If $y \leq 2m$ then $P = F_m$.*

Proof. The intersection a_2c occurs before the intersection a_1b_1. It follows that the partition $(y|b_1,\dots,b_n|-y)$ is Ulrich of type $(1,n,1)$. Recalling the classification of such partitions, the only possibility is that $n = y-1$, a_1 meets c at time $2y$, and $(b_1,\dots,b_n) = (y-1,\dots,1)$. □

On the other hand, if P is too large to be treated by Lemma 7.3.54 then we show that it is an elongation of a smaller partition. The next lemma completes the proof of Theorem 7.3.51.

Lemma 7.3.55. *Let $P = (a_1,a_2|b_1,\dots,b_n|c) = (y+2m,y|b_1,\dots,b_n|-y)$ be an Ulrich partition. If $y > 2m$ then there is some Ulrich partition P' of type $(2,n',1)$ with $E(P') = P$.*

Proof. Inducting on n, by Lemma 7.3.54, we may assume that any Ulrich partition of the form

$$P' = (y'+2m,y'|b_1',\dots,b_{n'}'|-y')$$

with $n' < n$ is equal to $E^k(F_m)$ for some k. In particular, P' satisfies the dimension formula

$$\dim P' = 2y' + m - 1$$

and the (b')'s begin with a contiguous block $y'-1,\dots,y'-l$ of length exactly m or $m-1$.

Claim 1. In the partition P, the time $t = 1$ intersection is a_2b_1, so $b_1 = y-1$. Suppose this is not the case. Then $b_n = -y+1$, and a_1 meets b_n at time $2y+2m-1$. By time $2y$, the a_2 and c_1 entries have already passed through the B-block, so all intersections for times $t \in [2y, 2y+2m)$ must occur between a_1 and the B-block. This gives that a contiguous block $B' = \{-y+2m,\dots,-y+1\} \subset B$ of length $2m$ occurs in the B-block. Since a_2 meets B' for times in $[2y-2m, 2y)$ and c meets B' for times in $[1,2m]$, it follows that the partition

$$P' = (y+m, y-m|B \setminus B'|-y+m)$$

is Ulrich. By induction, $B \setminus B'$ starts with a contiguous block $\{y-m-1,\dots,y-m-l\}$ of length exactly $l \in \{m, m-1\}$. In P' the intersection at time $l+1$ must be between c and some $b_0 \in B \setminus B'$. This is a contradiction, since in P we find that a_1 meets b_0 at the same time as a_2 meets an entry of B'. Therefore $b_1 = y-1$.

Having established the claim, let $m_1 \geq 1$ be the largest integer such that the contiguous block $B' = \{y-1,\dots,y-m_1\} \subset B$. Clearly, $m_1 \leq 2m$, since otherwise a_1 and a_2 would both intersect B' at the same time. An argument similar to the previous paragraph shows that, in fact, we must have $m_1 < 2m$. At time m_1+1 the intersection must be b_nc; let $m_2 \geq 1$ be the largest integer such that the contiguous block $B'' = \{-y+m_1+m_2,\dots,-y+m_1+1\} \subset B$. Observe that

$$\dim P = 2y + 2m - m_1 - 1$$

since the last intersection is a_1b_n, so P satisfies the dimension formula if and only if $m_1 = m$; our eventual goal is to show that $m_1 = m_2 = m$.

Claim 2. $m_1 + m_2 = 2m$. Let $t \in (m_1, 2m]$. When a_1 is at position $-y + t$, both a_2 and c have finished intersecting the B-block. Thus $-y + t \in B''$ for all $t \in (m_1, 2m]$, and so $m_1 + m_2 \geq 2m$. On the other hand, at time $t = 2m + 1$ we have an intersection a_1b_1, so $-y + 2m + 1 \notin B''$. Thus $m_1 + m_2 = 2m$.

Claim 3. $y > 2m + m_1$. By assumption $y > 2m$. If $t \in (2m, 2m + m_1]$ then a_1 meets B' at time t, so it is not possible for the intersection a_2c to happen at such a time. Thus $y > 2m + m_1$.

Claim 4. $m_1 = m_2 = m$. Consider the partition

$$P' = (y - m_1, y - 2m - m_1 | B \setminus (B' \cup B'') | -y + 2m + m_1)$$

obtained by looking at the time $2m + m_1$ evolution $P(2m + m_1)$ and removing the contiguous blocks B', B''. This makes sense since $y > 2m + m_1$, and P' is Ulrich because P is Ulrich. By induction, P' satisfies the dimension formula

$$\dim P' = 2(y - 2m - m_1) + m - 1.$$

On the other hand, the type of P' is $(2, n - 2m, 1)$, so

$$\dim P = \dim P' + 6m = 2y + 3m - 2m_1 - 1.$$

Comparing this with our earlier expression for $\dim P$ gives $m_1 = m$, and so $m_2 = m$ as well. This implies $P = E(P')$. □

The partitions $E^k(F_m)$ are all distinct, so Theorem 7.3.51 implies the following result.

Theorem 7.3.56. *There is a bijection between Ulrich partitions of type* $(1, n, 2)$ *and decompositions*

$$n = 2km + m - 1 \quad (k \geq 0, m > 0)$$

of the integer n.

Notice that according to this correspondence, the fundamental example of Ulrich partition of type $(1, n, 2)$ given in Example 7.3.46 corresponds to the trivial decomposition of n given by $k = 0$ and $m = n + 1$. On the other hand, when $n = 4$, $k = 2$, and $m = 1$ we will obtain a partition of type $(1, 4, 2)$ different from the fundamental example.

Finally, we will classify all Ulrich partitions of type $(2, n, 2)$. Therefore, we consider Ulrich partitions of the form $(a_1, a_2 | b_1, \ldots, b_n | c_1, c_2)$, and normalize the evolution of the partition to subtract 1, 0, −1 from the blocks. By symmetry, we may as well assume $a_1 - a_2 > c_1 - c_2$. We further normalize the positions

$$a_1 = y + s + r, \quad a_2 = y, \quad c_1 = -y, \quad \text{and} \quad c_2 = -y - s.$$

The intersection a_2c_1 occurs at time y, the gap between the a-entries is $s + r$, and the gap between the c-entries is s.

Definition 7.3.57. Let P_u be the partition of type $(2, 2u, 2)$ given by

$$P_u := (6u + 5, 2u + 1 | 2u, 2u - 2, \ldots, 4, 2, -1, -3, \ldots, -2u + 3, -2u + 1 | -2u - 1, -6u - 3).$$

We observe that the subpartition $(a_2|B|c_1)$ is the Ulrich partition of type $(1, 2u, 1)$ corresponding to the subset of $[2u]$ consisting of even numbers in the proof of Theorem 7.3.16.

Example 7.3.58. We have

$$P_1 = (11, 3|2, -1|-3, -9),$$
$$P_2 = (17, 5|4, 2, -1, -3|-5, -15).$$

The time evolution diagram of P_1 was given in Example 7.3.32. The time diagram for P_2 is Figure 7.11.

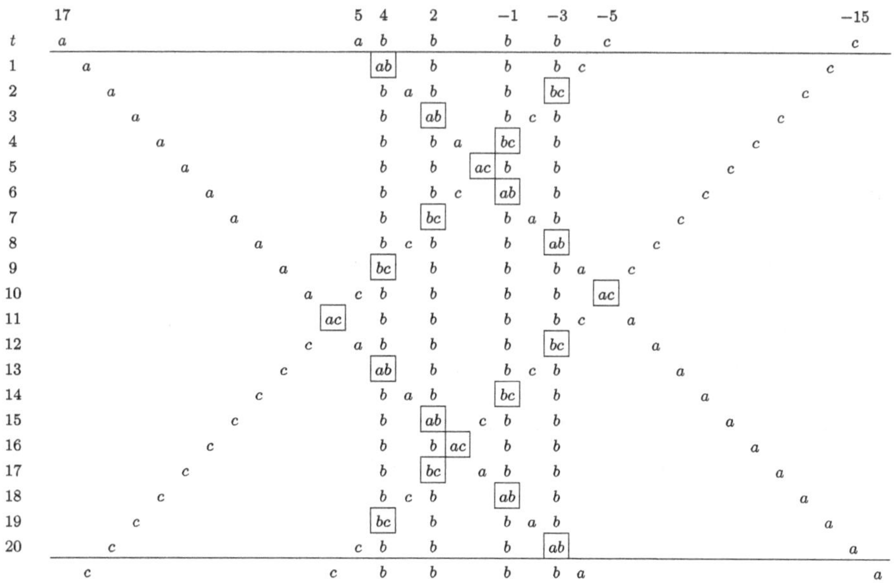

Figure 7.11: Time evolution diagram of the partition P_2 of type $(2, 4, 2)$.

We first show that these examples are, in fact, Ulrich.

Lemma 7.3.59. *The partition P_u is Ulrich. In particular, every flag variety* $(\mathrm{F}(2, 2n+2; 2n+4), \mathcal{O}_{\mathrm{F}}(1))$ *admits a Schur Ulrich bundle.*

Proof. The dimension of P_u is $N = 8u + 4$. For times $t \in [1, 4u + 1]$, there are intersections from the Ulrich subpartition $(a_2|B|c_1)$ of type $(1, 2u, 1)$. At time $4u + 2$ we have the intersection a_2c_2. As P_u is its own symmetric dual, times in $[4u + 3, 8u + 4]$ also have intersections. □

The following theorem classify all Ulrich partitions of type $(2, n, 2)$ and it asserts that P_u are the only Ulrich partitions of type $(2, n, 2)$, up to symmetry.

Theorem 7.3.60. *If P is an Ulrich partition of type $(2, n, 2)$, then $n = 2u$ is even and up to symmetry $P = P_u$.*

The plan of the proof of Theorem 7.3.60 is similar to the classification of partitions of type $(2, n, 1)$. We will first show that if P is an Ulrich partition with $y \leq s$ then P is a known example. We next show that if instead $y > s$ then P can be obtained from a shorter example by a process of elongation. However, we will finally show that elongations of the known examples are never Ulrich; this final step is the primary difference from the strategy in the $(2, n, 1)$ case, where such elongations were possible.

Before beginning the proof in earnest, we establish a couple of lemmas which are true for arbitrary Ulrich partitions of type $(2, n, 2)$. Let P be an Ulrich partition of the form $(a_1, a_2|b_1, \ldots, b_n|c_1, c_2)$ and normalize the evolution of the partition to subtract $1, 0, -1$ from the blocks. In particular, recall that $s + r = a_1 - a_2 > c_1 - c_2 = s$.

Lemma 7.3.61. *Let $P = (a_1, a_2|b_1, \ldots, b_n|c_1, c_2)$ be an Ulrich partition of type $(2, n, 2)$. Then, the intersection at time $t = 1$ is a_2b_1.*

Proof. If not, then the intersection at time $t = 1$ is of the form bc_1. Let $m \geq 1$ be the maximal number so that $\{-y + 1, \ldots, -y + m\} \subset B$, so $-y + m + 1 \notin B$. This implies that $y - i \notin B$ for $i \in [m]$. Note that $m \leq s$. At the time $t_0 = 2y + s + r - m - 1$ when $a_1(t_0) = -y + m + 1$ we have $c_2(t_0) = y - m - 1 + r \geq y - m$, so $c_2(t_0) \notin B$. Furthermore, $c_1(t_0) = y + s + r - m - 1 > y$ and $a_1(t_0) = -y - s - r + m + 1 < -y$. We conclude that there is no intersection at time t_0 even though a_2 has not yet passed through the B-block. □

Lemma 7.3.61 applied to the symmetric dual implies that the last intersection must be a_1b_n. From now on, we let $m \geq 1$ be the largest number such that $\{y-1, \ldots, y-m\} \subset B$.

Lemma 7.3.62. *According to the above notation, we have $m < \min\{r, y - 1\}$.*

Proof. First we show that $m < y-1$. Clearly, $m \leq y-1$ since the intersection a_2c_1 happens at time y. If $m = y - 1$ then B is the contiguous block $\{y - 1, \ldots, 1\}$. By Lemma 7.3.61, the last intersection in P is between a_1 and $1 \in B$, so $\dim P = s + r + y - 1 = 4y$. At time $2y$, we have $a_2(2y) = -y$ and $c_1(2y) = y$, so the intersection must be of the form ac; since

$a_2 - c_2 < a_1 - c_2$, it must be a_2c_2. Thus $s = 2y$, $r = y + 1$, and

$$P = (4y + 1, y|y - 1, \dots, 1|-y, -3y).$$

But then a_1 and c_2 meet at position $(y + 1)/2$, which is in B if $y \geq 3$. Parity is violated if $y = 2$, so we conclude $m < y - 1$.

Next we show $m < r$. If $m > r$ then the last intersection in P is b_1c_2, contradicting Lemma 7.3.61. Suppose $m = r$. At time $m+1$, the intersection is one of bc_1, a_1b, or a_2c_1. If it is a_2c_1 then $m = y - 1$ and we are done by the previous paragraph. If the intersection is bc_1 then $-y + r + 1 \in B$ and c_2 meets $y - 1 \in B$ when a_1 meets $-y + r + 1 \in B$, a contradiction. Finally, if the intersection is a_1b then again the last intersection is of type bc, violating Lemma 7.3.61. □

Remark 7.3.63. Combining Lemmas 7.3.61 and 7.3.62, we have $b_n = -y + m + 1$ for any Ulrich partition P of type $(2, n, 2)$, and thus we have the dimension formula

$$\dim P = 2y + s + r - m - 1.$$

Next we establish the fact that if the intersection a_2c_1 happens before a_1 or c_2 meet the B-block then P is a known example.

Lemma 7.3.64. *Let* $P = (a_1, a_2|b_1, \dots, b_n|c_1, c_2)$ *be an Ulrich partition of type* $(2, n, 2)$ *normalized as above. If* $y \leq s + m$ *then* $P = P_{\frac{s-2}{4}}$.

Proof. If $y \leq s + m$, then every intersection until time y is of the form a_2b or bc_1. Hence, there must be a b either at position p or $-p$ for every $1 \leq p < y$. Consequently, the total number of b entries is $y - 1$, and for $t \in (y, 2y)$ every intersection is also of the form a_2b or bc_1. By the dimension formula, $4y = 2y + s + r - m - 1$, so $2y = s + r - m - 1$.

At time $t = 2y$, a_2 and c_1 are at positions $-y$ and y, respectively. They cannot intersect at b. The intersection a_2c_2 happens before a_1c_1, so at time $t = 2y$ the intersection is a_2c_2 and $s = 2y$. At time $2y + 1$, c_2 cannot intersect at b, hence the intersection must be a_1c_1. We conclude that $r = 2$, so $m = 1$.

We now inductively determine the B-block. We have $y - 1, -y + 2 \in B$. When a_2 is at position $-y + k$ with $3 \leq k < y$ the entry c_1 is at position $y - k + 2$, while c_2 and a_1 have already intersected all other entries. By induction, we may assume $y - k + 2 \in B$ if and only if k is odd; therefore $-y + k \notin B$ if and only if k is odd and $y - k \in B$ if and only if k is odd.

Finally, note that y is odd. Equivalently, $2 \in B$. Otherwise, at time $t = 3y$, c_2 is at position 0 and a_2 is at position 2 and there is no intersection. Therefore $s \equiv 2 \pmod 4$, and $P = P_{\frac{s-2}{4}}$. □

We next argue that any Ulrich partition of type $(2, n, 2)$ must be obtained from some P_u by a process of elongation. Finally, we will see that P_u cannot be elongated.

Given a partition P at time $t = 0$, we will view it as three sequences ABC, where A is the sequence of a's and blank spaces $a_1 \times \cdots \times a_2$, C is the sequence of c's and blank spaces $c_1 \times \cdots \times c_2$ and B is the sequence of b's and blank spaces in between a_2 and c_1. The partition P is a concatenation of these three. Given a contiguous pattern Ψ of b's and blank spaces at positive positions, let Ψ^c be the complementary pattern which has a b at position p if Ψ does not have a b at position $-p$, and vice versa. Let $\ell(\Psi)$ be the length of Ψ. Let X_ℓ denote a contiguous sequence of blank spaces of length ℓ.

Example 7.3.65. If $\Psi = b \times b\, b \times$, then $\Psi^c = b \times \times\, b \times$. Also, $X_3 = \times \times \times$.

Definition 7.3.66. Let $P = ABC$ be a partition and Ψ a pattern of b's and blanks. The *Ψ-elongation of P* is the partition corresponding to the concatenation

$$A\Psi X_{\ell(\Psi)} B X_{\ell(\Psi)} \Psi^c C$$

of patterns.

Example 7.3.67. Continuing the previous example, if

$$P = a_1 \times \times \times \;\; a_2\, b_1\, b_2 \;\; \times \times\, b_3 \times \times\, c_1 \times c_2,$$

then the Ψ-elongation of P is

$$a_1 \times \times \times\, a_2\, b \times b\, b \times \times \times \times \times \times\, b_1\, b_2 \times \times\, b_3 \times \times \times \times \times \times \times\, b \times \times\, b \times c_1 \times c_2.$$

For integers $q, r > 0$, we let $K(q, r, m)$ be the pattern of b's consisting of q iterations of a contiguous block of m b's followed by $r - m$ blanks.

Example 7.3.68. $K(2, 6, 2) = b\, b \times \times \times \times\, b\, b \times \times \times \times$.

Lemma 7.3.69. *Let P be an Ulrich partition of type $(2, n, 2)$, normalized as above. If $y > s + m$, then r divides $s + m$; let the quotient be q. Then P is the $K(q, r, m)$-elongation of an Ulrich partition P' of type $(2, n - s - m, 2)$. Furthermore, the initial block of b's in P' also has length m.*

Proof. We can determine the pattern of intersections inductively until the time $s + m < y$ immediately before c_2 first meets the B-block. Until this time, all intersections are of the form $a_1 b$ or bc_2.

For the base of the induction, we first recall $\{y - 1, \ldots, y - m\} \subset B$. We know $-y + m + 1 \in B$ by Lemma 7.3.62, and $m < r$. Suppose $v \in (m, r]$. The only possible intersection at time t_0 when $a_1(t_0) = -y+v$ is $a_1 b$ since $a_2(t_0) = -y+v-r-s < -y$ and $c_2(t_0) = y-v+r \geq y$. Therefore $-y + v \in B$ for $v \in (m, r]$.

We now continue by induction until we reach time $s+m$. The positions $y-(h-1)r-v$ have a b for $1 \leq v \leq m$ by induction. Consequently, the positions $-y + hr + v$ cannot have b's for $1 \leq v \leq m$. Otherwise, when c_2 is at position $y - (h - 1)r - v$, a_1 would be at position $-y+hr+v$ giving a coincident intersection. Thus there is a contiguous B-block of length m at positions $y - hr - v$ for $1 \leq v \leq m$. Similarly, there are no b entries in positions $y - (h - 1)r - v$ for $m + 1 \leq v \leq r$ by induction. When c_2 is at these positions

a_1 is at the positions $-y + hr + v$. Since $a_1 b$ is the only possible intersection, there must be a contiguous B-block of length $r - m$ at these positions.

Claim. $r|s + m$. Write $s + m = qr + j$ with $0 \le j < r$ as in the division algorithm. What we have shown so far is that the pattern of b's and blanks in the interval $B \cap (y, y - s - m]$ is the truncation of $K(q, r, m)$ to a sequence of length $s + m$. Similarly, the pattern of b's and blanks in $B \cap [-y + s + m, -y)$ is the truncation of $K(q, r, m)^c$ to a sequence of length $s + m$. As $\ell(K(q, r, m)) = qr$, the claim is that no truncation actually takes place. There are two cases to consider depending on the remainder j.

Case 1. $1 \le j \le m$. In this case, $y - s - m \in B$. When c_2 is at position $y - s - m$, we find that c_1 is at position $y - m \in B$, a contradiction.

Case 2. $m < j < r$. Consider the time t_0 when $a_1(t_0) = -y + s + m + 1$. Then $c_2(t_0) = y - s - m - 1 + r \notin B$, and $c_1(t_0) \ge y$, $a_2(t_0) \le -y$ hold. Furthermore, $-y + s + m + 1 \notin B$, since c_1 is at this position when c_2 is at $-y + m + 1 \in B$. Thus there is no intersection at time t_0.

Therefore $r|s + m$. Let us analyze the known intersections. For times $t \in [1, s + m]$, the intersections are of type $a_2 b$ or bc_1. When $t \in (s + m, 2s + 2m]$, the intersections are all $a_1 b$ or bc_2. This gives $y - t, -y + t \notin B$ for all such times and $y > 2s + 2m$. Dually, a symmetric description holds for the last $2s + 2(r - m)$ times. Evolving P to time $2s + 2m$ and throwing out all the b's which have already met a's and c's, we arrive at an Ulrich partition

$$P' = (y - s + r - 2m, y - 2s - 2m|B'|-y + 2s + 2m, -y + s + 2m)$$

such that P is the $K(q, r, m)$-elongation of P'; it is easy to see that B' is nonempty.

Finally, we analyze the length m' of the initial block of b's in P'. The dimension formula of Remark 7.3.63 gives equalities

$$\begin{aligned}
\dim P' &= 2(y - 2s - 2m) + r + s - m' - 1,\\
\dim P &= 2y + r + s - m - 1,\\
\dim P &= \dim P' + 4(s + m),
\end{aligned}$$

from which $m = m'$ follows immediately. □

The above lemma easily implies Theorem 7.3.60.

Proof of Theorem 7.3.60. We have already classified the Ulrich partitions of type $(2, 2, 2)$ in Theorem 7.3.45. By induction on n, suppose that for $n < n_0$ the only Ulrich partitions of type $(2, n, 2)$ are P_u with $n = 2u$. Let P be an Ulrich partition of type $(2, n_0, 2)$. By Lemma 7.3.64, we may assume $y > s+m$. By Lemma 7.3.69, P is the $K(q, r, m)$-elongation of a smaller Ulrich partition P'. By induction and Lemma 7.3.69, we must have $r = 2$ and $m = 1$. However, s is even, so $r = 2$ does not divide $s + m = s + 1$. This contradiction proves the theorem. □

The classification of Ulrich partitions for two-step flag varieties is incomplete. However, we pose the following conjecture (see [61, Conjectures 5.8 and 5.9]).

Conjecture 7.3.70. *If (α, β, γ) is the type of an Ulrich partition, then up to symmetry this type is one of*

$$(1, n, 1), \quad (1, n, 2), \quad (2, 2n, 2), \quad (2, 2, 3), \quad (2, 1, k_1), \quad (1, 2, k_1), \quad \textit{or} \quad (1, 2, k_1 + k_2)$$

where n denotes a positive integer and k_i is a number of the form $(4^{m_i+1} - 1)/3$ for some $m_i \geq 0$.

The partial classification of Ulrich partitions for two-step flag varieties implies that Conjecture 7.3.70 is equivalent to the following a priori weaker statement.

Conjecture 7.3.71. *There are no Ulrich partitions of type (α, β, γ) if $\beta \geq 3$ and $\gamma \geq 3$.*

Indeed, suppose Conjecture 7.3.71 is true. If $\beta \leq 2$, then Theorems 7.3.44 and 7.3.45 completely classify Ulrich partitions of type (α, β, γ). So suppose $\beta \geq 3$. If $\alpha \leq 2$ and $\gamma \leq 2$, then Theorems 7.3.16, 7.3.56, and 7.3.60 completely classify Ulrich partitions of type (α, β, γ). If instead either $\alpha \geq 3$ or $\gamma \geq 3$, then by symmetry we may assume $\gamma \geq 3$, and therefore there are no Ulrich partitions of type (α, β, γ). Therefore Conjecture 7.3.71 implies Conjecture 7.3.70.

Remark 7.3.72. As evidence for Conjecture 7.3.71, we performed a brute-force computer search to verify that if there is an Ulrich partition of type (α, β, γ) with $\beta \geq 3$ and $\gamma \geq 3$, then $\alpha + \beta + \gamma \geq 13$.

We note that Conjecture 7.3.71 is open even in the special case $\alpha = 1$. Finally, we want to point out that the case $\beta = 3$ has been recently solved affirmatively in [64, Theorem 1.2].

7.4 Final comments and additional reading

This chapter developed the study of Ulrich bundles on some classical families of varieties of arbitrarily large dimension. More concretely, we have essentially addressed Conjecture 3.4.8 for Segre, Grassmann, and flag varieties. In Section 7.1 we have followed the paper [72] to construct families of arbitrarily large dimension of indecomposable Ulrich bundles on any Segre variety $\mathbb{P}^{n_1} \times \cdots \times \mathbb{P}^{n_s} \subseteq \mathbb{P}^N$, $N = \prod_{i=1}^{s}(n_i + 1) - 1$. Obviously, we had to exclude from the statement the case of the quadric surface $\mathbb{P}^1 \times \mathbb{P}^1$, for which we know, from Theorem 2.3.4, that it only supports the two Ulrich line bundles $\mathcal{O}_{\mathbb{P}^1 \times \mathbb{P}^1}(1, 0)$ and $\mathcal{O}_{\mathbb{P}^1 \times \mathbb{P}^1}(0, 1)$. As observed in Remark 7.1.18, the Segre variety $\mathbb{P}^1 \times \mathbb{P}^d$ is a particular case of rational normal scroll. Rational normal scrolls, jointly with quadric hypersurface, comprise the full list of *smooth varieties of minimal degree*. The proof of this fact can be found in [84]. In [156], the study of Ulrich bundles on rational normal scrolls was carried out. In [10], the authors gave a classification of Ulrich bundles on varieties of minimal degree in terms of some filtrations by twists of wedge product of the relative cotangent bundle.

In the remaining part of this chapter, we focused our attention on the study of irreducible homogeneous Ulrich bundles on Grassmann varieties (see Section 7.2) and, more generally, flag varieties (see Section 7.3).

The results from those sections should be framed in the following setting: flag varieties X belong to the larger and well-studied family of *rational homogenous varieties*. They are defined as the quotient of $X = G/P$ where G is a simple algebraic group and $P \subset G$ is a parabolic subgroup. Rational homogenous varieties are completely classified in terms of the list of simple algebraic groups. They consist of four families A_n, B_n, C_n, and D_n, and only 5 exceptional cases that are called E_6, E_7, E_8, F_4, G_2. In [186, pp. 45–55] it was proved that the category of homogeneous vector bundles on a rational homogeneous variety is equivalent to the category of representations of the parabolic group P. Thanks to this equivalence, Borel–Bott–Weil theorem (see [173, Theorem 11.4]) gives an explicit construction for any cohomology group of an irreducible homogeneous bundle on X. The reader can get a complete account and further references from [173].

In this setting, flag varieties (and, in particular, Grassmann varieties) are rational homogeneous varieties of type A_n with underlying simple group $\mathrm{SL}(n)$. In Section 7.2 we refined the results from Section 2.4, giving a characterization of irreducible homogeneous Ulrich bundles on Grassmann varieties. We have followed the approach of Costa and Miró-Roig in [68]. Section 7.3 comes from [61]. Theorem 7.3.5, which plays a crucial role in the development of the section, is a particular instance of the aforementioned Borel–Bott–Weil theorem.

Recently there have been further contributions to the classification of irreducible homogeneous bundles on other rational homogeneous varieties. Indeed, in [97] the author completely classifies the irreducible homogeneous Ulrich bundles on isotropic Grassmann varieties (namely, those of type B_n, C_n, and D_n). Finally, in [142] it was proved that the only rational homogeneous varieties with Picard number 1 from the list of exceptional algebraic groups admitting an irreducible homogeneous Ulrich bundle are the Cayley plane E_6/P_1 and the adjoint variety E_7/P_1. Furthermore, in both cases there is only one irreducible equivariant Ulrich bundle.

The existence of Ulrich bundles on higher-dimensional varieties in general is completely open and contributions from many different areas are expected. For instance, in [167], S. Novaković proved the existence of Ulrich bundles on any Brauer–Severi variety. The case of linear determinantal varieties is settled in [131].

A Categories and derived categories

This appendix contains the basic results on categories and derived categories used in the text. For more results, and a very detailed and careful treatment, see [146, 184] and [118].

Definition A.0.1. A *category* $\mathcal{C}$ is a collection of objects $\mathrm{Obj}(\mathcal{C})$ and, for each pair of objects $A, B \in \mathrm{Obj}(\mathcal{C})$, a set $\mathrm{Hom}_{\mathcal{C}}(A, B)$ of morphisms with a composition law

$$\mathrm{Hom}_{\mathcal{C}}(A,B) \times \mathrm{Hom}_{\mathcal{C}}(B,C) \longrightarrow \mathrm{Hom}_{\mathcal{C}}(A,C) \quad (f,g) \mapsto gf$$

and a distinguished element $1_A \in \mathrm{Hom}_{\mathcal{C}}(A,A)$ for each object $A \in \mathrm{Obj}(\mathcal{C})$ such that:
(1) The composition is associative in the sense that $h(gf) = (hg)f$ whenever both sides are defined.
(2) $f1_A = f$ and $1_A g = g$ whenever the compositions are defined.

We usually write $f : A \longrightarrow B$ to mean $f \in \mathrm{Hom}_{\mathcal{C}}(A, B)$ and say that certain diagrams commute to indicate that certain compositions are equal.

In this book, the category that appears most frequently is the category of coherent sheaves of $\mathcal{O}_X$-modules where X is a smooth projective variety and also the derived category $\mathcal{D}^b(X)$ of bounded complexes of coherent sheaves on X. Categories were defined by Eilenberg and Mac Lane to unify ideas from group theory and topology. Nowadays, the language of category theory is used very widely. Mac Lane in [146] gives a useful overview.

Definition A.0.2. Let $\mathcal{C}$ and $\mathcal{D}$ be two categories. A *functor* $F : \mathcal{C} \longrightarrow \mathcal{D}$ is a map such that
(1) For every object $A \in \mathrm{Obj}(\mathcal{C})$ assigns an object $F(A) \in \mathrm{Obj}(\mathcal{D})$.
(2) For every morphism $f : A \longrightarrow B$ assigns a morphism $F(f) : F(A) \longrightarrow F(B)$ preserving composition and identity elements.

There are also morphisms of functors.

Definition A.0.3. If $F, G : \mathcal{C} \longrightarrow \mathcal{D}$ are functors, then a morphism $\alpha : F \longrightarrow G$, called a *natural transformation*, is, for each object $A \in \mathrm{Obj}(\mathcal{C})$, a morphism $\alpha_A : F(A) \longrightarrow G(A)$ such that whenever $f : A \longrightarrow B$ is a morphism in $\mathcal{C}$, the diagram

$$\begin{array}{ccc} F(A) & \longrightarrow & G(A) \\ F(f)\downarrow & & \downarrow G(f) \\ F(B) & \longrightarrow & G(B) \end{array}$$

commutes. We say that F and G are isomorphic, written $F \cong G$, if there are natural transformations $\alpha : F \longrightarrow G$ and $\beta : G \longrightarrow F$ whose compositions $\alpha\beta$ and $\beta\alpha$ are the identity natural transformation (that is, $\beta_A\alpha_A = 1_{F(A)}$ and $\alpha_A\beta_A = 1_{G(A)}$ for all objects $A \in \mathrm{Obj}(\mathcal{C})$).

https://doi.org/10.1515/9783110647686-008

Definition A.0.4. We define the functor of points of a scheme X as the functor

$$h_X : (\text{schemes})^0 \longrightarrow (\text{sets})$$

where $(\text{schemes})^0$ and (sets) represent the category of schemes with the arrows reversed and the category of sets; h_X takes each scheme Y to the set $h_X(Y) = \text{Hom}(Y, X)$ and each morphism $f : Y \longrightarrow Z$ to the map of sets $h_X(Z) \longrightarrow h_X(Y)$ defined by sending an element $g \in h_X(Z) = \text{Hom}(Z, X)$ to the composition $gf \in \text{Hom}(Y, X)$. We say that a functor $F : (\text{schemes})^0 \longrightarrow (\text{sets})$ is *representable* if it is of the form h_X for some scheme X.

Definition A.0.5. Let $\mathcal{C}$ and $\mathcal{D}$ be two categories. A functor $F : \mathcal{C} \longrightarrow \mathcal{D}$ is said to be
(1) *Faithful* if the induced map $F_{X,Y} : \text{Hom}_{\mathcal{C}}(X, Y) \longrightarrow \text{Hom}_{\mathcal{D}}(F(X), F(Y))$ is injective for all $X, Y \in \text{Obj}(\mathcal{C})$.
(2) *Full* if the induced map $F_{X,Y} : \text{Hom}_{\mathcal{C}}(X, Y) \longrightarrow \text{Hom}_{\mathcal{D}}(F(X), F(Y))$ is surjective for all $X, Y \in \text{Obj}(\mathcal{C})$.
(3) *Fully faithful* if it is full and faithful, i. e., the induced map $F_{X,Y} : \text{Hom}_{\mathcal{C}}(X, Y) \longrightarrow \text{Hom}_{\mathcal{D}}(F(X), F(Y))$ is bijective for all $X, Y \in \text{Obj}(\mathcal{C})$.

One of the most useful notions from category theory is that of an adjoint of a functor which we recall now.

Definition A.0.6. Let $F : \mathcal{A} \longrightarrow \mathcal{B}$ and $G : \mathcal{B} \longrightarrow \mathcal{A}$ be functors. We say that F is a *left adjoint* for G (or, equivalently, G is a *right adjoint* for F) if there is a natural isomorphism $\alpha : \text{Hom}_{\mathcal{A}}(-, G(-)) \cong \text{Hom}_{\mathcal{B}}(F(-), -)$.

This means that for every pair of objects A of $\mathcal{A}$ and B of $\mathcal{B}$ there is an isomorphism $\alpha_{A,B} : \text{Hom}_{\mathcal{A}}(A, G(B)) \longrightarrow \text{Hom}_{\mathcal{B}}(F(A), B)$ such that for every morphism of objects $A \longrightarrow A'$ in $\mathcal{A}$ and $B \longrightarrow B'$ in $\mathcal{B}$, the diagram

$$\begin{array}{ccc} \text{Hom}_{\mathcal{A}}(A', G(B)) & \cong & \text{Hom}_{\mathcal{B}}(F(A'), B) \\ \downarrow & & \downarrow \\ \text{Hom}_{\mathcal{A}}(A, G(B')) & \cong & \text{Hom}_{\mathcal{B}}(F(A), B') \end{array}$$

commutes. We shall sometimes say that (F, G) is an *adjoint pair of functors*.

Any two left adjoints of a functor G are naturally isomorphic and, dually, two right adjoints of a functor F are naturally isomorphic.

Example A.0.7. Let $f : X \to Y$ be a proper morphism between two noetherian schemes X and Y. The pullback functor $f^* : \text{Coh}(Y) \to \text{Coh}(X)$ is left adjoint for the direct image functor $f_* : \text{Coh}(X) \to \text{Coh}(Y)$. So, for any coherent sheaf $\mathcal{F} \in \text{Coh}(X)$ and any coherent sheaf $\mathcal{G} \in \text{Coh}(Y)$ we have

$$\text{Hom}_{\mathcal{O}_X}(f^*\mathcal{G}, \mathcal{F}) \cong \text{Hom}_{\mathcal{O}_Y}(\mathcal{G}, f_*\mathcal{F}).$$

Definition A.0.8. An *additive category* is a category $\mathcal{A}$ such that for any pair of objects $A, B \in \text{Obj}(\mathcal{A})$, $\text{Hom}_{\mathcal{A}}(A, B)$ has a structure of an abelian group, the composition law is bilinear and finite direct sums and zero object exist. An *abelian category* is an additive category $\mathcal{A}$ such that every morphism has a kernel and a cokernel, every monomorphism is the kernel of its cokernel, every epimorphism is the cokernel of its kernel; and finally, every morphism can be factored into an epimorphism followed by a monomorphism.

As examples of abelian categories, we have the category of abelian groups, the category of modules over a commutative ring with identity, and the category of coherent sheaves of $\mathcal{O}_X$-modules on a scheme X.

Definition A.0.9. A *translation functor* on a category $\mathcal{D}$ is an automorphism (or for some authors, an autoequivalence) T from $\mathcal{D}$ to $\mathcal{D}$. For objects $X, Y \in \text{Obj}(\mathcal{D})$ one usually uses the notation $X[n] = T^n X$ and likewise for morphisms from X to Y.

Definition A.0.10. A *triangle* (X, Y, Z, u, v, w) in a category $\mathcal{D}$ consists of 3 objects X, Y, and Z, together with morphisms $u : X \longrightarrow Y$, $v : Y \longrightarrow Z$, and $w : Z \longrightarrow X[1]$. Triangles are generally written in the unravelled form:

$$X \xrightarrow{u} Y \xrightarrow{v} Z \xrightarrow{w} X[1].$$

Definition A.0.11. A triangulated category is an additive category $\mathcal{D}$ with a translation functor and a class of triangles, called *distinguished triangles*, satisfying the following properties:

TR1
- For any object $X \in \mathcal{D}$, the following triangle is distinguished: $X \xrightarrow{\text{Id}} X \longrightarrow 0 \longrightarrow X[1]$.
- For any morphism $u : X \longrightarrow Y$, there is an object $Z \in \mathcal{D}$ (called a *mapping cone of the morphism* u) fitting into a distinguished triangle $X \xrightarrow{u} Y \longrightarrow Z \longrightarrow X[1]$.
- Any triangle isomorphic to a distinguished triangle is distinguished. This means that if $X \xrightarrow{u} Y \xrightarrow{v} Z \xrightarrow{w} X[1]$ is a distinguished triangle, and $f : X \longrightarrow X'$, $g : Y \longrightarrow Y'$, and $h : Z \longrightarrow Z'$ are isomorphisms, then $X' \xrightarrow{guf^{-1}} Y' \xrightarrow{hvg^{-1}} Z' \xrightarrow{f[1]wh^{-1}} X'[1]$ is also a distinguished triangle.

TR2
- If $X \xrightarrow{u} Y \xrightarrow{v} Z \xrightarrow{w} X[1]$ is a distinguished triangle, then so are the two rotated triangles $Y \xrightarrow{v} Z \xrightarrow{w} X[1] \xrightarrow{-u[1]} Y[1]$ and $Z[-1] \xrightarrow{-w[-1]} X \xrightarrow{u} Y \xrightarrow{v} Z$.

TR3
- Given two distinguished triangles and a map between the first morphisms in each triangle, there exists a morphism between the third objects in each of the two triangles that makes everything commute. This means that in the following diagram

(where the two rows are distinguished triangles and f and g form the map of morphisms such that $gu = u'f$) there exists some map h (not necessarily unique) making all the squares commute:

$$\begin{array}{ccccccc} X & \overset{u}{\longrightarrow} & Y & \overset{v}{\longrightarrow} & Z & \overset{w}{\longrightarrow} & X[1] \\ \downarrow f & & \downarrow g & & \downarrow h & & \downarrow f[1] \\ X' & \overset{u'}{\longrightarrow} & Y' & \overset{v'}{\longrightarrow} & Z' & \overset{w'}{\longrightarrow} & X'[1] \end{array} .$$

TR4 (The *octahedral axiom*)

- Suppose we have morphisms $u : X \longrightarrow Y$ and $v : Y \longrightarrow Z$, so that we also have a composed morphism $vu : X \longrightarrow Z$. Form distinguished triangles for each of these three morphisms according to TR1. The octahedral axiom states that the three mapping cones can be made into the vertices of a distinguished triangle so that everything commutes. More formally, given distinguished triangles $X \overset{u}{\longrightarrow} Y \overset{j}{\longrightarrow} Z' \overset{k}{\longrightarrow} X[1]$, $Y \overset{v}{\longrightarrow} Z \overset{l}{\longrightarrow} X' \overset{i}{\longrightarrow} Y[1]$, and $X \overset{vu}{\longrightarrow} Z \overset{m}{\longrightarrow} Y' \overset{n}{\longrightarrow} X[1]$, there exists a distinguished triangle $Z' \overset{f}{\longrightarrow} Y' \overset{g}{\longrightarrow} X' \overset{h}{\longrightarrow} Z'[1]$ such that $l = gm$, $k = nf$, $h = j[1]i$, $ig = u[1]n$, and $fj = mv$.

Remark A.0.12. These axioms are not entirely independent, since (TR3) can be derived from the others.

As examples of triangulated categories, we have the category of vector spaces (over a field). Indeed, vector spaces form an elementary triangulated category in which $V[1] = V$ for all V. The homotopy category of chain complexes $K(\mathcal{A})$ in an abelian category $\mathcal{A}$ (i. e., the category of chain complexes modulo chain homotopy) is also a triangulated category. The bounded derived category of complexes of coherent sheaves on a smooth projective variety is also a triangulated category, among others.

We now give a very brief sketch of the construction of a derived category. The first complete formulation is due to Verdier [202]. Given an abelian category $\mathcal{A}$, we denote by $\mathrm{Kom}(\mathcal{A})$ the category of complexes of $\mathcal{A}$. It is well known that $\mathrm{Kom}(\mathcal{A})$ is again an abelian category. Given a complex $A^\bullet$ in $\mathrm{Kom}(\mathcal{A})$ as

$$\cdots \to A^{i-1} \overset{d_{i-1}}{\longrightarrow} A^i \overset{d_i}{\longrightarrow} A^{i+1} \to \cdots,$$

we define the ith cohomology as $\mathrm{H}^i(A^\bullet) := \ker(d_i)/\operatorname{im}(d_{i-1})$.

For any abelian category $\mathcal{A}$ and for any object $A^\bullet$ in $\mathrm{Kom}(\mathcal{A})$, we denote by $A^\bullet[1]$ the complex $(A^\bullet[1])^i := A^{i+1}$ and differential $d^i_{A^\bullet[1]} := -d^{i+1}_{A^\bullet}$.

The shift $f[1]$ of a morphism of complexes $f : A^\bullet \longrightarrow B^\bullet$ is the complex morphism $A^\bullet[1] \longrightarrow B^\bullet[1]$ given by $f[1]^i := f^{i+1}$.

The shift functor

$$T : \mathrm{Kom}(\mathcal{A}) \longrightarrow \mathrm{Kom}(\mathcal{A}), \quad A^\bullet \longmapsto A^\bullet[1]$$

defines an equivalence of abelian categories. Nevertheless, the category $\mathrm{Kom}(\mathcal{A})$ endowed with the shift functor T does not define a triangulated category.

Definition A.0.13. A morphism of complexes $f : A^\bullet \longrightarrow B^\bullet$ is a *quasiisomorphism* if for all $i \in \mathbb{Z}$ the induced map $\mathrm{H}^i(f) : \mathrm{H}^i(A^\bullet) \longrightarrow \mathrm{H}^i(B^\bullet)$ is an isomorphism.

Theorem A.0.14. *Let $\mathcal{A}$ be an abelian category and let $\mathrm{Kom}(\mathcal{A})$ be its category of complexes. Then there exist a category $D(\mathcal{A})$, the derived category of $\mathcal{A}$, and a functor*

$$Q : \mathrm{Kom}(\mathcal{A}) \longrightarrow D(\mathcal{A})$$

such that
1. *If $f : A^\bullet \longrightarrow B^\bullet$ is a quasiisomorphism, then $Q(f)$ is an isomorphism in $D(\mathcal{A})$.*
2. *Any functor $F : \mathrm{Kom}(\mathcal{A}) \longrightarrow \mathcal{D}$ satisfying property (1) factorizes uniquely over $Q : \mathrm{Kom}(\mathcal{A}) \longrightarrow D(\mathcal{A})$. This means that there exists a unique functor (up to isomorphism) $G : D(\mathcal{A}) \longrightarrow \mathcal{D}$ such that $F \cong G \circ Q$.*

The derived category $D(\mathcal{A})$ of $\mathcal{A}$, in general, is not abelian but it is always triangulated.

From now, until the end of this Appendix, we will deal with the derived category of bounded complexes of coherent sheaves of $\mathcal{O}_X$-modules on a projective variety X. Let X be a smooth projective variety defined over $\mathbf{k}$ and let $\mathcal{D} = \mathcal{D}^b(X) = D^b(\mathcal{O}_X\text{-mod})$ be the derived category of bounded complexes of coherent sheaves of $\mathcal{O}_X$-modules. If $\mathcal{F}$ is a coherent sheaf on X, $\mathcal{F}[i] \in \mathcal{D}$ denotes the bounded complex concentrated in degree i. In particular, $\mathcal{F} \in \mathcal{D}$ is seen as a bounded complex concentrated in degree 0. For any pair of objects $A, B \in \mathrm{Obj}(\mathcal{D})$, we introduce the following notation:

$$\mathrm{Hom}^\bullet_{\mathcal{D}}(A,B) := \bigoplus_{k\in\mathbb{Z}} \mathrm{Ext}^k_{\mathcal{D}}(A,B).$$

Definition A.0.15. A class of objects $\mathcal{C}$ *generates* a triangulated category $\mathcal{D}$ if the smallest full triangulated subcategory containing the objects of $\mathcal{C}$ is equivalent to $\mathcal{D}$. Equivalently, $\mathcal{C}$ spans $\mathcal{D}$ if for all $B \in \mathcal{D}$ the following holds:
(a) $\mathrm{Hom}(A, B[i]) = 0$ for all $A \in \mathcal{C}$ and all $i \in \mathbb{Z}$, then $B \cong 0$.
(b) $\mathrm{Hom}(B[i], A) = 0$ for all $A \in \mathcal{C}$ and all $i \in \mathbb{Z}$, then $B \cong 0$.

Unravelling this definition, one finds that this is equivalent to saying that, up to isomorphism, every object in $\mathcal{D}$ can be obtained by successively enlarging $\mathcal{C}$ through the following operations: taking finite direct sums, shifting in $\mathcal{D}$, and taking a cone $\mathcal{E}$ of a triangle $u : \mathcal{F} \longrightarrow \mathcal{G} \longrightarrow \mathcal{E} \longrightarrow \mathcal{F}[1]$ with $\mathcal{F}, \mathcal{G} \in \mathcal{C}$.

Definition A.0.16. Let X be a smooth projective variety.
1. An object $\mathcal{F} \in \mathcal{D}$ is *exceptional* if $\mathrm{Hom}^\bullet_{\mathcal{D}}(\mathcal{F}, \mathcal{F})$ is a 1-dimensional algebra generated by the identity.
2. An ordered collection $(\mathcal{F}_0, \mathcal{F}_1, \dots, \mathcal{F}_m)$ of objects of $\mathcal{D}$ is an *exceptional collection* if each object $\mathcal{F}_i$ is exceptional and $\mathrm{Hom}^\bullet_{\mathcal{D}}(\mathcal{F}_k, \mathcal{F}_j) = 0$ for $j < k$.
3. An exceptional collection $(\mathcal{F}_0, \mathcal{F}_1, \dots, \mathcal{F}_m)$ of objects of $\mathcal{D}$ is a *strongly exceptional collection* if, in addition, $\mathrm{Ext}^i_{\mathcal{D}}(\mathcal{F}_k, \mathcal{F}_j) = 0$ for $i \neq 0$ and $j \geq k$.

(4) An ordered collection of objects of $\mathcal{D}$, $(\mathcal{F}_0, \mathcal{F}_1, \dots, \mathcal{F}_m)$, is a *full (strongly) exceptional collection* if $(\mathcal{F}_0, \mathcal{F}_1, \dots, \mathcal{F}_m)$ is a (strongly) exceptional collection and $\mathcal{F}_0$, $\mathcal{F}_1, \dots, \mathcal{F}_m$ generate the bounded derived category $\mathcal{D}$.

Example A.0.17. (1) $(\mathcal{O}_{\mathbb{P}^r}, \mathcal{O}_{\mathbb{P}^r}(1), \mathcal{O}_{\mathbb{P}^r}(2), \dots, \mathcal{O}_{\mathbb{P}^r}(r))$ and $(\Omega^r_{\mathbb{P}^r}(r), \dots, \Omega^2_{\mathbb{P}^r}(2), \Omega^1_{\mathbb{P}^r}(1), \mathcal{O}_{\mathbb{P}^r})$ are full strongly exceptional collections of vector bundles on $\mathbb{P}^r$.

(2) Let $\mathrm{Gr}^a(k,n)$ be the Grassmann variety of k-dimensional subspaces of an n-dimensional vector space and let $\mathcal{Q}$ be the rank k universal quotient bundle on $\mathrm{Gr}^a(k,n)$. Denote by $\Sigma^\alpha \mathcal{Q}$ the space of the irreducible representations of the group $\mathrm{GL}(\mathcal{Q})$ with highest weight $\alpha = (\alpha_1, \dots, \alpha_s)$ and $|\alpha| = \sum_{i=1}^s \alpha_i$. Denote by $A(k,n)$ the set of locally free sheaves $\Sigma^\alpha \mathcal{Q}$ on $\mathrm{Gr}^a(k,n)$ where α runs over Young diagrams fitting inside a $k \times (n-k)$ rectangle. Set $\rho(k,n) := \#A(k,n)$. By [124, Propositions 2.2(a) and 1.4], $A(k,n)$ can be totally ordered in such a way that we obtain a full strongly exceptional collection $(\mathcal{E}_1, \dots, \mathcal{E}_{\rho(k,n)})$ of locally free sheaves on $\mathrm{Gr}^a(k,n)$.

(3) Let $Q_n \subset \mathbb{P}^{n+1}$, $n \geq 2$, be the smooth quadric hypersurface. By [123, Proposition 4.9], if n is even and $\mathcal{S}'$, $\mathcal{S}''$ are the spinor bundles on Q_n, then

$$(\mathcal{S}'(-n), \mathcal{S}''(-n), \mathcal{O}_{Q_n}(-n+1), \dots, \mathcal{O}_{Q_n}(-1), \mathcal{O}_{Q_n})$$

is a full strongly exceptional collection of vector bundles on Q_n; and if n is odd and $\mathcal{S}$ is the spinor bundle on Q_n, then

$$(\mathcal{S}(-n), \mathcal{O}_{Q_n}(-n+1), \dots, \mathcal{O}_{Q_n}(-1), \mathcal{O}_{Q_n})$$

is a full strongly exceptional collection of vector bundles on Q_n.

Remark A.0.18. The existence of a full strongly exceptional collection $(\mathcal{F}_0, \mathcal{F}_1, \dots, \mathcal{F}_m)$ of coherent sheaves on a smooth projective variety X imposes a rather strong restriction on X, namely that the Grothendieck group $K_0(X) = K_0(\mathcal{O}_X\text{-mod})$ is isomorphic to $\mathbb{Z}^{m+1}$.

Definition A.0.19. Let X be a smooth projective variety and let (A,B) be an exceptional pair of objects of $\mathcal{D}$. We define the *left mutation* of B, $L_A B$, and the *right mutation* of A, $R_B A$, with the aid of the following distinguished triangles in the category $\mathcal{D}$:

$$L_A B \longrightarrow \mathrm{Hom}^\bullet(A,B) \otimes A \longrightarrow B \longrightarrow L_A B[1], \tag{A.1}$$

$$R_B A[-1] \longrightarrow A \longrightarrow \mathrm{Hom}^\bullet(A,B)^\vee \otimes B \longrightarrow R_B A. \tag{A.2}$$

Remark A.0.20. If we apply $\mathrm{Hom}^\bullet(A,-)$ to the triangle (A.1) and apply $\mathrm{Hom}^\bullet(-,B)$ to the triangle (A.2), we get the following orthogonality relations:

$$\mathrm{Hom}^\bullet(A, L_A B) = 0 \quad \text{and} \quad \mathrm{Hom}^\bullet(R_B A, B) = 0.$$

Therefore, $(L_A B, A)$ and $(B, R_B A)$ are also exceptional pairs.

Definition A.0.21. Let X be a smooth projective variety and let $\sigma = (\mathcal{E}_0, \dots, \mathcal{E}_n)$ be an exceptional collection of objects of $\mathcal{D}$. A *left mutation* (respectively *right mutation*) of σ is defined as follows. For any integer i, $1 \le i \le n$, a left mutation L_i replaces the ith pair of consecutive elements $(\mathcal{E}_{i-1}, \mathcal{E}_i)$ by $(L_{\mathcal{E}_{i-1}}\mathcal{E}_i, \mathcal{E}_{i-1})$ and a right mutation R_i replaces the same pair of consecutive elements $(\mathcal{E}_{i-1}, \mathcal{E}_i)$ by $(\mathcal{E}_i, R_{\mathcal{E}_i}\mathcal{E}_{i-1})$:

$$L_i\sigma = L_{\mathcal{E}_{i-1}}\sigma = (\mathcal{E}_0, \dots, L_{\mathcal{E}_{i-1}}\mathcal{E}_i, \mathcal{E}_{i-1}, \dots, \mathcal{E}_n),$$
$$R_i\sigma = R_{\mathcal{E}_{i-1}}\sigma = (\mathcal{E}_0, \dots, \mathcal{E}_i, R_{\mathcal{E}_i}\mathcal{E}_{i-1}, \dots, \mathcal{E}_n).$$

It is convenient to agree that

$$R^{(j)}\mathcal{E}_i = R_{\mathcal{E}_{i+j}} \cdots R_{\mathcal{E}_{i+2}} R_{\mathcal{E}_{i+1}} \mathcal{E}_i$$

and similar notation for compositions of left mutations. According to these notations, mutations satisfy the following relations:

$$\begin{aligned} L_iR_i &= R_iL_i = \mathrm{Id}, & \\ L_iL_j &= L_jL_i & \text{for } |i-j| > 1, \\ L_{i+1}L_iL_{i+1} &= L_iL_{i+1}L_i & \text{for } 1 < i < n. \end{aligned}$$

Therefore, there exists a natural action of the *braid group* on $n+1$ strands on the set of left (respectively, right) mutations on σ.

Proposition A.0.22. *Let X be a smooth projective variety and let $\sigma = (\mathcal{E}_0, \dots, \mathcal{E}_n)$ be an exceptional collection of objects of $\mathcal{D}$. Then any mutation of σ is also an exceptional collection of objects of $\mathcal{D}$. Moreover, if σ generates the category $\mathcal{D}$, then the mutated collection also generates $\mathcal{D}$.*

Proof. See [29, Assertion 2.1 and Lemma 2.2]. □

Remark A.0.23. In general, a mutation of a strongly exceptional collection is not a strongly exceptional collection. For instance, let $X = \mathbb{P}^1 \times \mathbb{P}^1$ be a smooth quadric surface in $\mathbb{P}^3$ and denote by $\mathcal{O}_X(a,b) = \mathcal{O}_{\mathbb{P}^1}(a) \boxtimes \mathcal{O}_{\mathbb{P}^1}(b)$. By [66, Proposition 4.16], (see also [67])

$$\sigma = (\mathcal{O}_X, \mathcal{O}_X(1,0), \mathcal{O}_X(0,1), \mathcal{O}_X(1,1))$$

is a full strongly exceptional collection of line bundles on X. Using the exact sequence

$$0 \longrightarrow \mathcal{O}_X(-1,1) \longrightarrow \mathrm{Hom}(\mathcal{O}_X(0,1), \mathcal{O}_X(1,1)) \otimes \mathcal{O}_X(0,1) \longrightarrow \mathcal{O}_X(1,1) \longrightarrow 0,$$

we get that $L_{\mathcal{O}_X(0,1)}\mathcal{O}_X(1,1) = \mathcal{O}_X(-1,1)$. But, since $\mathrm{Ext}^1(\mathcal{O}_X(1,0), \mathcal{O}_X(-1,1)) = \mathbf{k}^2$, the mutated exceptional collection of line bundles on X given by

$$L\sigma = (\mathcal{O}_X, \mathcal{O}_X(1,0), \mathcal{O}_X(-1,1), \mathcal{O}_X(0,1))$$

is no more a strongly exceptional collection of line bundles on X.

Definition A.0.24. Let X be a smooth projective variety. Given any full exceptional collection $\sigma = (\mathcal{E}_0, \dots, \mathcal{E}_n)$, the collection

$$(L^{(n)}\mathcal{E}_n, L^{(n-1)}\mathcal{E}_{n-1}, \dots, L^{(1)}\mathcal{E}_1, \mathcal{E}_0)$$

will be called the *left dual base of* σ and the collection

$$(\mathcal{E}_n, R^{(1)}\mathcal{E}_{n-1}, \dots, R^{(n)}\mathcal{E}_0)$$

will be called the *right dual base of* σ.

Remark A.0.25. Given a full exceptional collection $\sigma = (\mathcal{E}_0, \dots, \mathcal{E}_n)$, its corresponding right and left dual bases are uniquely determined up to a unique isomorphism and they satisfy the following orthogonality conditions:

$$\mathrm{Hom}^{\alpha}(R^{(j)}\mathcal{E}_i, \mathcal{E}_k) = 0, \quad \mathrm{Hom}^{\alpha}(\mathcal{E}_k, L^{(j)}\mathcal{E}_i) = 0$$

for all α, i, j, and k, except for

$$\mathrm{Hom}^{k}(R^{(k)}\mathcal{E}_{n-k}, \mathcal{E}_{n-k}) = \mathrm{Hom}^{n-k}(\mathcal{E}_{n-k}, L^{(n-k)}\mathcal{E}_{n-k}) = \mathbf{k}.$$

Example A.0.26. (1) The right dual collection of $(\mathcal{O}_{\mathbb{P}^r}, \mathcal{O}_{\mathbb{P}^r}(1), \mathcal{O}_{\mathbb{P}^r}(2), \dots, \mathcal{O}_{\mathbb{P}^r}(r))$ is $(\mathcal{O}_{\mathbb{P}^r}(r), T_{\mathbb{P}^r}(r-1), \dots, \bigwedge^j T_{\mathbb{P}^r}(r-j), \dots, \bigwedge^r T_{\mathbb{P}^r})$ and its left dual collection is $(\Omega^r_{\mathbb{P}^r}(r), \dots, \Omega^1_{\mathbb{P}^r}(1), \mathcal{O}_{\mathbb{P}^r})$.

(2) Let $\mathbb{F}_e$ be a Hirzebruch surface and denote by C_0 (respectively f) the section with nonpositive self-intersection $C_0^2 = -e$ (respectively the fiber of the structural map). These C_0 and f generate of $\mathrm{Pic}(\mathbb{F}_e)$. The collection of line bundles

$$(\mathcal{O}_{\mathbb{F}_e}(-C_0 - f), \mathcal{O}_{\mathbb{F}_e}(-C_0), \mathcal{O}_{\mathbb{F}_e}(-f), \mathcal{O}_{\mathbb{F}_e}) \tag{A.3}$$

is a full exceptional collection on $\mathbb{F}_e$ and its left dual collection is given by

$$(\mathcal{O}_{\mathbb{F}_e}(-C_0 - (e+1)f), \mathcal{O}_{\mathbb{F}_e}(-C_0 - ef), \mathcal{O}_{\mathbb{F}_e}(-f), \mathcal{O}_{\mathbb{F}_e}).$$

In fact, it follows from [171, Corollary 2.7] that (A.3) is a full exceptional collection and its left dual can be computed using the orthogonality conditions of Remark A.0.25.

Theorem A.0.27. *Let $(X, \mathcal{O}_X(1))$ be a smooth projective variety with a full exceptional collection $(\mathcal{E}_0^{\vee}[-k_0], \dots, \mathcal{E}_n^{\vee}[-k_n])$ where each $\mathcal{E}_i$ is a vector bundle and $(k_0, \dots, k_n) \in \mathbb{Z}^{n+1}$ is such that there exists a sequence $(\mathcal{F}_n, \dots, \mathcal{F}_0)$ of vector bundles satisfying*

$$\mathrm{Ext}^{k}(\mathcal{E}_i^{\vee}[-k_i], \mathcal{F}_j) = \mathrm{H}^{k+k_i}(\mathcal{E}_i \otimes \mathcal{F}_j) = \begin{cases} \mathbf{k} & \text{if } i = j = k, \\ 0 & \text{otherwise.} \end{cases} \tag{A.4}$$

Equivalently, the collection $(\mathcal{F}_n, \dots, \mathcal{F}_0)$ is the left dual collection of $(\mathcal{E}_0^{\vee}[-k_0], \dots, \mathcal{E}_n^{\vee}[-k_n])$ and $(k_0, \dots, k_n) \in \mathbb{Z}^{n+1}$ is determined by the orthogonality condition. Then, *for any coherent sheaf $\mathcal{G}$ on X, there is a spectral sequence in the square $-n \le p \le 0$, $0 \le q \le n$ with E_1-term*

$$E_1^{p,q} = \mathrm{H}^{q+k_{-p}}(\mathcal{E}_{-p} \otimes \mathcal{G}) \otimes \mathcal{F}_{-p}$$

which is functorial in $\mathcal{G}$ and converges to

$$E^{p+q} = \begin{cases} \mathcal{G} & \text{if } p+q=0, \\ 0 & \text{otherwise.} \end{cases}$$

If, in addition, $\mathrm{Ext}^k(\mathcal{F}_i, \mathcal{F}_j) = 0$ for $k > 0$ and all i, j, then there exists a complex of vector bundles $L^\bullet$ such that

(1) $\mathrm{H}^k(L^\bullet) = \begin{cases} \mathcal{G} & \text{if } k=0, \\ 0 & \text{otherwise.} \end{cases}$

(2) $L^k = \bigoplus_{p+q=k} \mathrm{H}^{q+k_{-p}}(\mathcal{G} \otimes \mathcal{E}_{-p}) \otimes \mathcal{F}_{-p}$ *for* $0 \le q \le n$ *and* $-n \le p \le 0$.

Proof. See [99, 2.7.3]. □

Example A.0.28. (1) It follows from Theorem A.0.27 and Example A.0.26 (1) that for any vector bundle $\mathcal{G}$ in $\mathbb{P}^n$ there is a spectral sequence in the square $-n \le p \le 0$, $0 \le q \le n$ with E_1-term

$$E_1^{p,q} = \mathrm{H}^q(\mathbb{P}^n, \mathcal{G}(p)) \otimes \Omega_{\mathbb{P}^n}^{-p}(-p)$$

which is functorial in $\mathcal{G}$ and converges to

$$E^{p+q} = \begin{cases} \mathcal{G} & \text{if } p+q=0, \\ 0 & \text{otherwise.} \end{cases}$$

This spectral sequence is known as the Beilinson's spectral sequence for vector bundles on $\mathbb{P}^n$.

(2) For the quadric hypersurface $Q_n \subset \mathbb{P}^{n+1}$, we set $\Omega^p := \Omega^p_{\mathbb{P}^{n+1}}$. We are going to define inductively a sequence of vector bundles ψ_i that will provide a spectral sequence for sheaves on quadrics analogous to Beilinson's spectral sequence for sheaves on $\mathbb{P}^n$. We set $\psi_0 := \mathcal{O}_{Q_n}$, $\psi_1 := \Omega^1(1)_{|Q_n}$ and, for $i \ge 2$, we define ψ_i as the middle term of the exact sequence

$$0 \longrightarrow \Omega^i(i)_{|Q_n} \longrightarrow \psi_i \longrightarrow \psi_{i-2} \longrightarrow 0. \tag{A.5}$$

This sequence of vector bundles is well-defined since by [4, Theorem 5.3], we have that

$$\mathrm{H}^s(\Omega^q(q)_{|Q_n} \otimes \psi_i^*) = \begin{cases} 0 & \text{for } s \ne 1 \text{ or } q \ne i+2, \\ \mathbf{k} & \text{for } s = 1 \text{ and } q = i+2. \end{cases}$$

In particular, the vector bundle ψ_{i+2} is defined as the unique nontrivial extension of ψ_i by $\Omega^{i+2}(i+2)_{|Q_n}$. Moreover, since $\Omega^i = 0$ for $i \geq n+2$, we have defined canonically ψ_i for all $i \in \mathbb{N}$. Indeed, it holds that $\psi_i \cong \psi_{i+2}$ for $i \geq n$. These bundles were originally defined via graded Clifford algebras in [123]. For n odd, they define a left dual collection for the exceptional collection $(\mathcal{S}(-n), \mathcal{O}_{Q_n}(-n+1), \dots, \mathcal{O}_{Q_n}(-1), \mathcal{O}_{Q_n})$, and for n even, for the collection $(\mathcal{S}'(-n), \mathcal{S}''(-n), \mathcal{O}_{Q_n}(-n+1), \dots, \mathcal{O}_{Q_n}(-1), \mathcal{O}_{Q_n})$, both given in Example A.0.17. In particular, by Theorem A.0.27 for a coherent sheaf $\mathcal{E}$ on Q_n, there exists a complex of vector bundles $L^\bullet$ on Q_n such that

$$\mathrm{H}^k(L^\bullet) = \begin{cases} \mathcal{E} & \text{if } k = 0, \\ 0 & \text{if } k \neq 0, \end{cases}$$

where $L^k = \bigoplus_{j+k=i} X_j^i$, with $X_j^i := \psi_j^{h^i(\mathcal{E}(-j))}$ for $j < n$, and

$$X_n^i = \begin{cases} \mathcal{S}^{h^i(\mathcal{E}\otimes\mathcal{S}^\vee(-n))} & \text{for } n \text{ odd}, \\ \mathcal{S}'^{\,h^i(\mathcal{E}\otimes\mathcal{S}'^\vee(-n))} \oplus \mathcal{S}''^{\,h^i(\mathcal{E}\otimes\mathcal{S}''^\vee(-n))} & \text{for } n \text{ even}, \end{cases}$$

and all other X_j^i are zero (see [123] and [4, Theorems 5.5 and 5.6] for all the details).

Now let X_1 and X_2 be two smooth projective varieties defined over $\mathbf{k}$ and let $\mathcal{D}_i = D^b(\mathcal{O}_{X_i}\text{-mod})$ be the derived category of bounded complexes of coherent sheaves of $\mathcal{O}_{X_i}$-modules for $i = 1, 2$. Denote by $\pi_i : X_1 \times X_2 \longrightarrow X_i$ $(i = 1, 2)$ the two natural projections. For an object $\mathcal{E} \in \mathrm{Obj}(D^b(X_1 \times X_2))$, let us define the exact functor

$$\begin{array}{rcl} \Phi_{\mathcal{E}} : \mathcal{D}_1 & \longrightarrow & \mathcal{D}_2, \\ \mathcal{F} & \mapsto & \Phi_{\mathcal{E}}(\mathcal{F}) := R\pi_{2*}(L\pi_1^*(\mathcal{F}) \otimes \mathcal{E}), \end{array}$$

where $R\pi_{2*}$ and $L\pi_1^*$ denote the corresponding derived functors. However, notice that, since π_1 is flat, $L\pi_1^*$ is the usual pullback.

Definition A.0.29. Let $F : \mathcal{D}_1 \longrightarrow \mathcal{D}_2$ be an exact functor. F is said a Fourier–Mukai functor (or of Fourier–Mukai type) if there exists an object $\mathcal{E} \in \mathrm{Obj}(D^b(X_1 \times X_2))$ such that $F \cong \Phi_{\mathcal{E}}$. In this case, $\mathcal{E}$ is called the kernel of F.

Fourier–Mukai functors always have right and left adjoints, as they were defined in Definition A.0.6. Indeed, the right and left adjoints of a Fourier–Mukai functor are again Fourier–Mukai functors. Their kernels can be defined as follows. For any object $\mathcal{E} \in D^b(X_1 \times X_2)$, we define

$$\mathcal{E}_R := \mathcal{E}^\vee \otimes \pi_1^* \omega_{X_1}[\dim X_1], \quad \mathcal{E}_L := \mathcal{E}^\vee \otimes \pi_2^* \omega_{X_2}[\dim X_2].$$

Proposition A.0.30. *Let $\Phi_{\mathcal{E}} : \mathcal{D}_1 \longrightarrow \mathcal{D}_2$ be the Fourier–Mukai functor with kernel $\mathcal{E}$. Then $\Phi_{\mathcal{E}_R} : \mathcal{D}_2 \longrightarrow \mathcal{D}_1$ (respectively $\Phi_{\mathcal{E}_L} : \mathcal{D}_2 \longrightarrow \mathcal{D}_1$) is the right (respectively left) adjoint to $\Phi_{\mathcal{E}}$.*

Proof. See [161, Theorem 2.2] and [118, Proposition 5.9]. □

Bibliography

[1] Allan Adler, Sundararaman Ramanan, *Moduli of Abelian varieties*, Lecture Notes in Math., **1644**, Springer-Verlag (1996).

[2] James Alexander and André Hirschowitz, *Polynomial interpolation in several variables*, J. Alg. Geom., **4** (1995), 201–222.

[3] Allen B. Altman and Steven L. Kleiman, *Compactifying the Picard scheme*, Adv. Math., **35** (1980), 50–112.

[4] Vicenzo Ancona and Giorgio Ottaviani, *Some applications of Beilinson's theorem to projective spaces and quadrics*, Forum Math., **3** (1991), 157–176.

[5] Vincenzo Antonelli, *Characterization of Ulrich bundles on Hirzabruch surfaces*, Revista Matemática Complutense (2020).

[6] Marian Aprodu, *Lazarsfeld–Mukai bundles and applications*, in Commutative Algebra, Springer (2013), 1–23.

[7] Marian Aprodu, Gianfranco Casnati, Laura Costa, Rosa Maria Miró-Roig and Montserrat Teixidor, *Theta divisors and Ulrich bundles on Geometrically ruled surfaces*, Annali di Matematica Pura ed Applicata (2020).

[8] Marian Aprodu, Laura Costa and Rosa Maria Miró-Roig, *Ulrich bundles on ruled surfaces*, J. Pure Appl. Algebra, **222** (2018), 131–138.

[9] Marian Aprodu, Gavril Farkas and Angela Ortega, *Minimal resolutions, Chow forms and Ulrich bundles on $K3$ surfaces*, J. Reine Angew. Math., **730** (2017), 225–249.

[10] Marian Aprodu, Sukmoon Huh, Francesco Malaspina and Joan Pons-Llopis, *Ulrich bundles on smooth projective varieties of minimal degree*, Proc. Amer. Math. Soc., **147** (2019), 5117–5129.

[11] Enrico Arbarello, Maurizio Cornalba, Phillip A. Griffiths and Joe Harris, *Geometry of algebraic curves. Vol. I*, Grundlehren der Mathematischen Wissenschaften [Fundamental Principles of Mathematical Sciences], **267**, Springer-Verlag (1985).

[12] Enrique Arrondo and Laura Costa, *Vector bundles on Fano 3-folds without intermediate cohomology*, Comm. Algebra, **28** (2000), 3899–3911.

[13] Igor Vadimovich Artamkin, *Deformation of torsion-free sheaves on an algebraic surface*, Izv. Akad. Nauk SSSR Ser. Mat., **54** (1990), 435–468.

[14] Michael Francis Atiyah, *Vector bundles over an elliptic curve*, Proc. London Math. Soc. **7** (1957), 414–452.

[15] Michael Francis Atiyah, *Geometry on Yang–Mills fields*, Scuola Normale Superiore Pisa, Pisa, 1979, 99 pp.

[16] Jörgen Backelin and Jürgen Herzog, *On Ulrich-modules over hypersurface rings*, in Commutative Algebra (Berkeley, CA, 1987), Math. Sci. Res. Inst. Publ., **15** (1989), 63–68.

[17] Jörgen Backelin, Jürgen Herzog and Herbert Sanders, *Matrix factorizations of homogeneous polynomials*, in Algebra some current trends (Varna, 1986), Lecture Notes in Math., **1352** (1988).

[18] Edoardo Ballico, Sukmoon Huh, Francesco Malaspina and Joan Pons-Llopis, *aCM sheaves on the double plane*, Trans. Amer. Math. Soc., **372** (2019), 1783–1816.

[19] Wolf P. Barth, Klaus Hulek, *Monads and moduli of vector bundles*, Manuscripta Math., **25** (1978), 323–347.

[20] Wolf P. Barth, Klaus Hulek, Chris A. M. Peters and Antonius Van de Ven, *Compact complex surfaces*, Ergebnisse der Mathematik und ihrer Grenzgebiete. 3. Folge. A Series of Modern Surveys in Mathematics, **4** (2004).

The original version of this chapter was revised: Ref. [63] on p. 251 has been corrected. An Erratum is available at DOI: https://doi.org/10.1515/9783110647686-014
https://doi.org/10.1515/9783110647686-009

[21] Arnaud Beauville, *Surfaces Algébriques complexes*, Astérisque, **54** (1978).

[22] Arnaud Beauville, *Determinantal hypersurfaces*, Michigan Math. J., **48** (2000), 39–64.

[23] Arnaud Beauville, *Ulrich bundles on abelian surfaces*, Proc. Amer. Math. Soc., **144** (2016), 4609–4611.

[24] Arnaud Beauville, *An introduction to Ulrich bundles*, Eur. J. Math., **4** (2018), 26–36.

[25] Christina Birkenhake and Herbert Lange, *Complex abelian varieties*, Grundlehren der mathematischen Wissenschaften, **302** (2003).

[26] Jishnu Biswas and Girivaru Venkata Ravindra, *Arithmetically Cohen–Macaulay bundles on complete intersection varieties of sufficiently high multidegree*, Math. Z., **265** (2010), 493–509.

[27] Markus Bläser, David Eisenbud and Frank-Olaf Schreyer, *Ulrich complexity*, Differential Geometry and its Applications, **55** (2017), 128–145.

[28] Guntram Bohnhorst and Heinz Spindler, *The stability of certain vector bundles on* $\mathbb{P}^n$, in Proc. Bayreuth Conference Complex Algebraic Varieties, Lecture Notes in Mathematics, **1507** (1992), 39–50.

[29] Alexey Bondal, *Representation of associative algebras and coherent sheaves*, Math. USSR Izvestiya, **34** (1990), 23–42.

[30] Alexey Bondal and Mikhail M. Kapranov, *Homogeneous bundles* in Helices and vector bundles, London Math. Soc. Lecture Note Ser., **148**, Cambridge Univ. Press, Cambridge (1990), 45–55.

[31] Alexey Bondal and Dmitri O. Orlov, *Semiorthogonal decompositions for algebraic varieties*, arXiv:alg-geom/9506012v1 (1995).

[32] Maria Chiara Brambilla and Daniele Faenzi, *Moduli spaces of rank 2 ACM bundles on prime Fano threefolds*, Michigan Math. J., **60** (2011), 113–148.

[33] Maria Chiara Brambilla and Daniele Faenzi, *Rank 2 stable sheaves with odd determinant on Fano threefolds of genus 9*, Math. Z., **275** (2013), no. 1–2, 185–210.

[34] Maria Chiara Brambilla and Daniele Faenzi, *Vector bundles and Fano threefolds of genus 7 and Brill–Noether loci*, Internat. J. Math., **25** (2014), 59 pp.

[35] Joseph P. Brennan, Jürgen Herzog and Bernd Ulrich, *Maximally generated Cohen–Macaulay modules*, Math. Scandinavica, **61** (1987), 181–203.

[36] Winfried Bruns and Jürgen Herzog, *Cohen–Macaulay rings*, Cambridge Studies in Advanced Mathematics, **39** (1993).

[37] David A. Buchsbaum and David Eisenbud, *Algebra structures for finite free resolutions, and some structure theorems for ideals of codimension* 3, Amer. J. Math., **99** (1977), 447–485.

[38] Ragnar-Olaf Buchweitz, Gert-Martin Greuel and Frank-Olaf Schreyer, *Cohen–Macaulay modules on hypersurface singularities, II*, Invent. Math., **88** (1987), 165–182.

[39] Marta Casanellas and Robin Hartshorne, *Gorenstein biliaison and ACM sheaves*, J. of Algebra, **278** (2004), 314–341.

[40] Marta Casanellas and Robin Hartshorne, *ACM bundles on cubic surfaces*, J. Eur. Math. Soc., **13** (2011), 709–731.

[41] Marta Casanellas, Robin Hartshorne, Florian Geiss and Frank-Olaf Schreyer, *Stable Ulrich bundles*, Int. J. Math., **23** (2012), 50 pp.

[42] Gianfranco Casnati, *Special Ulrich bundles on non-special surfaces with* $p_g = q = 0$, Int. J. Math., **28** (2017).

[43] Gianfranco Casnati, *Erratum to: "Special Ulrich bundles on non-special surfaces with* $p_g = q = 0$", Int. J. Math., **29** (2018).

[44] Gianfranco Casnati, *Examples of surfaces which are Ulrich-wild*, Proc. Amer. Math. Soc. (2020).

[45] Gianfranco Casnati and Yeongrak Kim, *Ulrich bundles on blowing up (and an erratum)*, C.R. Acad. Sci. Paris, Ser. I, **355** (2017), 1291–1297.

[46] Fabrizio Catanese, *Chow varieties, Hilbert schemes and moduli spaces of surfaces of general type*, Journal of Algebraic Geometry, **1(4)** (1992), 561–595.

[47] Fabrizio Catanese, *Cayley forms and self-dual varieties*, Proc. of the Edinbourgh Mathematical Society **57(1)** (2014) 89–109.

[48] Arthur Cayley, *On the triple tangent planes of surfaces of the third order*, Cambridge and Dublin Math. J., **4** (1849), 118–138.

[49] Arthur Cayley, *On a new analytical representation of curves in space*, Quarterly J. of Pure and Appl. Math., **III** (1860), 225–236.

[50] Arthur Cayley, *On a new analytical representation of curves in space*, Quarterly J. of Pure and Appl. Math., **V** (1862), 81–86.

[51] Arthur Cayley, *A memoir on quartic surfaces*, Proc. London Math. Soc., **3** (1869–1871), 19–69.

[52] Luca Chiantini and Carlo Madonna, *ACM bundles on a general quintic threefold*, Le Matematiche, **55** (2000), 239–258.

[53] Luca Chiantini and Carlo Madonna, *A splitting criterion for rank 2 bundles on a general sextic threefold*, Internat. J. Math., **15** (2004), 341–359.

[54] Luca Chiantini and Carlo Madonna, *ACM bundles on general hypersurfaces in $\mathbb{P}^5$ of low degree*, Collect. Math., **56** (2005), 85–96.

[55] Luca Chiantini and Rosa Maria Miró-Roig, *Arithmetically Cohen–Macaulay bundles and sets of points on general surfaces*, J. of Algebra, **451** (2016), 268–292.

[56] Yonghwa Cho, Yeongrak Kim and Kyoung-Seog Lee, *Ulrich bundles on intersections of two 4-dimensional quadrics*, Int. Math. Res. Notices (2019).

[57] Arthur Byron Coble, *Algebraic geometry and theta functions* (reprint of the 1929 edition), A.M.S. Coll. Publ., **10**, A.M.S., Providence R.I. (1982).

[58] Roger J. Cook and Alan D. Thomas, *Line bundles and homogeneous matrices*, Quart. J. Math. Oxford, **30** (1979), 423–429.

[59] Emre Coskun, *A survey of Ulrich bundles*, Analytic and algebraic geometry, 85–106, Hindustan Book Agency (2017).

[60] Emre Coskun and Ozhan Genc, *Ulrich bundles on Veronese Surfaces*, Proc. Amer. Math. Soc., **145** (2017), 4687–4701.

[61] Izzet Coskun, Laura Costa, Jack Huizenga, Rosa Maria Miró-Roig and Matthew Woolf, *Ulrich Schur bundles on flag varieties*, J. of Algebra, **474** (2017), 49–96.

[62] Izzet Coskun and Jack Huizenga, *Brill–Noether theorems and globally generated vector bundles on Hirzebruch surfaces*, Nagoya Mathematical Journal, (2018), DOI 10.1017/nmj.2018.17.

[63] Izzet Coskun and Jack Huizenga, *Brill–Noether problems, Ulrich bundles and the cohomology of moduli spaces of sheaves*, Matemática Contemporânea, **47** (2020), 21–72.

[64] Izzet Coskun and Luke Jaskowiak, *Ulrich partitions for two-step flag varieties*, Involve, **10** (2017), 531–539. DOI 10.2140/involve.2017.10.531.

[65] Emre Coskun, Rajesh S. Kulkarni and Yusuf Mustopa, *The geometry of Ulrich bundles on del Pezzo surfaces*, J. of Algebra, **375** (2013), 280–301.

[66] Laura Costa and Rosa Maria Miró-Roig, *Tilting sheaves on toric varieties*, Math. Z., **248** (2004), 849–865.

[67] Laura Costa and Rosa Maria Miró-Roig, *Derived categories of projective bundles*, Proc. Amer. Math. Soc., **133** (2005), 2533–2537.

[68] Laura Costa and Rosa Maria Miró-Roig, GL(V)*-invariant Ulrich bundles on Grassmannians*, Math. Ann., **361** (2015), 443–457.

[69] Laura Costa and Rosa Maria Miró-Roig, *Homogeneous ACM bundles on a Grassmannian*, Advances in Mathematics, **289**(5) (2016), 95–113.

[70] Laura Costa and Rosa Maria Miró-Roig, *Ulrich bundles on Veronese surfaces*, in Singularities, algebraic geometry, commutative algebra, and related topics, 375–381, Springer, Cham (2018).

[71] Laura Costa and Rosa Maria Miró-Roig, *Instanton bundles vs Ulrich bundles on projective spaces*, Beitr. Algebra Geom. (2020), 10.1007/s13366-020-00506-7.

[72] Laura Costa, Rosa Maria Miró-Roig and Joan Pons-Llopis, *The representation type of Segre varieties*, Adv. Math., **230** (2012), 1995–2013.

[73] David A. Cox, *The Noether–Lefschetz locus of regular elliptic surfaces with section and $p_g \geq 2$*, Am. J. Math., **112**(2) (1990), 289–329.

[74] Luigi Cremona, *Sulle superficie gobbe di quatro grado*, Mem. Accad. Scienze Ist. Bologna, **8** (1868), 235–250.

[75] Richard Dedekind and Heinrich Weber, *Theorie der algebraischen Functionen einer Veränderlichen*, J. Reine Angew. Math., **92** (1882), 181–290.

[76] Michele Demazure, *A very simple proof of Bott theorem*, Inv. Math., **33** (1976), 271–272.

[77] Leonard Eugene Dickson, *Determination of all general homogeneous polynomials expressible as determinants with linear elements*, Trans. Amer. Math. Soc., **22** (1921), 167–179.

[78] Alfred Dixon, *Note on the reduction of a ternary quartic to a symmetrical determinant*, Proc. Cambridge Philos. Soc., **11** (1900–1902), 350–351.

[79] Yuri A. Drozd and Gert-Martin Greuel, *Tame and wild projective curves and classification of vector bundles*, J. of Algebra, **246** (2001), 1–84.

[80] David Eisenbud, *Homological algebra with an application to group representations*, Trans. Amer. Math. Soc., **260** (1980), 35–64.

[81] David Eisenbud, *Commutative Algebra. With a view toward algebraic geometry*, Graduate Texts in Mathematics, **150** (1995).

[82] David Eisenbud, Gunnar Fløystad and Frank-Olaf Schreyer, *Sheaf cohomology and free resolutions over exterior algebra*, Transactions of the AMS, **355** (2003), 4397–4426.

[83] David Eisenbud, Mark Green and Joe Harris, *Cayley–Bacharach theorems and conjectures*, Bulletin of the AMS, **33** (1996), 295–324.

[84] David Eisenbud and Joe Harris, *On varieties of minimal degree (a centennial account)*, in Algebraic geometry, Bowdoin, 1985 (Brunswick, Maine, 1985), 3–13, Proc. Sympos. Pure Math., **46**, Part 1, Amer. Math. Soc., Providence, RI, 1987.

[85] David Eisenbud and Jürgen Herzog, *The classification of homogeneous Cohen– Macaulay rings of finite representation type*, Math. Ann., **280** (1988), 347–352.

[86] David Eisenbud and Irena Peeva, *Minimal free resolutions over complete intersections*, Lecture Notes in Mathematics, **2152** (2016).

[87] David Eisenbud and Frank-Olaf Schreyer, *Boij–Söderberg Theory*, in Combinatorial Aspects of Commutative Algebra and Algebraic Geometry, Abel Symposia, **6** (2009), 35–48.

[88] David Eisenbud, Frank-Olaf Schreyer and Jerzy Weyman, *Resultants and Chow forms via exterior syzygies*, J. Amer. Math. Soc., **16** (2003), 537–579.

[89] Phillipe Ellia and André Hirschowitz, *Generation de certain fibres sur les spaces projectifs et application*, J. Algebraic Geom., **1** (1992) 531–547.

[90] Federigo Enriques, *Memorie scelte di geometria, vol. I*, Zanichelli, Bologna (1956), 541 pp.

[91] Graham E. Evans and Phillip A. Griffiths, *The syzygy problem*, Ann. of Math., **114** (1981), 323–333.

[92] Daniele Faenzi, *Ulrich bundles on K3 surfaces*, Algebra Number Theory, **13** (2019), 1443–1454.

[93] Daniele Faenzi and Joan Pons-Llopis *The Cohen–Macaulay representation type of arithmetically Cohen–Macaulay varieties*, ArXiv e-print: arXiv:1504.03819.

[94] Maria Lucia Fania, Margherita Lelli-Chiesa, and Joan Pons-Llopis, *Ulrich bundles on three dimensional scrolls*, Int. Math. Res. Notices, rnz288, DOI 10.1093/imrn/rnz288, Published: 18 November 2019.

[95] Gunnar Fløystad, Joe Kileel and Giorgio Ottaviani, *The Chow form of the essential variety in computer vision*, J. of Symbolic Compt., **86**, 97–119.

[96] John Fogarty, *Algebraic families on an algebraic surface*, Amer. J. Math., **10** (1968), 511–521.

[97] Anton Fonarev, *Irreducible Ulrich bundles on isotropic Grassmannians*, Mosc. Math. J., **16** (2016), 711–726.

[98] William Fulton and Joe Harris, *Representation theory: a first course*, GTM **129**, Springer (2004).

[99] Alexey L. Gorodentsev and Sergej A. Kuleshov, *Helix theory*, Moscow Mathematical Journal, **4(2)** (2004), 377–440.

[100] Hermann Grassmann, *Die stereometrischen Gleichungen dritten Grades, und die dadurch erzeugten Oberflächen*, J. Reine Angew. Math., **49** (1885), 47–65.

[101] Hans Grauert and Reinhold Remmert, *Zur Spaltung lokal-freier Garben über Riemannschen Flächen*, Math. Z., **144** (1975), 35–43.

[102] Daniel Grayson and Michael Stillman, *Macaulay2, a software system for research in algebraic geometry*, http://www.math.uiuc.edu/Macaulay2/.

[103] Phillip A. Griffiths and Joseph Harris, *Principles of algebraic geometry*, Wiley & Sons, New York (1978).

[104] Benedict H. Gross and Nolan R. Wallach, *On the Hilbert polynomials and Hilbert series of homogeneous projective varieties*, in Arithmetic geometry and automorphic forms, Adv. Lect. Math., **19**, 253–263.

[105] Alexander Grothendieck, *La théorie des classes de Chern*, Bull. Soc. Math. de France, **86** (1958), 137–154.

[106] Douglas Hanes, *On the Cohen–Macaulay modules of graded subrings*, Trans. Amer. Math. Soc., **357** (2005), 735–756.

[107] Robin Hartshorne, *Varieties of small codimension in projective space*, Bulletin of AMS, **80** (1974), 1017–1032.

[108] Robin Hartshorne, *Algebraic geometry*, Graduate Texts in Mathematics, **52** (1977).

[109] Robin Hartshorne, *Generalized divisors on Gorenstein curves and a theorem of Noether*, J. Mat. Kyoto Univ., **26** (1983), 375–386.

[110] Robin Hartshorne, *Generalized divisors on Gorenstein schemes*, K-Theory, **8** (1994), 287–339.

[111] Robin Hartshorne and André Hirschowitz, *Smoothing algebraic space curves*, Lecture Notes in Math., **1124** (1985), 98–131.

[112] Georg Hein, *Raynaud's vector bundles and base points of the generalized theta divisor*, Mathematische Zeitschrift, **257 (3)** (2007), 597–611.

[113] Jürgen Herzog, Bernd Ulrich and Jörgen Backelin, *Linear maximal Cohen–Macaulay modules over strict complete intersections*, J. Pure Appl. Algebra, **71** (1991), 187–202.

[114] Otto Hesse, *Ueber determinanten und ihre Anwendung in der Geometrie, insbesondere auf Curven vierter Ordnung*, J. Reine Angew. Math., **49** (1855), 243–264.

[115] Otto Hesse, *Ueber die Doppeltangenten der Curven vierter Ordnung*, J. Reine Angew. Math., **49** (1855), 279–332.

[116] Friedrich Hirzebruch, *Topological methods in algebraic geometry*, Grundlehre der Math. Wissenschaften, **131**, Springer-Verlag, (1966).

[117] Geoffrey Horrocks, *Vector bundles on the punctured spectrum of a local ring*, Proc. London Math. Soc., **3** (1964), 689–713.

[118] Daniel Huybrechts, *Fourier–Mukai transforms in algebraic geometry*, Oxford Mathematical Monographs (2006).

[119] Daniel Huybrechts and Manfred Lehn, *The geometry of moduli spaces of sheaves*, Aspects of Mathematics, **E31** (1997).

[120] Atanas Iliev and Laurent Manivel, *On cubic hypersurfaces of dimensions 7 and 8*, Proc. Lond. Math. Soc., **108**, (2014), 517–540.

[121] Vasilii A. Iskovskikh and Yuri G. Prokhorov, *Fano varieties, Algebraic Geometry V*, Encyclopedia Math. Sci., **47**, Springer, Berlin (1999).

[122] Victor G. Kac, *Infinite root systems, representations of graphs and invariant theory*, Inv. Math., **56** (1980), 57–92.

[123] Mikhail M. Kapranov, *On the derived categories of coherent sheaves on some homogeneous spaces*, Invent. Math., **92** (1988), 479–508.

[124] Mikhail M. Kapranov, *On the derived category of coherent sheaves on Grassmann manifolds*, Math. USSR Izvestiya, **24** (1985), 183–192.

[125] Dmitry Kerner and Victor Vinnikov, *Determinantal representations of singular hypersurfaces in* $\mathbb{P}^n$, Advances in Mathematics, **231** (2012), 1619–1654.

[126] Yeongrak Kim, *Ulrich bundles on blowing ups*, C.R. Acad. Sci. Paris, Ser. I, **354** (2016), 1215–1218.

[127] Yeongrak Kim and Frank-Olaf Schreyer, *An explicit matrix factorization of cubic hypersurfaces of small dimension*, Journal of Pure and Applied Algebra, **224(8)** (2020), 106346.

[128] Steven L. Kleiman and John Landolfi *Geometry and deformation of special Schubert varieties*, Compositio Mathematica, **23(4)** (1971), 407–434.

[129] Hans Kleppe, *Deformation of schemes defined by vanishing of Pfaffians*, J. of Algebra, **53** (1978), 84–92.

[130] Jan Oddvar Kleppe and Rosa Maria Miró-Roig, *The dimension of the Hilbert scheme of Gorenstein codimension* 3 *subschemes*, J. Pure Appl. Algebra, **127** (1998), 73–82.

[131] Jan Oddvar Kleppe and Rosa Maria Miró-Roig, *The Representation type of determinantal varieties*, Algebras and Representation Theory, **20** (2017), 1029–1059.

[132] Horst Knörrer, *Cohen-Macaulay modules on hypersurface singularities, I*, Invent. Math., **88** (1987), 153–164.

[133] Janos Kollár, *Rational curves on algebraic varieties*, Springer (1996).

[134] Neithalath Mohan Kumar, Aroor Prabhakar Rao and Girivaru Venkata Ravindra, *Arithmetically Cohen–Macaulay bundles on hypersurfaces*, Comment. Math. Helv., **82** (2007), 829–843.

[135] Neithalath Mohan Kumar, Aroor Prabhakar Rao and Girivaru Venkata Ravindra, *Arithmetically Cohen–Macaulay bundles on three dimensional hypersurfaces*, International Math Research Notices **2007**, rnm025 (2007), 11 pp.

[136] Mario Kummer and Cynthia Vinzant, *The Chow form of a reciprocal linear space*, Michigan Math. J., **68** (2019), 831–858.

[137] Alexander Kuznetsov, *Instanton bundles on Fano threefolds*, Cent. Eur. J. Math., **10** (2012), 1198–1231.

[138] Martí Lahoz, Emanuele Macrí and Paolo Stellari, *ACM sheaves on cubic threefolds*, Algebr. Geom., **2** (2015), 231–269.

[139] Adrian Langer, *D-affinity and C-Frobenius morphism on quadrics*, Int. Math. Res. Not. (2008), 26 pp.

[140] Robert Lazarsfeld, *Lectures on linear series*, Complex Algebraic Geometry, **3**, 161–219 (1993).

[141] Kyoung-Seog Lee and Kyeong-Dong Park, *Moduli spaces of Ulrich bundles on the Fano 3-fold* V_5, Preprint arXiv:1711.08305.

[142] Kyoung-Seog Lee and Kyeong-Dong Park, *Equivariant Ulrich bundles on exceptional homogeneous varieties*, Advances in Geometry (2020).

[143] Zhiming Lin, *Ulrich bundles on projective spaces*, Preprint arXiv:1703.06424.

[144] Angelo Felice Lopez, *On the existence of Ulrich vector bundles on some irregular surfaces*, Proc. Amer. Math. Soc., **149** (2021), no. 1, 13–26.

[145] Angelo Felice Lopez, Roberto Muñoz and José Carlos Sierra, *On the extendability of elliptic surfaces of rank two and higher*, Ann. Inst. Fourier, **59** (2009), 311–346.

[146] Saunders Mac Lane, *Categories for the working mathematician*, Graduate Texts in Mathematics, **5**, Springer-Verlag, New York–Berlin (1971), ix+262 pp.

[147] Marco Maggesi, *Some results on holomorphic vector bundles over projective spaces*, Tesi di Dottorato, Università di Firenze (1999).

[148] Yuri Manin, *Cubic forms*, North Holland (1986).

[149] Laurent Manivel, *On the syzygies of flag manifolds*, Proc. Amer. Math. Soc., **124** (1996), no. 8, 2293–2299.
[150] Laurent Manivel, *Ulrich and aCM bundles from invariant theory*, Communications in Algebra, **47** (2019), 706–718.
[151] Masaki Maruyama, *Moduli of stable sheaves, I*, J. Math. Kyoto Univ., **17** (1977), 91–126.
[152] Masaki Maruyama, *Moduli of stable sheaves, II*, J. Math. Kyoto Univ., **18** (1978), 557–614.
[153] Hideyuki Matsumura, *Commutative algebra*, Ed. Benjamin (1970).
[154] Rick Miranda, *The basic theory of elliptic surfaces*, PhD Thesis (1989).
[155] Rosa Maria Miró-Roig, *Determinantal ideals*, Progress in Mathematics, **264**, Birkhäuser (2008).
[156] Rosa Maria Miró-Roig, *The representation type of rational normal scrolls*, Rendiconti del Circolo Matematico di Palermo, **62** (2013), 153–164.
[157] Rosa Maria Miró-Roig and Joan Pons-Llopis, *Representation type of rational ACM surfaces* $X \subseteq \mathbb{P}^4$, Algebr. Represent. Theory, **16** (2013), 1135–1157.
[158] Rosa Maria Miró-Roig and Joan Pons-Llopis, *n-dimensional Fano varieties of wild representation type*, J. Pure Appl. Algebra, **218** (2014), 1867–1884.
[159] Rosa Maria Miró-Roig and Joan Pons-Llopis, *Special Ulrich bundles on regular Weierstrass fibrations*, Math. Z., **293** (2019), 1431–1441.
[160] Shigefumi Mori, *Projective manifolds with ample tangent bundles*, Annals of Mathematics, **110** (1979), 593–606.
[161] Shigeru Mukai, *Duality between* $D(X)$ *and* $D(\hat{X})$ *with its applications to Picard sheaves*, Nagoya Math. J., **81** (1981), 153–175.
[162] Shigeru Mukai, *Curves and Grassmannians*, in *Algebraic geometry and related topics*, Conf. Proc. Lect. Notes Alg. Geom. I, 19–40 (1993).
[163] David Mumford, *Projective invariants of projective structures and applications*, in Proc. Int. Cong. Math. Stockholm, 526–530 (1962).
[164] Masayoshi Nagata, *On self-intersection numbers of a section on a ruled surface*, Nagoya Math. J., **37** (1970), 191–196.
[165] Mudumbai S. Narasimhan and Sundararaman Ramanan, *Moduli of vector bundles on a compact Riemann surface*, Ann. of Math., **89** (1969) 14–51.
[166] Peter E. Newstead, *Stable bundles of rank 2 and odd degree over a curve of genus 2*, Topology, **7** (1968), 205–215.
[167] Saša Novaković, *Ulrich bundles on Brauer–Severi varieties*, Proc. Amer. Math. Soc. (2020).
[168] Kieran G. O'Grady, *Moduli of vector bundles on projective surfaces: Some basic results*, Invent. Math., **123** (1996), 141–207.
[169] Christian Okonek, Michael Schneider and Heinz Spindler, *Vector bundles on complex projective spaces*, Progress in Mathematics, **3** (1980).
[170] Michael A. van Opstall and Răzvan Veliche, *Variation of hyperplane sections* Contemp. Math., **448** (2007), 255–260.
[171] Dmitri Olegovich Orlov, *Projective bundles, monoidal transformations and derived categories of coherent sheaves*, Russian Acad. Sci. Ivz. Math., **41** (1993), 133–141.
[172] Giorgio Ottaviani, *Spinor bundles on quadrics*, Trans. Amer. Math. Soc., **307** (1988), 301–316.
[173] Giorgio Ottaviani, *Rational Homogeneous Varieties*, SMI course in Algebraic Geometry, Cortona (1995).
[174] Joan Pons-Llopis, *Ulrich bundles and varieties of wild representation type*, PhD Thesis (2011).
[175] Joan Pons-Llopis and Fabio Tonini, *ACM bundles on del Pezzo surfaces*, Le Matematiche, **64** (2009), 177–211.
[176] Mircea Popa, *On the base locus of the generalized theta divisor*, C.R. Acad. Sci. Paris, **329** (1999), 507–512.

[177] Mircea Popa, *Generalized theta linear series on moduli spaces of vector bundles on curves*, in Handbook of moduli, vol. III, Adv. Lect. Math. (ALM), **26**, 219–255, Int. Press, Somerville, MA, 2013.

[178] Sundararaman Ramanan, *Holomorphic vector bundles on homogeneous spaces*, Topology, **5** (1966), 159–177.

[179] Annamalai Ramanathan, *Schubert varieties are arithmetically Cohen-Macaulay*, Invent. Math., **80** (1985), 283–294.

[180] Girivaru Venkata Ravindra and Vasudevan Srivinas, *The Noether–Lefschetz theorem for the divisor class group*, J. of Algebra, **322** (2009), 3373–3391.

[181] Girivaru Venkata Ravindra and Amit Tripathi, *Rank 3 ACM bundles on general hypersurfaces in* $\mathbb{P}^5$, Advances in Mathematics, **355** (2019) 106780.

[182] Michel Raynaud, *Sections des fibrés vectoriels sur une courbe*. Bull. Soc. Math. France, **110** (1982), 103–125.

[183] Miles Reid, *The complete intersection of two or more quadrics*, PhD Thesis (1982).

[184] Joseph J. Rotman, *An introduction to homological algebra*, Universitext, Springer (2009).

[185] Elena Rubei, *Stability of homogeneous bundles on* $\mathbb{P}^3$, Geometriae Dedicata, **158** (2012), 1–21.

[186] Aleksei N. Rudakov et al., *Helices and vector bundles*, London Math. Soc. Lect. Note Series, **148**, Cambridge University Press (1990).

[187] Francesco Russo, *On the geometry of special varieties*, Lectures Notes of the Unione Matematica Italiana, **18** (2016).

[188] Barbara Russo and Montserrat Teixidor, *On a conjecture of Lange*, J. Algebraic Geom., **8** (1999), 483–496.

[189] George Salmon, *On the triple tangent planes to a surface of the third order*, Cambridge and Dublin Math. J., **4** (1849), 252–260.

[190] Tadakazu Sawada, *A sufficient condition for splitting of arithmetically Cohen–Macaulay bundles on general hypersurfaces*, Comm. Algebra, **38** (2010), 1633–1639.

[191] Olivier Schneider, *Sur la dimension de l'ensemble des points base du fibré déterminant sur SUC(r)*, Ann. Inst. Fourier, **57**(2) (2007), 481–490.

[192] Friedrich Schur, *Über die durch collineare Grundgebilde erzeugten Curven und Flächen*, Math. Annalen, **18** (1881), 1–32.

[193] Rolph L. E. Schwarzenberger, *Vector bundles on the projective plane*, Proc. London Math. Soc., **(3)11** (1961), 623–640.

[194] Saverio Andrea Secci, *On the existence of Ulrich bundles on blow-up varieties at a point*, Bolletino dell'Unione Matematica Italiana, **13** (2020), 131–135.

[195] Jean-Pierre Serre, *Faisceaux Algebriques Coherents*, Annals of Mathematics, Second Series, **61**(2) (Mar., 1955), 197–278.

[196] Montserrat Teixidor, *Stable extensions by line bundles*, J. Reine Angew. Math., **502** (1998), 163–173.

[197] Douglas A. Torrance, *Generic forms of low Chow rank*, J. Algebra Appl., **16** (2017), 10 pp.

[198] Douglas A. Torrance and Nick Vannieuwenhoven, *All secant varieties of the Chow variety are nondefective for cubics and quaternary forms*, arXiv:2005.12436.

[199] Amit Tripathi, *Splitting of low-rank ACM bundles on hypersurfaces of high dimension*, Comm. Algebra, **44** (2016), 1011–1017.

[200] Bernd Ulrich, *Gorenstein rings and modules with high numbers of generators*, Math. Z., **188** (1984), 23–32.

[201] Hiroshi Umemura, *On a theorem of Ramanan*, Nagoya Math. J., **69** (1978), 131–138.

[202] Jean-Louis Verdier, *Derived categories*, Thesis. Reprinted in SGA 4 l/2, Lect. Notes in Math., **269**.

[203] Bartel L. van der Waerden, *Einführung in die Algebraische Geometrie*, Grundlehren der Math. B. XXXIII–XXIV (1930).

[204] Jerzy Weyman, *Cohomology of vector bundles and syzygies*, Cambridge Tracts in Mathematics, **149** (2003).
[205] Yuji Yoshino, *Cohen–Macaulay modules over Cohen–Macaulay rings*, London Math. Soc. Lecture Notes, **146** (1990).
[206] Fyodor L. Zak, *Tangent and secants of algebraic varieties*, Amer. Math. Soc. (1993).

Index

Laura Costa, Rosa María Miró-Roig, and Joan Pons-Llopis

Erratum to: Introduction

published in: Laura Costa, Rosa María Miró-Roig, Joan Pons-Llopis, Ulrich Bundles, 978-3-11-064540-8

Erratum

Despite careful production of our books, sometimes mistakes happen. We apologize sincerely for the following mistakes contained in the original version of this chapter of the printed book:

Page	Line	Printed	Should Read
XIII	39	bundles.	bundles. The material in Section 5.6 is based on the paper *Brill–Noether Problems, Ulrich Bundles and the Cohomology of Moduli Spaces of Sheaves* by Izzet Coskun and Jack Huizenga, Matemática Contemporânea 47 (2020), 21–72 [63].
XIV	21	this problem for most of the flag varieties and propose a conjecture for the rest of them.	this problem for most of the flag varieties and propose a conjecture for the rest of them. The authors wish to acknowledge the contribution of Professor Izzet Coskun, Professor Jack Huizenga and Dr. Matthew Woolf for their collaboration on the paper *Ulrich Schur bundles on flag varieties*, J. of Algebra, 474 (2017), 49–96 [61]. The material in Section 7.3 is a reprint of this material. In addition, the authors thank Professor Jack Huizenga for creating all of the original illlustrations. We gratefully acknowledge the permission from the publisher Elsevier for including this material.

The updated original chapter is available at DOI: https://doi.org/10.1515/9783110647686-201
https://doi.org/10.1515/9783110647686-011

Laura Costa, Rosa María Miró-Roig, and Joan Pons-Llopis

Erratum to: Chapter 5 Ulrich bundles on surfaces

published in: Laura Costa, Rosa María Miró-Roig, Joan Pons-Llopis, Ulrich Bundles, 978-3-11-064540-8

Erratum

Despite careful production of our books, sometimes mistakes happen. We apologize sincerely for the following mistakes contained in the original version of this chapter of the printed book:

Page	Line	Printed	Should Read
151	1	5.6 Ulrichness and change of very ample line bundles	5.6 Ulrichness and change of very ample line bundles* [[with footnote]] Footnote *: The material in Section 5.6 is based on the paper I. Coskun and J. Huizenga, *Brill–Noether Problems, Ulrich bundles and the Cohomology of Moduli Spaces of Sheaves*, Matemática Contemporânea 47 (2020), 21–72 [63].

The updated original chapter is available at DOI: https://doi.org/10.1515/9783110647686-005
https://doi.org/10.1515/9783110647686-012

Laura Costa, Rosa María Miró-Roig, and Joan Pons-Llopis

Erratum to: Chapter 7 Ulrich bundles on higher-dimensional varieties

published in: Laura Costa, Rosa María Miró-Roig, Joan Pons-Llopis, Ulrich Bundles, 978-3-11-064540-8

Erratum

Despite careful production of our books, sometimes mistakes happen. We apologize sincerely for the following mistakes contained in the original version of this chapter of the printed book:

Page	Line	Printed	Should Read
197	1	7.3 Ulrich bundles on flag varieties	7.3 Ulrich bundles on flag varieties* [[with footnote]] Footnote *: The authors wish to acknowledge the contribution of Professor Izzet Coskun, Professor Jack Huizenga and Dr. Matthew Woolf for their collaboration on the paper *Ulrich Schur bundles on flag varieties*, J. of Algebra, 474 (2017), 49–96 [61]. The material in Section 7.3 is a reprint of this material. In addition, the authors thank Professor Jack Huizenga for creating all of the original illlustrations. We gratefully acknowledge the permission from the publisher Elsevier for including this material.

The updated original chapter is available at DOI: https://doi.org/10.1515/9783110647686-007
https://doi.org/10.1515/9783110647686-013

Laura Costa, Rosa María Miró-Roig, and Joan Pons-Llopis

Erratum to: Bibliography

published in: Laura Costa, Rosa María Miró-Roig, Joan Pons-Llopis, Ulrich Bundles, 978-3-11-064540-8

Erratum

Despite careful production of our books, sometimes mistakes happen. We apologize sincerely for the following mistakes contained in the original version of this chapter of the printed book:

Page	Line	Printed	Should Read
251	31–33	[63] Izzet Coskun and Jack Huizenga, *Brill-Noether theorems, Ulrich bundles and the cohomology of moduli spaces of sheaves*, in Proceedings of the ICM 2018 Satellite Meeting "Moduli spaces in algebraic geometry and applications". Preprint (2020).	[63] Izzet Coskun and Jack Huizenga, *Brill–Noether problems, Ulrich bundles and the cohomology of moduli spaces of sheaves*, Matemática Contemporânea **47** (2020), 21–72.

The updated original chapter is available at DOI: https://doi.org/10.1515/9783110647686-009
https://doi.org/10.1515/9783110647686-014

De Gruyter Studies in Mathematics

Volume 76
Anton Zettl
Recent Developments in Sturm-Liouville Theory, 2021
ISBN 978-3-11-071884-3, e-ISBN 978-3-11-071900-0, e-ISBN (ePUB) 978-3-11-071938-3

Volume 75
Elena Cordero, Luigi Rodino
Time-Frequency Analysis of Operators, 2020
ISBN 978-3-11-053035-3, e-ISBN 978-3-11-053245-6, e-ISBN (ePUB) 978-3-11-053060-5

Volume 25
Karl H. Hofmann, Sidney A. Morris
The Structure of Compact Groups, 4th Edition, 2020
ISBN 978-3-11-069595-3, e-ISBN 978-3-11-069599-1, e-ISBN (ePUB) 978-3-11-069601-1

Volume 74
Alessandro Carbotti, Serena Dipierro, Enrico Valdinoci
Local Density of Solutions to Fractional Equations, 2019
ISBN 978-3-11-066069-2, e-ISBN 978-3-11-066435-5, e-ISBN (ePUB) 978-3-11-066131-6

Volume 41
Ulrich Knauer, Kolja Knauer
Algebraic Graph Theory. Morphisms, Monoids and Matrices, 2nd Edition, 2019
ISBN 978-3-11-061612-5, e-ISBN 978-3-11-061736-8, e-ISBN (ePUB) 978-3-11-061628-6

Volume 43
Mark M. Meerschaert, Alla Sikorskii
Stochastic Models for Fractional Calculus, 2nd Edition, 2019
ISBN 978-3-11-055907-1, e-ISBN 978-3-11-056024-4, e-ISBN (ePUB) 978-3-11-055914-9

Volume 52
Miroslav Pavlović
Function Classes on the Unit Disc, 2nd Edition, 2019
ISBN 978-3-11-062844-9, e-ISBN 978-3-11-063085-5, e-ISBN (ePUB) 978-3-11-062865-4